Advanced Nanomaterials for Electrochemical Energy Conversion and Storage

Advanced Nanomaterials for Electrochemical Energy Conversion and Storage

Editors

Rongming Wang
Shuhui Sun

MDPI • Basel • Beijing • Wuhan • Barcelona • Belgrade • Manchester • Tokyo • Cluj • Tianjin

Editors
Rongming Wang
University of Science and
Technology Beijing
China

Shuhui Sun
Institut National de la
Recherche Scientifique (INRS)
Canada

Editorial Office
MDPI
St. Alban-Anlage 66
4052 Basel, Switzerland

This is a reprint of articles from the Special Issue published online in the open access journal *Nanomaterials* (ISSN 2079-4991) (available at: https://www.mdpi.com/journal/nanomaterials/special_issues/Energy_Electrochemical).

For citation purposes, cite each article independently as indicated on the article page online and as indicated below:

LastName, A.A.; LastName, B.B.; LastName, C.C. Article Title. *Journal Name* **Year**, *Volume Number*, Page Range.

ISBN 978-3-0365-5839-4 (Hbk)
ISBN 978-3-0365-5840-0 (PDF)

Cover image courtesy of Rongming Wang.

© 2022 by the authors. Articles in this book are Open Access and distributed under the Creative Commons Attribution (CC BY) license, which allows users to download, copy and build upon published articles, as long as the author and publisher are properly credited, which ensures maximum dissemination and a wider impact of our publications.

The book as a whole is distributed by MDPI under the terms and conditions of the Creative Commons license CC BY-NC-ND.

Contents

About the Editors vii

Zhihong Zhang and Rongming Wang
Editorial for the Special Issue: "Advanced Nanomaterials for Electrochemical Energy Conversion and Storage"
Reprinted from: *Nanomaterials* 2022, 12, 3579, doi:10.3390/nano12203579 1

Xinxing Zhan, Xin Tong, Manqi Gu, Juan Tian, Zijian Gao, Liying Ma, Yadian Xie, Zhangsen Chen, Hariprasad Ranganathan, Gaixia Zhang and Shuhui Sun
Phosphorus-Doped Graphene Electrocatalysts for Oxygen Reduction Reaction
Reprinted from: *Nanomaterials* 2022, 12, 1141, doi:10.3390/nano12071141 5

Juzhe Liu, Rui Hao, Binbin Jia, Hewei Zhao and Lin Guo
Manipulation on Two-Dimensional Amorphous Nanomaterials for Enhanced Electrochemical Energy Storage and Conversion
Reprinted from: *Nanomaterials* 2021, 11, 3246, doi:10.3390/nano11123246 27

Haoxian Chen, Jiayi Wang, Yan Zhao, Qindan Zeng, Guofu Zhou and Mingliang Jin
Three-Dimensionally Ordered Macro/Mesoporous Nb_2O_5/Nb_4N_5 Heterostructure as Sulfur Host for High-Performance Lithium/Sulfur Batteries
Reprinted from: *Nanomaterials* 2021, 11, 1531, doi:10.3390/nano11061531 47

Chong Wang, Jian-Hao Lu, Zi-Long Wang, An-Bang Wang, Hao Zhang, Wei-Kun Wang, Zhao-Qing Jin and Li-Zhen Fan
Synergistic Adsorption-Catalytic Sites TiN/Ta_2O_5 with Multidimensional Carbon Structure to Enable High-Performance Li-S Batteries
Reprinted from: *Nanomaterials* 2021, 11, 2882, doi:10.3390/nano11112882 55

Xiaoya Li, Yajun Zhao, Lei Ding, Deqiang Wang, Qi Guo, Zhiwei Li, Hao Luo, Dawei Zhang and Yan Yu
Enhancing the Capacity and Stability by $CoFe_2O_4$ Modified $g-C_3N_4$ Composite for Lithium-Oxygen Batteries
Reprinted from: *Nanomaterials* 2021, 11, 1088, doi:10.3390/nano11051088 65

Ndeye F. Sylla, Samba Sarr, Ndeye M. Ndiaye, Bridget K. Mutuma, Astou Seck, Balla D. Ngom, Mohamed Chaker and Ncholu Manyala
Enhanced Electrochemical Behavior of Peanut-Shell Activated Carbon/Molybdenum Oxide/Molybdenum Carbide Ternary Composites
Reprinted from: *Nanomaterials* 2021, 11, 1056, doi:10.3390/nano11041056 75

Yan Rong and Siping Huang
Self-Templating Synthesis of N/P/Fe Co-Doped 3D Porous Carbon for Oxygen Reduction Reaction Electrocatalysts in Alkaline Media
Reprinted from: *Nanomaterials* 2022, 12, 2106, doi:10.3390/nano12122106 95

Heng Luo, Xiaoxu Wang, Chubin Wan, Lu Xie, Minhui Song and Ping Qian
A Theoretical Study of Fe Adsorbed on Pure and Nonmetal (N, F, P, S, Cl)-Doped $Ti_3C_2O_2$ for Electrocatalytic Nitrogen Reduction
Reprinted from: *Nanomaterials* 2022, 12, 1081, doi:10.3390/nano12071081 105

Shutao Zhao, Xiao Tang, Jingli Li, Jing Zhang, Di Yuan, Dongwei Ma and Lin Ju
Improving the Energetic Stability and Electrocatalytic Performance of Au/WSSe Single-Atom Catalyst with Tensile Strain
Reprinted from: *Nanomaterials* **2022**, 12, 2793, doi:10.3390/nano12162793 . . . . **119**

Yongsheng Zhu, Fengxin Sun, Changjun Jia, Tianming Zhao and Yupeng Mao
A Stretchable and Self-Healing Hybrid Nano-Generator for Human Motion Monitoring
Reprinted from: *Nanomaterials* **2022**, 12, 104, doi:10.3390/nano12010104 . . . . **129**

Penglin Gao, Yuanhua Xia, Jian Gong and Xin Ju
Structure and Magnetic Properties of ErFe$_x$Mn$_{12-x}$ ($7.0 \leq x \leq 9.0$, $\Delta x = 0.2$)
Reprinted from: *Nanomaterials* **2022**, 12, 1586, doi:10.3390/nano12091586 . . . . **145**

Hongyu Du, Min Zhang, Ke Yang, Baohe Li and Zhenhui Ma
A Self-Assembly of Single Layer of Co Nanorods to Reveal the Magnetostatic Interaction Mechanism
Reprinted from: *Nanomaterials* **2022**, 12, 2499, doi:10.3390/nano12142499 . . . . **155**

Panpan Gao, Minhui Song, Xiaoxu Wang, Qing Liu, Shizhen He, Ye Su and Ping Qian
Theoretical Study on the Electronic Structure and Magnetic Properties Regulation of Janus Structure of M'MCO$_2$ 2D MXenes
Reprinted from: *Nanomaterials* **2022**, 12, 556, doi:10.3390/nano12030556 . . . . **167**

Baoshan Cui, Zengtai Zhu, Chuangwen Wu, Xiaobin Guo, Zhuyang Nie, Hao Wu, Tengyu Guo, Peng Chen, Dongfeng Zheng, Tian Yu, Li Xi, Zhongming Zeng, Shiheng Liang, Guangyu Zhang, Guoqiang Yu and Kang L. Wang
Comprehensive Study of the Current-Induced Spin–Orbit Torque Perpendicular Effective Field in Asymmetric Multilayers
Reprinted from: *Nanomaterials* **2022**, 12, 1887, doi:10.3390/nano12111887 . . . . **179**

About the Editors

Rongming Wang

Rongming Wang, Ph.D., is a professor at the University of Science and Technology Beijing, Director of the Beijing Key Laboratory for Magneto-Photoelectrical Composite and Interface Science, and Team Leader for Advanced Functional Material, Beijing Advanced Innovation Center of Materials Genome Engineering. His research interests include magnetic nanomaterials, transmission electron microscopy, interface science, and materials genome engineering. He aims to figure out the physical mechanism at the atomic scale by which microstructures influence the properties. He has published more than 260 SCI-indexed papers in reputable scientific journals with over 13,000 citations and a H-factor of 61. He has been recognized as a Highly Cited Chinese Researcher by Elsevier for exceptional research performance in the field of Physics.

Shuhui Sun

Shuhui Sun, Ph.D., is a professor in the Center for Energy, Materials and Telecommunications, Institut National de la Recherche Scientifique (INRS), Canada. His research interests focus on the rational design of advanced nanomaterials for clean energy conversion and storage, and environmental applications. He is a Fellow of the Canadian Academy of Engineering (CAE, 2022) and a Member of the Royal Society of Canada (College of New Scholars, 2020), and Vice President of the International Academy of Electrochemical Energy Science (IAOEES). He serves as the Executive Editor-in-Chief of Electrochemical Energy Reviews (EER, IF=32.8, Springer-Nature), Associate Editor of SusMat (Wiley), and is an editorial board member of over 10 journals

Editorial

Editorial for the Special Issue: "Advanced Nanomaterials for Electrochemical Energy Conversion and Storage"

Zhihong Zhang and Rongming Wang *

Beijing Advanced Innovation Center for Materials Genome Engineering, Beijing Key Laboratory for Magneto-Photoelectrical Composite and Interface Science, Institute for Multidisciplinary Innovation, School of Mathematics and Physics, University of Science and Technology Beijing, Beijing 100083, China
* Correspondence: rmwang@ustb.edu.cn

Developing efficient and low-cost energy conversion and storage devices and technologies is all-important issue in order to achieve a low-carbon society, whose performance essentially depends on the properties of materials. Nanomaterials have been extensively demonstrated to have great potential applications in energy conversion and storage devices and technologies, i.e., batteries, capacitors, electrocatalysis, and nanogenerators. On the other hand, to realize their practical use, further optimization of the nanomaterials' structures and properties is still needed.

This Special Issue aims to communicate the recent advances of advanced nanomaterials for energy conversion and storage. It covers the design, synthesis, properties, and applications of advanced nanomaterials for energy conversion and storage. Twelve research works focus on various nanomaterials for batteries, capacitors, electrocatalysis, nanogenerators, and magnetic nanomaterials. Furthermore, two reviews present two kinds of two-dimensional (2D) materials, namely 2D amorphous nanomaterials and phosphorus-doped graphene in electrochemical energy conversion and storage. A brief overview of the published articles is presented in the following, and we hope to provide useful information for potential readers.

Lithium–sulfur (Li-S) batteries are receiving increasing attention as next-generation high-energy-density storage systems. However, currently, Li-S batteries suffer from low volumetric energy density and poor cycling stability due to the intrinsic low conductivity of sulfur and its discharge product, lithium polysulfide shuttle effect, and so on. Many strategies have been proposed to improve the performance of Li-S batteries, and one effective way is to design the "adsorptive-catalytic" cathode. Chen et al. [1] developed a three-dimensionally ordered macro/mesoporous Nb_2O_5/Nb_4N_5 through in situ nitridation to serve as a multi-functional sulfur host. The strong adsorption of Nb_2O_5 and high conductivity and catalytic activity of Nb_4N_5, combined with the porous structure, enable batteries with the $S/Nb_2O_5/Nb_4N_5$ cathode to exhibit excellent cycling stability and higher discharge capacity. Wang et al. [2] adopted similar concept to construct composite cathodes. They synthesized $TiN@C/S/Ta_2O_5$ with high sulfur fraction through a simple and low-cost co-precipitation method. Benefiting from the high conductivity of TiN, the strong adsorption of Ta_2O_5, and the micro- and mesoporous structure of the multidimensional carbon structure, the batteries with such cathode showed superior cycle stability and high areal capacity with a high sulfur utilization.

Lithium–oxygen (Li-O_2) batteries also have high theoretical capacity, but in practice their energy density is less than half the theoretical one, due to the great energy loss during the charging and discharging process. Designing highly active catalyst with low cost and great stability is one efficient way to tackle the problem and promote the practical use of Li-O_2 batteries. Li et al. [3] fabricated a $CoFe_2O_4/g-C_3N_4$ composite catalyst with $CoFe_2O_4$ particles supported on the flaky $g-C_3N_4$ using a scalable facile method. Both $CoFe_2O_4$ and $g-C_3N_4$ can provide reactive sites for the discharge–charge reaction, and the flaky

Citation: Zhang, Z.; Wang, R. Editorial for the Special Issue: "Advanced Nanomaterials for Electrochemical Energy Conversion and Storage". *Nanomaterials* **2022**, *12*, 3579. https://doi.org/10.3390/nano12203579

Received: 20 September 2022
Accepted: 28 September 2022
Published: 12 October 2022

Publisher's Note: MDPI stays neutral with regard to jurisdictional claims in published maps and institutional affiliations.

Copyright: © 2022 by the authors. Licensee MDPI, Basel, Switzerland. This article is an open access article distributed under the terms and conditions of the Creative Commons Attribution (CC BY) license (https://creativecommons.org/licenses/by/4.0/).

g-C_3N_4 with a high specific area and high chemical stability enabled a strong mass transfer ability and stable support for restraining the aggregation of $CoFe_2O_4$ particles. Under the synergistic effect of $CoFe_2O_4$ and g-C_3N_4, the Li-O_2 batteries exhibited improved capacity and stability.

Superconductors are another important energy storage devices with advantages of high-power density and long lifetime. One type of superconductor is the pseudo capacitor, where reversible faradic-type redox reactions occur at the electrode surface. Sylla et al. [4] prepared a composite electrode material for pseudo capacitors, where MoO_2 and Mo_2C nanostructures were incorporated into a peanut-shell-activated carbon (PAC) network via a one-step pyrolysis route. The composite combined the high specific area of PAC, the pseudocapacitive effect of MoO_2, and the superior conductivity and stability of the Mo_2C, thus delivering excellent capacitive performance when used as electrodes in a symmetric supercapacitor.

In the energy conversion devices, the properties of the catalyst determine their performance, and thus developing low-cost, highly active, and stable catalysts is of great significance to realize high-efficiency and high-selectivity energy conversion. Many nanomaterials exhibited various advantages as catalysts due to their high specific area which can offer abundant active sites. Carbon-based doped nanomaterials are some of most widely studied catalysts. Rong et al. [5] developed a self-template-assisted pyrolysis route to prepare a three-dimensional nanoporous carbon structure co-doped with N/P/Fe. The synthesis process is simple, avoiding the extra process of template removal, and the distribution and content of N/P/Fe can be well controlled. The prepared catalysts exhibited superior oxygen reduction reaction catalytic activity and durability compared to the currently advanced Pt/C catalysts.

Two-dimensional materials have also attracted extensive attention in the catalysis field. Graphene has a high specific surface area, high stability, and excellent conductivity, making it a promising candidate for catalyst. However, the pristine graphene is semimetal with no bandgap and shows poor catalytic activity. Heteroatom doping could greatly alter the electronic properties and improve the catalytic activity of graphene. Zhan et al. [6] reviewed recent advances in P-doped graphene electrocatalysts for the oxygen reduction reaction, including the synthesis and performance of the materials and catalytic mechanism. MXenes, as a new member of 2D materials, have also been demonstrated to have great potential in the electrocatalysis process. Luo et al. [7] studied how the Fe atom adsorption on pure or doped $Ti_3C_2O_2$ affected the catalytic performance of the nitrogen reduction reaction using the density functional theory (DFT)They found the charge transfer of the adsorbed Fe atoms to N_2 could promote the hydrogenation of N_2 and thus improve the catalytic performance. Two-dimensional materials supported single-atom catalysts which exhibit unique catalytic performance, represent one of the hot topics in the catalysis field. Zhao et al. [8] designed a single-atom catalyst Au/WSSe by filling the single Au atom at the S vacancy site in the Janus WSSe monolayer. By DFT calculation, they found the strong binding between the single Au atom and the WSSe resulted from the electron transfer and orbital hybridization between Au and W. Moreover, the tensile strain in the support could further improve the electrocatalytic performance in the hydrogen evolution reaction of the Au/WSSe catalyst. In fact, due to their unique structure and excellent properties, 2D materials have been widely investigated in electrochemical fields, not only the electrocatalyst, but also the batteries and supercapacitor. Liu et al. [9] focused on the 2D amorphous nanomaterials and summarized various regulation strategies, including composition and structure design, to enhance the electrochemical performance for the batteries, supercapacitors, and electrocatalysts.

The nanogenerator is a kind of mirco-nano device that can realize energy conversion. Since the first piezoelectric nano-generator (PENG) came onto the scene, the concept, mechanism, and applications of nanogenerators have been widely studied and made great progress. Zhu et al. [10] developed a self-powered sport sensor which was composed of a triboelectric nanogenerator (TENG), a PENG, and a flexible transparent stretchable self-healing hydrogel electrode. The prepared sensor is stretchable, wearable, and transparent, and can be used to monitor human three-dimensional motions.

Magnetic materials provide an alternate source of clean and renewable energy, and also play an irreplaceable role in energy conversion and storage. Researches on the physics and properties of magnetic materials are essential to realize their optimization for energy application. Gao et al. [11] prepared the ErFe$_x$Mn$_{12-x}$ series alloy samples with $\Delta x = 0.2$ using the arc melting method, and achieved a detailed magnetic phase diagram of the samples. They also studied the exchange bias effect and magnetocaloric effect in such a magnetic alloy. Nano-magnetic materials with small size and unique magnetic properties have played vital roles in many fields and their properties can be modulated by distinct strategies. Du et al. [12] developed a self-assembly method to prepare a single layer of aligned Co nanorods to obtain improved magnetic performance. They also studied the magnetic interaction of Co nanorods with different shapes, offering guidance to the magnetostatic interaction of shape anisotropic magnetic nanostructures. Gao et al. [13] investigated the magnetic and electronic properties of Janus MXene (M'MCO$_2$, M' and M = V, Cr, and Mn) via first-principles calculations and they found that the transition metal and configuration could tune the band gap, the magnetic ground state, and the net output magnetic moments of the Janus MXene materials. The work points out a new path to design and regulate novel magnetic materials. Cui et al. [14] synthesized laterally asymmetric heavy metal (HM)/ferromagnetic metal (FM) multilayers by growing the FM layer in a wedge shape, and studied field-free spin–orbit torques (SOTs) switching in the asymmetric multilayers. They found that the switching efficiency strongly depended on the HM/FM interface and the FM layer thickness.

In this Special Issue, some original research works and high-quality reviews on the advanced nanomaterials for energy conversion and storage are presented, and we hope that these articles prove informative and instructive for readers. The research in this field is booming and great advances in the development of efficient, clean, and sustainable energy devices and technologies are expected.

Funding: This work was supported by the Beijing Natural Science Foundation (No. 2212034), the National Natural Science Foundation of China (Nos. 51971025 and 12034002), and the Fundamental Research Funds for the Central Universities (Nos. 06108248 and 06500235).

Conflicts of Interest: The authors declare no conflict of interest.

References

1. Chen, H.; Wang, J.; Zhao, Y.; Zeng, Q.; Zhou, G.; Jin, M. Three-Dimensionally Ordered Macro/Mesoporous Nb$_2$O$_5$/Nb$_4$N$_5$ Heterostructure as Sulfur Host for High-Performance Lithium/Sulfur Batteries. *Nanomaterials* **2021**, *11*, 1531. [CrossRef] [PubMed]
2. Wang, C.; Lu, J.H.; Wang, Z.L.; Wang, A.B.; Zhang, H.; Wang, W.K.; Jin, Z.Q.; Fan, L.Z. Synergistic Adsorption-Catalytic Sites TiN/Ta$_2$O$_3$ with Multidimensional Carbon Structure to Enable High-Performance Li-S Batteries. *Nanomaterials* **2021**, *11*, 2882. [CrossRef] [PubMed]
3. Li, X.; Zhao, Y.; Ding, L.; Wang, D.; Guo, Q.; Li, Z.; Luo, H.; Zhang, D.; Yu, Y. Enhancing the Capacity and Stability by CoFe$_2$O$_4$ Modified g-C$_3$N$_4$ Composite for Lithium-Oxygen Batteries. *Nanomaterials* **2021**, *11*, 1088. [CrossRef] [PubMed]
4. Sylla, N.F.; Sarr, S.; Ndiaye, N.M.; Mutuma, B.K.; Seck, A.; Ngom, B.D.; Chaker, M.; Manyala, N. Enhanced Electrochemical Behavior of Peanut-Shell Activated Carbon/Molybdenum Oxide/Molybdenum Carbide Ternary Composites. *Nanomaterials* **2021**, *11*, 1056. [CrossRef] [PubMed]
5. Rong, Y., Huang, S. Self Templating Synthesis of N/P/Fe Co-Doped 3D Porous Carbon for Oxygen Reduction Reaction Electrocatalysts in Alkaline Media. *Nanomaterials* **2022**, *12*, 2106. [CrossRef] [PubMed]
6. Zhan, X.; Tong, X.; Gu, M.; Tian, J.; Gao, Z.; Ma, L.; Xie, Y.; Chen, Z.; Ranganathan, H.; Zhang, G.; et al. Phosphorus-Doped Graphene Electrocatalysts for Oxygen Reduction Reaction. *Nanomaterials* **2022**, *12*, 1141. [CrossRef] [PubMed]
7. Luo, H.; Wang, X.; Wan, C.; Xie, L.; Song, M.; Qian, P. A Theoretical Study of Fe Adsorbed on Pure and Nonmetal (N, F, P, S, Cl)-Doped Ti$_3$C$_2$O$_2$ for Electrocatalytic Nitrogen Reduction. *Nanomaterials* **2022**, *12*, 1081. [CrossRef] [PubMed]
8. Zhao, S.; Tang, X.; Li, J.; Zhang, J.; Yuan, D.; Ma, D.; Ju, L. Improving the Energetic Stability and Electrocatalytic Performance of Au/WSSe Single-Atom Catalyst with Tensile Strain. *Nanomaterials* **2022**, *12*, 2793. [CrossRef] [PubMed]
9. Liu, J.; Hao, R.; Jia, B.; Zhao, H.; Guo, L. Manipulation on Two-Dimensional Amorphous Nanomaterials for Enhanced Electrochemical Energy Storage and Conversion. *Nanomaterials* **2021**, *11*, 3246. [CrossRef] [PubMed]

10. Zhu, Y.; Sun, F.; Jia, C.; Zhao, T.; Mao, Y. A Stretchable and Self-Healing Hybrid Nano-Generator for Human Motion Monitoring. *Nanomaterials* **2021**, *12*, 104. [CrossRef] [PubMed]
11. Gao, P.; Xia, Y.; Gong, J.; Ju, X. Structure and Magnetic Properties of ErFe$_x$Mn$_{12-x}$ ($7.0 \leq x \leq 9.0$, $\Delta x = 0.2$). *Nanomaterials* **2022**, *12*, 1586. [CrossRef] [PubMed]
12. Du, H.; Zhang, M.; Yang, K.; Li, B.; Ma, Z. A Self-Assembly of Single Layer of Co Nanorods to Reveal the Magnetostatic Interaction Mechanism. *Nanomaterials* **2022**, *12*, 2499.
13. Gao, P.; Song, M.; Wang, X.; Liu, Q.; He, S.; Su, Y.; Qian, P. Theoretical Study on the Electronic Structure and Magnetic Properties Regulation of Janus Structure of M′MCO$_2$ 2D MXenes. *Nanomaterials* **2022**, *12*, 556. [CrossRef] [PubMed]
14. Cui, B.; Zhu, Z.; Wu, C.; Guo, X.; Nie, Z.; Wu, H.; Guo, T.; Chen, P.; Zheng, D.; Yu, T.; et al. Comprehensive Study of the Current-Induced Spin-Orbit Torque Perpendicular Effective Field in Asymmetric Multilayers. *Nanomaterials* **2022**, *12*, 1887. [CrossRef] [PubMed]

Review

Phosphorus-Doped Graphene Electrocatalysts for Oxygen Reduction Reaction

Xinxing Zhan [1], Xin Tong [1,2,*], Manqi Gu [1], Juan Tian [1], Zijian Gao [1], Liying Ma [1], Yadian Xie [2], Zhangsen Chen [3], Hariprasad Ranganathan [3], Gaixia Zhang [3] and Shuhui Sun [3,*]

[1] School of Chemistry and Material Science, Guizhou Normal University, Guiyang 550001, China; zhanxinxing@gznu.edu.cn (X.Z.); gumanqi@gznu.edu.cn (M.G.); tianjuan@gznu.edu.cn (J.T.); 20010080258@gznu.edu.cn (Z.G.); maliying@gznu.edu.cn (L.M.)

[2] Key Laboratory of Low-Dimensional Materials and Big data, Guizhou Minzu University, Guiyang 550025, China; xieyadian@gzmu.edu.cn

[3] Centre Énergie, Matériaux et Télécommunications, Institut National de la Recherche Scientifique (INRS), 1650 Boulevard Lionel-Boulet, Varennes, QC J3X 1P7, Canada; zhangsen.chen@inrs.ca (Z.C.); hariprasad.ranganathan@inrs.ca (H.R.); gaixia.zhang@inrs.ca (G.Z.)

* Correspondence: tongxin@gznu.edu.cn (X.T.); shuhui.sun@inrs.ca (S.S.)

Abstract: Developing cheap and earth-abundant electrocatalysts with high activity and stability for oxygen reduction reactions (ORRs) is highly desired for the commercial implementation of fuel cells and metal-air batteries. Tremendous efforts have been made on doped-graphene catalysts. However, the progress of phosphorus-doped graphene (P-graphene) for ORRs has rarely been summarized until now. This review focuses on the recent development of P-graphene-based materials, including the various synthesis methods, ORR performance, and ORR mechanism. The applications of single phosphorus atom-doped graphene, phosphorus, nitrogen-codoped graphene (P, N-graphene), as well as phosphorus, multi-atoms codoped graphene (P, X-graphene) as catalysts, supporting materials, and coating materials for ORR are discussed thoroughly. Additionally, the current issues and perspectives for the development of P-graphene materials are proposed.

Keywords: doped graphene; oxygen reduction reaction; phosphorus-doped; codoped

1. Introduction

The energy conversion technique is one of the most important parts that significantly influence the development of human society. The conversion efficiency of converting chemical energy into electrical energy is many folds higher than that of conversion of chemical energy into thermal (e.g., combustion) or kinetic energies (e.g., modern IC engines). Accordingly, great attention has been given to the study of fuel cells and metal-air batteries as the next-generation energy conversion techniques because these devices can directly convert chemical energy into electric energy with nearly zero carbon footprint [1–9]. In fuel cells and metal-air batteries, the electrochemical reactions, in particular, the oxygen reduction reaction (ORR) in the cathode, determine the performance [10,11]. Hence, a suitable catalyst is highly needed to boost the kinetics and reduce the overpotential. Thus far, the noble metal platinum (Pt)-based materials have proven to be state-of-the-art catalysts for ORR [12,13]. However, the commercial application of Pt-based catalysts is hampered by their prohibitive cost, limited supply, and unsatisfactory durability. Therefore, developing cheap and earth-abundant alternatives with outstanding activity and excellent durability is of significance [14–18].

Graphene-based materials have attracted extensive interest in the field of electrocatalysis due to their fascinating mechanical, electronic, and thermal properties [19–23]. In particular, the high surface and the high conductivity are beneficial to the mass transfer and the electron transfer, thus increasing the activity of heterogeneous catalysis [17].

However, the pristine graphene has no intrinsic bandgap and shows poor electrocatalytic activity [24]. It is necessary to change the electronic density of the graphene sheet to modulate its electrochemical properties. When the carbon atoms in the graphene lattice are partially substituted by heteroatoms, the structure and electronic properties of the heteroatoms-doped graphene can be dramatically altered by the specific characters of the dopants. Usually, the neighboring elements of the carbon, such as nitrogen(N), boron(B), oxygen(O), phosphorus(P), sulfur(S), and fluorine(F), are preferred as the dopants for graphene [25,26]. Depending on their atom size, electronegativity, and doping level, the adsorption behavior, the chemisorption energy, and the active sites of the doped graphene can be modulated.

Among these elements, nitrogen is by far the most studied dopant because of its similar atomic size to carbon and slightly higher electronegativity than carbon. The ORR performance can be remarkably enhanced in nitrogen-doped graphene (N-graphene), which has been demonstrated and reviewed in previous works [27–29]. Besides, many previous articles reviewed the synthesis methods and the application of different heteroatoms-doped graphene (X-graphene) [25,30,31]. However, a limited number of articles focused on phosphorus-doped graphene (P-graphene). Therefore, to figure out the internal decisive factor of doped-graphene to affect the ORR performance, intensive research on P-graphene is indispensable.

Phosphorus is located on the periodic table in Period 3 and Group 5. Like its congener, nitrogen, phosphorous has five electrons in the valence shell. The geometric and electronic effects of phosphorus doping differ from nitrogen doping because of its large atom size and low electronegativity compared to both nitrogen and carbon. As seen from the STEM images in Figure 1, there is a protrusion of the phosphorus atom above the plane of the graphene lattice in P-graphene because the atom size of phosphorus (98 pm) is larger than that of carbon (67 pm) [32]. The C–P bond and the nearest C–C bond are elongated, compared to the pristine C–C bond. Although nitrogen has the same valence number as phosphorus, the difference between P-graphene and N-graphene is distinct. In graphene, the pyramidal-like bonding phosphorus can lead to the transformation of sp^2 hybridized carbon to sp^3 state. These lattice distortion-induced defects can serve as the active site for ORR. Moreover, the phosphorus in P-graphene could act as a donor and show an n-type nature because of its weak attraction with the valence electrons in the M-shell. To further confirm the electronic structure difference between P-graphene and pristine graphene, the density functional theory (DFT) simulations are conducted. As in the electrostatic potential map in Figure 2a, the pristine graphene shows electroneutrality except for the unsaturated carbon atoms at the edge. In this case, the undoped graphene affords limited active sites and thus poor activity towards ORR. When the phosphorus is doped into graphene, as shown in Figure 2b, it could cause the redistribution of the surface charge. The carbon atoms near phosphorus are negatively charged. As the coordination environment changed, the adsorption behavior of the oxygen molecule and the binding strength of the intermediate species are changed [33]. Thus, it is interesting and worthwhile to carry out an in-depth study on how phosphorus impacts the catalytic performance toward ORR.

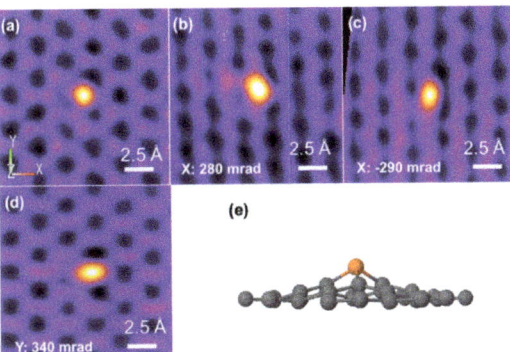

Figure 1. (**a**–**d**) Atomically resolved STEM of P-graphene with different angles along the X-axis and Y-axis (**e**) The experimental three-dimensional model of P-graphene. Reproduced with permission from [32]. Copyright © 2022, American Chemical Society.

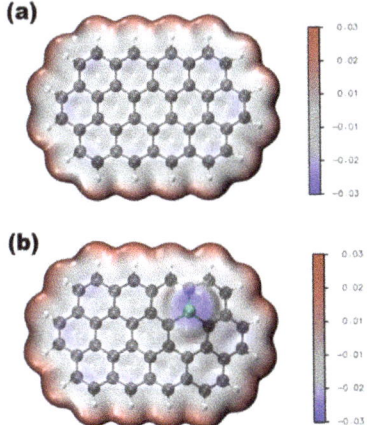

Figure 2. Electrostatic potential map of (**a**) pristine graphene, (**b**) P-graphene.

Accordingly, in this review, we discuss the recent studies of phosphorus-doped graphene, including phosphorus mono-doped (P-) and phosphorus with other elements codoped (P, X-) graphene. Herein, the works reported on phosphorus-doped graphene served as metal-free catalysts, supporting materials, and coating materials are reviewed. The synthesis methods, catalytic mechanism, and ORR performance are systematically summarized. The synergistic effect of inducing defects and electron distribution on the ORR activity and durability are also discussed. Finally, the current challenges and future perspectives in the field of P-graphene ORR catalysts are proposed.

2. The Synthesis of Phosphorus-Doped Graphene

It is well known that there are a large number of synthesis strategies to obtain graphene. According to the formation mechanism and the source of the carbon (small carbon molecule or graphite), the synthesis strategies can be divided into two types: top-down and bottom-up. The intrinsic properties and electronic structure of as-prepared doped graphene largely depend on the pristine graphene. Therefore, we divide these doping methods into two categories, as shown in Figure 3. In those strategies, if the heteroatoms partially replace some of the carbon atoms in the carbon skeleton simultaneously during the formation of graphene, it can be called in situ doping. If the doping is achieved based on the as-prepared graphene with dopant-containing precursors, it can be called post-treatment doping.

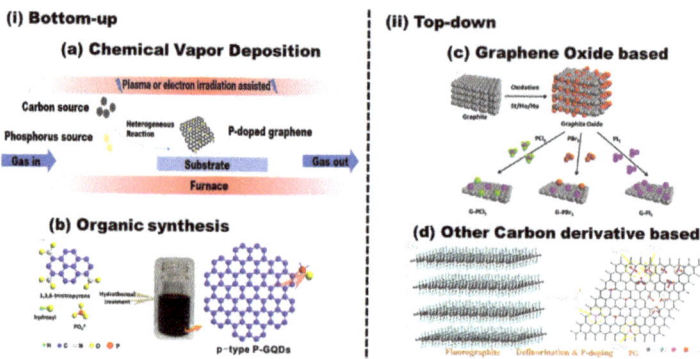

Figure 3. The synthesis routes of phosphorus-doped graphene: (**i**) bottom-up: (**a**) chemical vapor deposition (CVD) and (**b**) organic synthesis; reproduced with permission from [34] © 2022, American Chemical Society. (**ii**) Top-down: (**c**) graphene oxide-based method; reproduced with permission from [35] © 2022 Wiley-VCH VerlagGmbH & Co. KGaA, Weinheim; (**d**) other carbon derivative-based method; reproduced with permission from [36]; copyright © 2022, American Chemical Society.

2.1. Bottom-Up Approaches

The in situ doping is mainly based on the bottom-up pathway to produce graphene from small carbon molecules. Theoretically, the structure of the doped graphene can be fabricated on an atomic level. The doped graphene with high purity and fine structure can be obtained. However, this kind of method suffers from low yield, high cost, and difficulties in scaling up. Moreover, phosphorus is not easy to be introduced into the carbon skeleton because of its larger atom size than that of carbon. Engineering-desired, doped structures remain a challenge and detailed theoretical and experimental investigations are warranted.

2.1.1. Chemical Vapor Deposition (CVD)

Chemical vapor deposition (CVD) is one of the most used technologies to produce low-dimensional materials, such as carbon nanotubes and graphene. It is based on the chemical reaction of the precursors in the vapor phase [37]. As shown in Figure 3a, the carbon source and phosphorus source in the gas phase are transported into the reactor to form thin P-graphene film on the heated substrate via homogeneous reactions. In L.G. Bulusheva's work [38], P-graphene was deposited on copper substrates under low pressure of a gaseous mixture of $CH_4/PH_3/H_2$, and the phosphorus content is about 0.1%. In the prepared P-graphene, the phosphorus atoms are located at the edge of graphene as phosphorus oxides, such as C–P=O or P–Ox. Triphenylphosphine (TPP) [39,40] can also be used as phosphorus sources. The nucleation, growth, and coalescence behavior of the P-graphene is largely determined by the temperature, the surface chemistry of the substrate, the pressure, and the precursor used.

After the graphene is grown on the catalyst surface, additional energy from external sources, including heating [41], ions implantation [42], electron irradiation [43], and laser direct writing [44], is essential to break the chemical bond to achieve doping. The phosphorus precursor can be triphenylphosphine [43], triphenylamine [41], P-ions [42], and phosphoric acid [44]. Depending on the doping method and the precursor used, a large range of P content can be obtained, 5.56% in [42], 4.96% in [41], 0.92% in [44], and 0.26% in [43] (based on XPS results). Interestingly, the phosphorus in those P-graphenes shows two kinds of bonding configurations, P–C and P–O [42,44].

Although high-quality doped graphene films can be obtained by CVD, complicated procedures, such as substrate pretreatment, or protection layer coating are needed. How to transfer the doped graphene from the catalyst surface onto arbitrary substrates without damaging the graphene is still a million-dollar question. Furthermore, the precise controlling of the P–C bond configuration is desired for further study.

2.1.2. Organic Synthesis

The P-doping can also be achieved simultaneously during the formation of graphene by stepwise solution chemistry of organic precursors. As shown in Figure 3b, Na_2HPO_4 was used as a phosphorus source [34]. After hydrothermal treatment, the phosphorus in graphene mainly exists as CPO_3/C_2PO_2 and C_3PO structures. Similar methods were used and the tetrakis(hydroxymethyl)phosphonium chloride [45], Na_2HPO_4 [46], and sodium phytate [47] were used as phosphorus sources. Although this kind of method can effectively control the morphology and structures of the as-prepared products, the sizes of the P-graphenes are usually very small, called P-graphene dots. Their applications mainly focus on luminescence-related fields, such as photodegradation and photocatalyst.

2.2. *Top-Down Approaches*

Top-down approaches mean the doped graphene comes from the graphite and graphite derivatives. To extract the graphene from graphite and graphite derivatives, different methods, such as sonication and hydrothermal, were used. The precursor mainly determines the structure and intrinsic properties of the as-prepared doped graphene. Graphite oxide (GO) is the most used precursor for doped graphene. Thus, we devote a separate chapter to this method. These methods are always cost-effective, easy to scale up, and have high yields. Yet, it is difficult to precisely control the configuration of the doped graphene.

2.2.1. Graphite Oxide-Based Routine

Graphite oxide (GO), which comes from the oxidation of layered graphite mainly by a modified hummers method, is a promising precursor and versatile building block for chemically converted graphene in large volumes owing to its tailorable surface chemistry [48]. After oxidation, the space between GO layers is increased, compared to that of the graphite, because of the repulsive force between hydrophilic oxygen-containing functional groups, such as peroxy, hydroxyl, epoxy, and carboxyl groups on the surface. The GO can be easily dispersed in water and conduct chemical modification further [49]. The larger number of oxygen functional groups and defects on graphene can also play a critical role in heteroatoms doping. Therefore, the GO-based routine is by far the most used method to get heteroatoms-doped graphene, such as N-graphene, P-graphene, B-graphene, and P, N-graphene [50–52].

P-graphene can be obtained by simply mixing the phosphorus source with GO in solution [53]. The XPS results show that the leading P components are the P–O bond without reduction. To prepare the doped graphene with less oxygen content, the thermal treatment was conducted. Different chemicals, such as triphenylphosphine [54],1-butyl-3-methylimidazolium hexafluorophosphate [55], phosphoric acid [56], phytic acid [57], and tetrabutylphosphonium bromide [58], were used as phosphorus precursors. Usually, the mixture of GO and phosphorus precursors was annealed in high temperatures (more than 800 °C) under an inert atmosphere. In addition, a hydrothermal or solvothermal procedure can also be added to build three-dimensional (3D) P-graphene hydrogel [58–61]. After mixing the GO with phosphorus precursors, the solution was transferred to a Teflon-lined autoclave at high temperatures (180–220 °C); then, the annealing process was carried out. Additionally, the reduced GO with less residual oxygen functionalities was also used to prepare P-graphene [62].

Moreover, different combinations of phosphorus sources and heteroatom sources are used to prepare phosphorus and heteroatoms codoped graphene, such as P, F-graphene, P, N-graphene, and P, S-graphene [35,63–68]. As shown in Figure 3c, different phosphorus-

and halogen-codoped graphene were obtained by refluxing the mixture containing phosphorus and halogen precursor. Three kinds of GO were synthesized by different methods [35]. The doping level of phosphorus and halogen depends on the type of GO. GO prepared by Hummers shows large amounts of dopant.

Because of the abundant oxygen-containing functional groups on GO, there is a considerable amount of oxygen content in the final P-graphene. Thus, there should be different P–O bonds and P–C bonds, such as C_3–P=O, C–P–O, C–O–P. The presence of these phosphorus configurations plays a key role in electrochemical performance. We will discuss these impacts later in this article.

2.2.2. Other Graphite Derivative-Based Routine

Graphite is thousands of graphene layers stacked together. The graphite was also used as the starting material to produce P-graphene by an electrochemical exfoliated method [69,70]. Because of the narrow space between graphite, the expanded graphite layer should be prepared by the intercalation of the solvent molecule. Then, the introduction of phosphorus can be achieved after the subsequent erosion/expansion process. Phosphoric acid (H_3PO_4) was used as a phosphorus source and electrolyte. Ball milling of graphene stack and red phosphorus were also used to produce P-graphene [71]. Generally, using graphite as the starting material, the phosphorus content in P-graphene can reach less than 1% (0.68% in [69], up to 0.7% in [70], 0.91% in [71]).

To enhance the phosphorus-doping content, fluorography was used [36]. As shown in Figure 3d, there are many vacancies in the graphene sheets, and the space between the graphene layer is large. Thus, it is easy to introduce phosphorus atoms, and the phosphorus content in P-graphene can reach 6.40% by thermal annealing in red phosphorus vapor.

3. Phosphorus-Doped Graphene for Metal-Free ORR Electrocatalyst

Metal-free catalyst based on carbon nanomaterials has been studied intensively as an emerging class of ORR catalysts since 2009 [72]. Recently, this kind of catalyst has been demonstrated to be a promising alternative ORR catalyst with low cost and high efficiency, as well as tolerance to methanol crossover and carbon-monoxide-poisoning effects. Although there is an argument that the metal-free catalyst strictly does not exists because of the residual metal on the catalyst, the active sites are mainly proven to be the adjacent area of the heteroatom. The doping-induced charge transfer in doped carbon material takes a crucial role to modulate the oxygen chemisorption mode and weaken the O–O bonding to facilitate the ORR [31,73,74]. Nitrogen or phosphorus single doping graphene was used in earlier research. Then, numerous works have been devoted to codoping graphene with a more complicated system.

3.1. Phosphorus-Doped Graphene

The P-doped graphene made its debut as an ORR catalyst in 2013. Wei's group simply annealed the mixture of GO and 1-butyl-3-methylimidazolium hexafluorophosphate to achieve in situ P-doping of reduced graphene oxide (RGO) [55]. The phosphorus atoms were incorporated into the carbon frame with the formation of the C–P bond and the P–O bond, with the amount of phosphorus in P-graphene being 1.16%. The oxygen atoms are bonded with phosphorus atoms as C_3PO, C_2PO_2, and CPO_3. Thus, the polarized phosphorus serves as a bridge between the oxygen atoms and the carbon atoms. The carbon atom that bonded with a phosphorus atom and oxygen atoms is positively charged. Therefore, the active sites are these positively charged carbon atoms. P-graphene shows a one-step 4 e$^-$ pathway and comparable ORR catalytic performance than that of the Pt/C catalyst. As shown in Figure 3, a similar method using GO and triphenylphosphine (TPP) as a precursor was used, and the phosphorus content is 1.81%. The TEM image in Figure 4a shows a typical crumpled surface. Besides, there are C–P bonds and the P–O bonds in P-graphene, as displayed in Figure 4b. The onset potential of P-graphene is 0.92 v vs. RHE and shows a combination of the 2 e$^-$ and 4 e$^-$ pathways (Figure 4c). Moreover, the prepared P-graphene (PG) maintains better stability than that of the Pt/C (Figure 4d). With the change of phosphorus source, similar methods thus show distinct results.

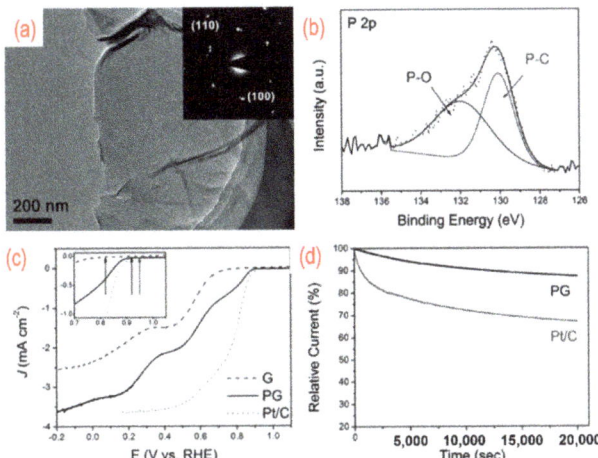

Figure 4. (**a**) The TEM image (the inset is the corresponding SAED pattern), (**b**) the high-resolution P 2p peaks, (**c**) the activity test (LSV curves recorded by RDE in O_2-saturated 0.1 M KOH solution with 1600 rpm at a scanning rate of 10 mV s^{-1}), (**d**) the stability test (current-time chronoamperometric in O_2-saturated 0.1 M KOH solution) of P-graphene. Reproduced with permission from [54]. Copyright © 2022 WILEY-VCH Verlag GmbH & Co. KGaA, Weinheim.

Further theoretical and experimental works were conducted (some typical experimental works about P-graphene for ORR have been listed in Table 1) [75–79]. In Yang's work, a phosphorus atom is believed to be the active site for ORR. The ORR could process by an indirect 2 e-pathway and the *OOH is intermediate [75]. Phosphorus-doped divacancy graphene was also used to study the reaction mechanism by Bai and co-workers. The ORR can take place on the phosphorus atom and its adjacent carbon atoms and demonstrates the 4 e-process [76]. As shown in Figure 5, four different models of P-doped graphene were considered to identify the relationships between chemical-bonding states and ORR performance by Wei's group [80]. In their research, the impact of the oxygen, which can be doped into graphene lattice during the synthesis and electrochemical process were studied in which two possible structures, such as the OPC3G and PC4G, show the best ORR activity. The following simulation results show that it is difficult to directly form the PC4G structure while the OPC3G structure is more energetically favorable. The active sites for P-graphene could be the negatively charged carbon atoms in OPC3G. However, some scholars asserted that the positively charged phosphorus atoms can be regarded as active sites [25]. In summary, the ORR performance of P-doped graphene is attributed to many complicated factors, depending on the chemistry state of the precursor and synthesis method. It is still a long process to identifying the real mechanism of phosphorus-doped graphene for ORR.

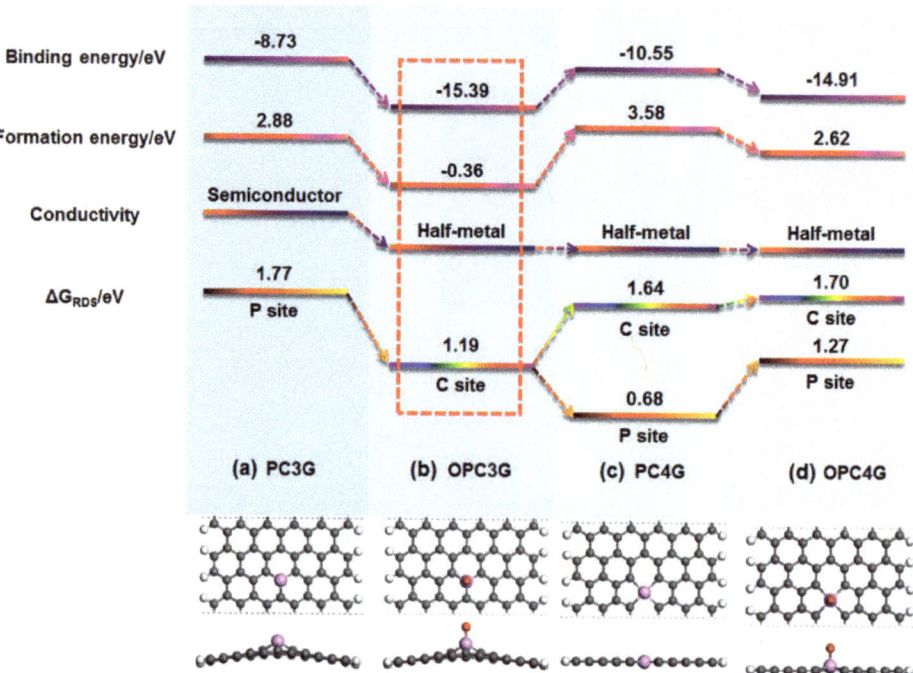

Figure 5. The simulation of different models of P-doped graphene for ORR performance. Reproduced with permission from [80]. Copyright © 2022, American Chemical Society.

Table 1. Summary of some typical works dedicated to P-graphene for ORR.

The Material [1]	Synthesis Method	P-Content (at.%) [2]	Onset Potential	Half-Wave Potential	Electron Transfer Number	Ref.
P-graphene	Annealing the mixture of GO and 1-butyl-3-methlyimidazolium	1.16%	−0.0261 V vs. SCE in 0.1 M KOH	~−0.2 V vs. SCE	3.9	[55]
P-graphene	Annealing the mixture of GO and triphenylphosphine	1.81%	0.92 V vs. RHE in 0.1 M KOH	-	3.0–3.8	[54]
P-graphene	Supercritical fluid processing of GO and triphenylphosphine	1.4–3.2% by EDX	0.12 V vs. MMO in 0.1 M KOH	-	-	[78]
P-graphene	Immersed into the mixture of GO and NaH$_2$PO$_4$	2.6%	~−0.3 V vs. Ag/AgCl in 0.1 M KOH	-	3.9	[79]
P, N-graphene	Pyrolysis of hexachlorocyclotriphosphazene (HCCP) and GO	1.08%	−0.20 V vs. Ag/AgCl	-	3.4–3.73	[81]

Table 1. Cont.

The Material [1]	Synthesis Method	P-Content (at.%) [2]	Onset Potential	Half-Wave Potential	Electron Transfer Number	Ref.
P, N-graphene	Twice pyrolysis treatment of GO and phosphoric acid	-	0.87 V vs. RHE in 0.1 M HClO4	0.64 V vs. RHE	-	[81]
P, N-graphene	Pyrolysis treatment of GO, polyaniline, and phytic acid	1.72%	1.01 V vs. RHE	~0.84 V vs. RHE	3.96	[82]
P, N-graphene	Two-step solution process using phytic acid and GO	0.6%	0.89 V vs. RHE	0.69 V vs. RHE	3.9	[83]
P, N-graphene	Pyrolysis treatment of GO and phytic acid	-	−0.11 V vs. Ag/AgCl	−0.34 V vs. Ag/AgCl	-	[84]
P, N-graphene	Hydrothermal and subsequent pyrolysis processes (GO and phytic acid)	1.22%	0.983 V vs. RHE	0.865 V vs. RHE	3.9–4.0	[85]
P, N-graphene	Pyrolysis treatment of phytic acid	0.67–0.71%	1.0 V vs. RHE	0.86 V vs. RHE	-	[86]
P, N-graphene	Pyrolysis treatment of GO and diammonium hydrogen phosphate	1.16%	~−0.2 V vs. Ag/AgCl	~−0.18 V vs. Ag/AgCl	3.66	[87]
P, N-graphene	Pyrolysis treatment of GO and diammonium phosphate	2.32%	~0.84 V vs. RHE	~0.87 V vs. RHE	3.99	[88]
P, Fe-graphene	Sol-gel polymerization and pyrolysis process	~2%	−0.139 V	-	3.74–3.89	[89]
P, Fe-graphene	Pyrolysis treatment of GO, phytic acid and FeCl$_2$	0.84%	−0.05 V vs. Ag/AgCl in 0.5 M H$_2$SO$_4$	-	3.84	[90]
P, Co-graphene	Pyrolysis treatment of GO, tetraphenylphosphonium bromide, Co(NO$_3$)$_2$	0.639%	0.89 V vs. RHE	~0.78 V vs. RHE	3.87–3.96	[91]
N, P, F-graphene	Pyrolysis of the mixture of GO, polyaniline, and ammonium hexafluorophosphate	0.37%	~0.83 V vs. RHE	~0.72 V vs. RHE	3.85	[92]
P, S, N-graphene	Pyrolysis treatment of GO and acephate	0.42%	−0.192 V vs. SCE	-	2.99	[93]
P, S, N-graphene	Pyrolysis treatment of GO and phosphoric acid	-	−0.052 V vs. SCE	0.015 V vs. SCE	3.67–3.97	[94]
Fe, B, N, S, P-graphene	Pyrolysis treatment of GO and triphenylphosphine	0.54%	1.06 V vs. RHE	0.9 V vs. RHE	3.98	[95]
P, B, N-graphene	Hydrothermal method (GO and boron phosphate)	-	−0.12 V vs. Ag/AgCl	-	3.7	[96]
P, B, N-graphene	Pyrolysis treatment of GO and phenylphosphine	0.43%	0.88 V vs. RHE	0.80 V vs. RHE	3.8	[97]
P, Fe, N-graphene	Pyrolysis treatment of aphytic acid	1.11%	~0.95 V vs. RHE	0.84 V vs. RHE	3.2	[98]
P, Ni, N-graphene	Pyrolysis treatment of GO and phytic acid	-	0.88 V vs. RHE	-	3.5	[99]
P, S, N-graphene	Pyrolysis treatment of ammonium monohydrogenphosphate	0.95%	0.856 V vs. RHE	0.74 V vs. RHE	3.07	[100]

Table 1. Cont.

The Material [1]	Synthesis Method	P-Content (at.%) [2]	Onset Potential	Half-Wave Potential	Electron Transfer Number	Ref.
P, S, N-graphene	Ball milling and pyrolysis treatment of phosphonitrilic chloride trimer	1.16%	~0.93 V vs. RHE	0.88 V vs. RHE	-	[101]
MoP_x @MnP_y /P, N-graphene	Annealing the mixture of GO and the desired chemical	4.12%	0.965 V vs. RHE in 0.1 M KOH	0.842 V vs. RHE	3.95–3.97	[102]
$CoMn_2O_4$/P, N-graphene	Hydrothermal method and soaking hypophosphorous acid	1.22%	−0.094 V vs. SCE in 0.1 M KOH	−0.2 V vs. SCE	3.64–3.70	[103]
Co/P, N-graphene	A hydrothermal method with the subsequent pyrolysis procedure	0.83% using elemental analysis	0.04 V vs. SCE	0.18 V vs. SCE	~4	[104]
Co_2P/Co, P, N-graphene	Supramolecular gel-assisted strategy and annealing method	2.86%	0.90 V vs. RHE	0.81 V vs. RHE	3.96	[105]
$Co_3(PO_4)_2$/P, N-graphene	Hydrothermal and annealing the mixture of desired chemical and phytic acid	-	0.95 V vs. RHE in 0.1 M KOH	0.81 V vs. RHE	3.7–3.84	[106]
Cu_3P@ P, N-graphene	Annealing the mixture of desired chemical and 1-hydroxyethylidene1,1-diphosphonic acid	-	-	0.78 V vs. RHE	3.96–4.0	[107]
Co@N, P, S -graphene	Thermal treatment of the mixture of desired chemical and kelp	-	0.90 V vs. RHE	0.74 V vs RHE	4.0	[108]
FeCo@P, N-graphene/N-CNTs	Thermal treatment of the mixture of desired chemical and polystyrene spheres	2.77%	0.95 V vs. RHE	-	3.67–3.82	[109]
FeP@P-graphene	Annealing of the mixture of hemin diammonium phosphate and melamine	1.1%	0.95 V vs. RHE in 0.1 M KOH	0.81 V vs. RHE	3.8	[110]

Notes: [1] X/Y means Y-supported X. For instance, Co/P, N-graphene means P, N-doped graphene-supported Co. X@Y means X coating with Y. For instance, Co@N, P, S -graphene means Co coating with N, P, S-graphene. [2] The content is based on the XPS results unless the exception is in the chart.

3.2. P, N Codoped Graphene

Codoping of the heteroatoms, such as nitrogen and boron, has been proven to improve the ORR performance. Among them, P, N-graphene has become a hotspot of research and shows superior electrocatalysts performances. As shown in Figure 6, L. Dong et al. reported on N, P- graphene that was prepared by the pyrolysis of hexachlorocyclotriphosphazene (HCCP) and GO [75]. Because the P–Cl in $N_3P_3Cl_6$ with exatomic ring structure is prone to break under high temperature, the residual active groups containing N and P atoms can be incorporated into the graphene. The morphology of P, N-graphene is similar to pristine graphene and the N and P atoms are homogeneously distributed on the graphene sheets, as shown in Figure 6b. In the XPS spectra of the as-obtained P, N-graphene, three chemical bonds, such as C–P, C–PO_3, and P–O, are found. The formation of P–O is related to the active oxygen released from GO under high temperatures. The content of C–P increases with the increase of the annealing temperature. The annealing temperature was adjusted to estimate the effect on ORR. The onset potential of optimized P, N-graphene (−0.20 V vs./AgCl) annealed under 1000 °C is close to that of the commercial Pt/C (−0.02 V vs./AgCl) and shows better electrochemical stability (Figure 6c,d). The best catalyst with the highest annealing temperature can result in a high degree of graphitization and excellent conductivity. Moreover, the synergistic effects of N, P-codoping and the suitable ratio of nitrogen and phosphorus can produce the asymmetrical spin and charge density, thus enhancing the catalytic activity. In an acidic environment, the P, N-graphene shows moderate ORR performance [111]. The onset potential of the P, N-graphene is 0.87 v vs. RHE, and the limiting current is only around 3.5 mA cm^{-2}. Lately, phytic acid [82–86], ammonium hydrogen phosphate [87], and diammonium phosphate [88] were also used as the phosphorus source to obtain N, P-graphene.

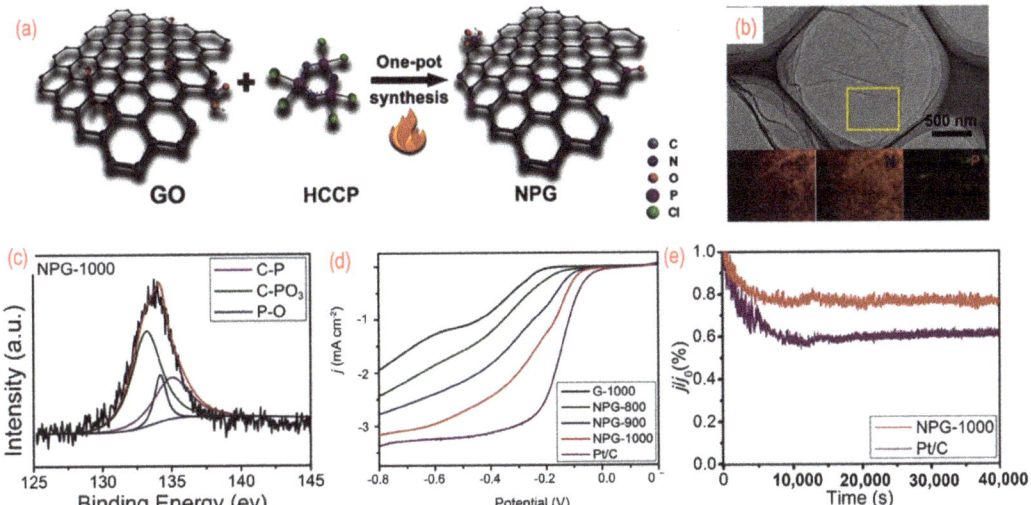

Figure 6. (**a**) The schematic of the synthesis process, (**b**) the TEM image (the inset is C-, N-, and P-elemental mappings), (**c**) the high-resolution P 2p peaks, (**d**) the activity test (LSV curves recorded by RDE in O$_2$-saturated 0.1 M KOH solution with 1600 rpm at a scanning rate of 10 mV s^{-1}), (**e**) the stability test (current-time chronoamperometric in O$_2$-saturated 0.1 M KOH solution) of P, N-graphene. Reproduced with permission from [81]. © 2022 WILEY-VCH Verlag GmbH & Co. KGaA, Weinheim.

To clarify the real active sites in P, N-graphene, the theoretical works were conducted [88,112,113]. As shown in Figure 7, the free energy of different P-containing structures for ORR are calculated by DFT simulation. The active sites of P, N- graphene are the edge carbon atoms next to the N–P bond as C2 in Figure 7b. To get the high concentration of P–N bonds, the dosage of the precursor was adjusted according to the P/N ratio. The optimal P, N-graphene with the highest P–N bond and lowest NH$_2$ group concentration shows the best ORR performance.

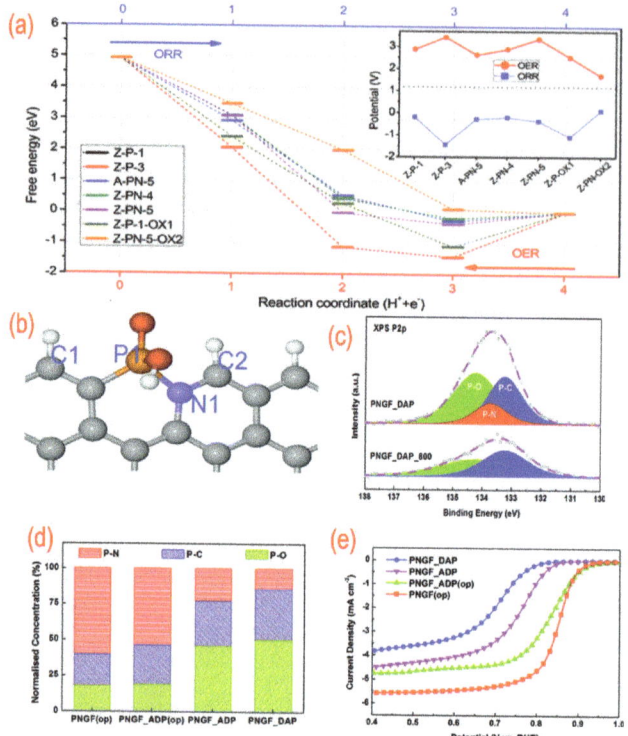

Figure 7. (**a**) The free energy variations of different P-containing structures for each ORR step. (**b**) The favorable N, P-containing structures. (**c**) The high-resolution P 2p peaks of N, P-graphene. (**d**) The relative ratio of XPS P2p binding configurations for different N, P-graphene materials. (**e**) The activity test (LSV curves recorded by RDE in O_2-saturated 0.1 M KOH solution with 1600 rpm at a scanning rate of 10 mV s^{-1}) for different N, P-graphene materials. Reproduced from [88] with permission from the Royal Society of Chemistry.

3.3. Phosphorus, X Codoped Graphene

Codoping other elements except nitrogen with phosphorus have been studied to figure out the relationship between the redistributed electronic state of graphene and the catalytical performance. Phosphorus and metal, such as Co, Fe codoped catalysts, were developed as ORR catalysts [89,90,114]. The metal P bond, such as Fe–Px or Co-P, can always be found and serve as active sites.

Furthermore, the third element can also be introduced into graphene lattice to get tridoped graphene. As shown in Figure 8, N-, P-, and F-tridoped graphenes are synthesized by pyrolysis of the mixture of GO, polyaniline, and ammonium hexafluorophosphate. The phosphorus, nitrogen, and fluorine are homogeneously doped into wrinkled graphene, thus introducing defects and changing the surface properties. There are P–C bonds and P–O bonds in N, P, F-graphene. The third element, fluorine, can induce strong charge redistribution for the carbon atom in the graphene. The N, P, F-graphene show a 4 e- pathway for ORR. The P, S, N-graphene [93,94,100,101], P, N, Fe-graphene [91,98], Fe, B, N, S, P-graphene [95], P, N, B-graphene [96,97], and Ni, N, P-graphene [99] are developed as well. Significantly, these materials always served as the multifunctional catalyst for ORR, OER, HER, and carbon dioxide reduction reaction (CO_2RR). This is mainly related to the different charged sites of the graphene due to the different heteroatoms doping.

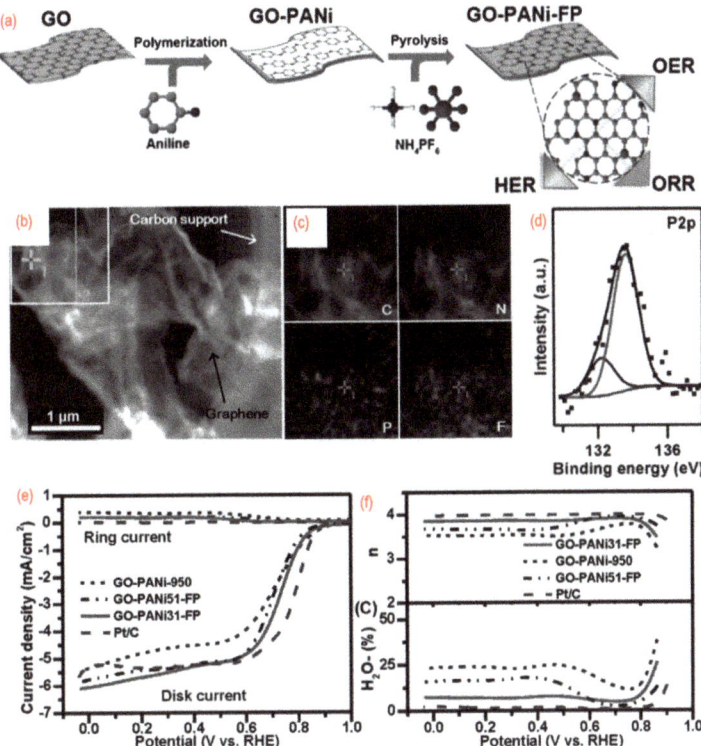

Figure 8. (a) The schematic of the synthesis process, (b) the TEM image, (c) C-, N-, F-, and P-elemental mappings, (d) the high-resolution P 2p peaks, (e) the activity test (LSV curves recorded by RRDE in O_2-saturated 0.1 M KOH solution with 1600 rpm at a scanning rate of 10 mV s^{-1}), (f) the electrons transfer number and H_2O_2 yield of N, P, F-graphene. Reproduced from [92]. © 2022 WILEY-VCH Verlag GmbH & Co. KGaA, Weinheim.

4. Phosphorus-Doped Graphene Composite for ORR

4.1. P-Graphene as Catalyst Support Material for ORR

The nanosized elemental metal, metal oxide, metal nitrides, and metal phosphides have potentially served as ORR catalysts. However, these nanoparticles are prone to agglomerate and aggregate during the synthesis and operation process because of the high surface energy. Besides, these nanosized catalysts always suffer from limited active surface areas and poor charge conductivity. Therefore, the substrate with a high surface area is generally used to disperse the nanoparticles. Doped graphene is a good candidate to accelerate the homogeneous dispersion. In particular, phosphorus atoms with multielectron orbital in P-graphene can lead to enhanced interactions between the catalyst and supports. The surface chemical state of P-graphene can not only affect the morphology but also modulate the electronic structure of the catalysts. A few research works have been devoted to using the P-graphene as catalyst support for ORR [102–106].

As shown in Figure 9, $MoP_x@MnP_y$ nanoparticles on P, N-graphene were obtained by annealing the mixture of GO and the desired chemicals. The phosphorus is $(NH_4)_2HPO_4$ and $Na_2HPO_2 \cdot H_2O$. The $MoP_x@MnP_y$ homogeneously disperse on the P, N-graphene due to the negatively charged state of nitrogen and phosphorus codoping. There are P–C, Mn–P, Mo–P, P–O, P–O–H, and PO_4 in the hybrid structure, indicating the enhanced interaction between $MoP_x@MnP_y$ and P, N-graphene. Hence, the composite shows good electrocatalytic activity and stability. Besides, different $CoMn_2O_4$, cobalt, Co_2P, and cobalt phosphate nanoparticles are also dispersed on the P, N-graphene [103–106]. These works are also summarized in Table 1.

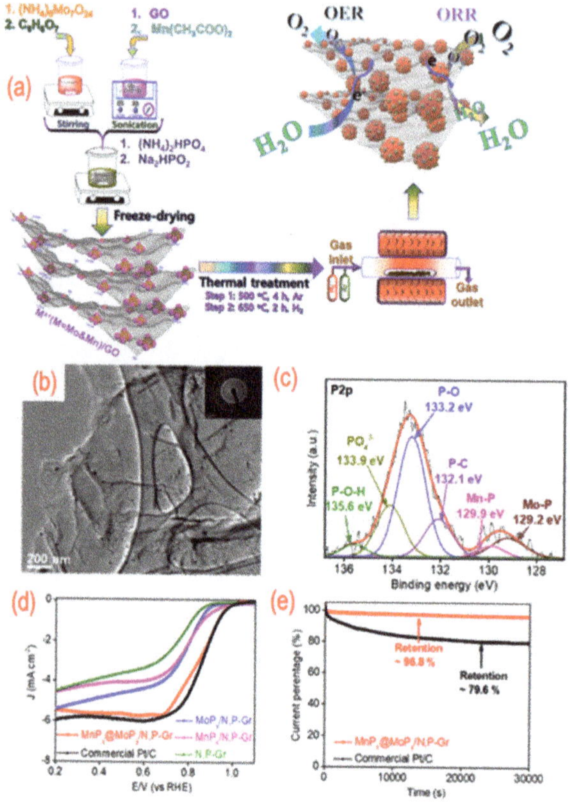

Figure 9. (**a**) The schematic of the synthesis process, (**b**) the TEM image (the inset is the corresponding SAED pattern), (**c**) the high-resolution P 2p peaks, (**d**) the activity test (LSV curves recorded by RDE in O2-saturated 0.1 M KOH solution with 1600 rpm at a scanning rate of 10 mV s^{-1}), (**e**) the stability test (current-time chronoamperometric in O2-saturated 0.1 M KOH solution) of MnPx@MoPy supported on P, N-graphene. Reproduced with permission from [102]. Copyright © 2022, American Chemical Society.

4.2. P-Graphene as a Coating Material for ORR

Encapsulating the active nanosized catalyst into carbon materials is another way to resist aggregation and agglomeration by dual chemical and physical protection. Furthermore, the structural and electronic properties of the active catalyst can be modulated with the contact of the doped carbon materials. The carbon materials outside the catalyst can be called graphene or carbon shells in different literature. When the thickness of this carbon layer is very thin, it is called graphene in this review.

As shown in Figure 10, the precursors of the Cu_3P, which were obtained by self-assembly, and 1-hydroxyethylidene-1,1-diphosphonic acid (HEDP) were annealed under N_2 atmosphere to obtain $Cu_3P@P$, N-graphene. The TEM image shows that the Cu_3P NPs are covered with a thin carbon shell. The N and P atoms in the precursor play a vital role in incorporating Cu_3P NPs into the carbon network and in situ doping nitrogen and phosphorus. A strong interaction can be found between the Cu_3P and P, N-graphene because of the chemical bonding of copper with phosphorus. The optimized $Cu_3P@P$, N-graphene shows comparable catalytic activity to that of the Pt/C catalyst because of the synergistic effect between the Cu$_3$P and P, N-graphene shell. Later, similar strategies were used by different groups to prepare Co@N, P, S -graphene, FeCo@P, N-graphene/N-CNTs, and FeP@P-graphene [108–110]. Interestingly, these materials are always used as bifunctional catalysts, such as OER, HER, and triiodide reduction reaction (TIRR). As a result, presently, this type of P-graphene encapsulating technology is gaining particular attention.

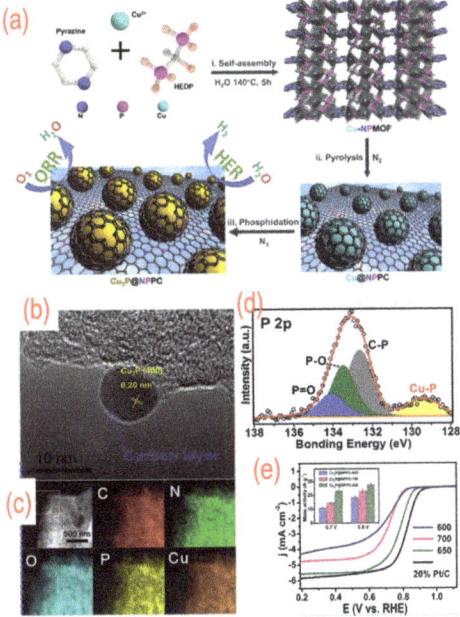

Figure 10. (**a**) The schematic of the synthesis process, (**b**) the TEM image, (**c**) C-, N-, O-, P-, and Cu-elemental mappings, (**d**) the high-resolution P 2p peaks, (**e**) the activity test (LSV curves recorded by RDE in O$_2$-saturated 0.1 M KOH solution with 1600 rpm at a scanning rate of 10 mV s^{-1}), of Cu$_3$P@ P, N-graphene. The inset of (**e**) is the mass activity of Cu$_3$P@ P, N-graphene catalysts. Reproduced with permission from [107]. © 2022 Wiley-VCH Verlag GmbH & Co. KGaA, Weinheim.

5. Perspectives

Regardless of abundant works done on the P-graphene, continuous efforts are needed to rationally design and fabricate the highly efficient ORR catalysts. Some current challenges and future perspectives are listed below.

First, the precisely controlled synthesis for heteroatom-doped graphene remains a key challenge and needs to be explored further. In present catalysts, there are many types of dopants and various defects. The electrochemical reaction is delicate, so it is difficult to identify the real factors that are responsible for enhancing the ORR activity. The synergistic mechanism is one of the most used explanations for the improvement of ORR. However, it is essential to clarify the effect of a specific chemical bond. Doping desired heteroatom(s) with a dominated specific type of chemical bond in a particular position is the first step to studying the doping effect. Currently, GO is being used as the carbon support in most of the research related to electrocatalysis. The conventional pyrolysis of GO cannot meet the demands. The changing structure and surface chemistry of GO during thermal treatment should be tuned. In addition, the controllable synthetic routes should be further developed. For example, magnetron sputtering is one of the promising physical vapor deposition techniques for the preparation of metal, alloy, and compound thin films. Under a high vacuum environment with a magnetically confined plasma, the atoms from the target can be sputtered onto the substrates. P, S, F, and N are commonly used as dopants by this technique [115–117]. Furthermore, the catalyst layer can be directly deposited on the membrane of the membrane electrode assembly (MEA) in fuel cells by magnetron sputtering.

Secondly, more efforts should be made on the physical property evaluations of P-graphene, such as thermal stability, wettability, and thermal conductivity, which also play important roles in ORR performance. It is also important for the synthesis routes. For example, the defects can be found in sing layer-graphene prepared by CVD at ~500 °C, while defects appear in bilayer graphene fabricated at ~600 °C [118]. More studies should be done on the physical property characterizations to achieve controlled production of P-graphene.

Thirdly, identifying the types and quantifying the content of dopants accurately is still a big challenge. XPS is a frequently used technology to study the information of chemical bonding. However, even with a similar catalyst, the analysis of XPS data varies between different research groups. Other advanced characterization techniques, such as X-ray absorption spectroscopy, should be involved.

Fourthly, the underlying catalytic mechanism, as well as the real active sites for ORR of doped graphene is not yet identified clearly. Although some theoretical simulations have been done, there is still a big gap between the theoretical model and the actual working state of the catalysts. Some well-designed experiments and in situ characterizations should be done to unveil the catalytic mechanism.

Last but not least, the building of the multiple doped structures or the composites will play a more important role. For example, we have recently synthesized P, N-graphene dots/N-3D-graphene [27]. We found that P–N bonded structures can serve as ORR active sites. Because of the edge defects and abundant functional groups, the catalyst shows excellent ORR performance. Also, the P-doped graphene can be used as the support for metal catalysts.

6. Conclusions

In summary, doping graphene with phosphorus and various heteroatoms has been proven as a promising way to enhance the ORR performance. More importantly, the experimental and theoretical studies of P-graphene are essential to study the catalytic mechanisms towards ORR, thus providing a general strategy to obtain the advanced non-noble metal catalysts for the application of energy conversion devices, such as fuel cells and metal-air batteries. In this review, the recent advances in P-doped graphene with an overview of the various synthesis methods, ORR performance, and ORR mechanism are discussed. Furthermore, we believe that these achievements can be used in other energy conversion and storage fields. Further efforts on the study of P-graphene are warranted.

Author Contributions: Conceptualization, X.T.; methodology, X.Z.; software, Z.G. and Y.X.; validation, M.G., J.T., L.M. and G.Z.; formal analysis, X.Z.; investigation, X.T.; resources, Y.X.; data curation, M.G. and J.T.; writing—original draft preparation, X.Z. and X.T.; writing—review and editing, Z.C., H.R., G.Z. and S.S.; visualization, X.T.; supervision, X.T. and S.S.; project administration, X.Z. and X.T.; funding acquisition, X.Z. and X.T. All authors have read and agreed to the published version of the manuscript.

Funding: This study has been accomplished with financial support by the Youth Science and Technology Talent Growth Program of Guizhou Provincial Education Department (No. QianjiaoheKY [2021]295, QianjiaoheKY [2022]163), the National Natural Science Foundation of China (NSFC, No. 22005072, 21506041), the Science and Technology Project of Guizhou Province ([2019]1226), the Natural Sciences and Engineering Research Council of Canada (NSERC), Fonds de la Recherche du Québec sur la Nature et les Technologies (FRQNT), Centre Québécois sur les Materiaux Fonctionnels (CQMF), and the Canada Research Chair program. S. Sun acknowledges the ECS-Toyota Young Investigator Fellowship.

Data Availability Statement: All raw data in this study can be provided by the corresponding authors on request.

Conflicts of Interest: The authors declare no conflict of interest.

References

1. Mu, Y.; Wang, T.; Zhang, J.; Meng, C.; Zhang, Y.; Kou, Z. Single-atom catalysts: Advances and challenges in metal-support interactions for enhanced electrocatalysis. *Electrochem. Energy Rev.* **2022**, *5*, 145–186. [CrossRef]
2. Ren, X.F.; Liu, B.H.; Liang, X.Y.; Wang, Y.R.; Lv, Q.Y.; Liu, A.M. Review-current progress of non-precious metal for ORR based electrocatalysts used for fuel cells. *J. Electrochem. Soc.* **2021**, *168*, 044521. [CrossRef]
3. Zhao, Z.H.; Li, M.T.; Zhang, L.P.; Dai, L.M.; Xia, Z.H. Design principles for heteroatom-doped carbon nanomaterials as highly efficient catalysts for fuel cells and metal-air batteries. *Adv. Mater.* **2015**, *27*, 6834–6840. [CrossRef]
4. Zhang, J.T.; Zhao, Z.H.; Xia, Z.H.; Dai, L.M. A metal-free bifunctional electrocatalyst for oxygen reduction and oxygen evolution reactions. *Nat. Nanotechnol.* **2015**, *10*, 444–452. [CrossRef] [PubMed]
5. Liu, J.; Song, P.; Ning, Z.G.; Xu, W.L. Recent advances in heteroatom-doped metal-free electrocatalysts for highly efficient oxygen reduction reaction. *Electrocatalysis* **2015**, *6*, 132–147. [CrossRef]
6. Bai, L.; Zhang, Y.; Tong, W.; Sun, L.; Huang, H.; An, Q.; Tian, N.; Chu, P.K. Graphene for energy storage and conversion: Synthesis and interdisciplinary applications. *Electrochem. Energy Rev.* **2020**, *3*, 395–430. [CrossRef]
7. Han, A.; Zhang, Z.; Yang, J.; Wang, D.; Li, Y. Carbon-supported single-atom catalysts for formic acid oxidation and oxygen reduction reactions. *Small* **2021**, *17*, 2004500. [CrossRef] [PubMed]
8. Xu, X.; Sun, H.; Jiang, S.P.; Shao, Z. Modulating metal–organic frameworks for catalyzing acidic oxygen evolution for proton exchange membrane water electrolysis. *SusMat* **2021**, *1*, 460–481. [CrossRef]
9. Xu, Z.; Deng, W.; Wang, X. 3D hierarchical carbon-rich micro-/nanomaterials for energy storage and catalysis. *Electrochem. Energy Rev.* **2021**, *4*, 269–335. [CrossRef]
10. Tang, L.; Xu, Q.; Zhang, Y.; Chen, W.; Wu, M. MOF/PCP-based electrocatalysts for the oxygen reduction reaction. *Electrochem. Energy Rev.* **2022**, *5*, 32–81. [CrossRef]
11. Cui, X.; Luo, Y.; Zhou, Y.; Dong, W.; Chen, W. Application of functionalized graphene in Li–O_2 batteries. *Nanotechnology* **2021**, *32*, 132003. [CrossRef] [PubMed]
12. Zhu, S.; Wang, X.; Luo, E.; Yang, L.; Chu, Y.; Gao, L.; Jin, Z.; Liu, C.; Ge, J.; Xing, W. Stabilized Pt cluster-based catalysts used as low-loading cathode in proton-exchange membrane fuel cells. *ACS Energy Lett.* **2020**, *5*, 3021–3028. [CrossRef]
13. Tong, X.; Wei, Q.; Zhan, X.; Zhang, G.; Sun, S. The new graphene family materials: Synthesis and applications in oxygen reduction reaction. *Catalysts* **2017**, *7*, 1. [CrossRef]
14. Khan, K.; Tareen, A.K.; Aslam, M.; Zhang, Y.; Wang, R.; Ouyang, Z.; Gou, Z.; Zhang, H. Recent advances in two-dimensional materials and their nanocomposites in sustainable energy conversion applications. *Nanoscale* **2019**, *11*, 21622–21678. [CrossRef] [PubMed]
15. Wang, Y.; Li, J.; Wei, Z. Recent progress of carbon-based materials in oxygen reduction reaction catalysis. *Chemelectrochem* **2018**, *5*, 1764–1774. [CrossRef]
16. Ji, Y.; Dong, H.; Liu, C.; Li, Y. The progress of metal-free catalysts for the oxygen reduction reaction based on theoretical simulations. *J. Mater. Chem. A* **2018**, *6*, 13489–13508. [CrossRef]
17. Tong, X.; Zhan, X.; Rawach, D.; Chen, Z.; Zhang, G.; Sun, S. Low-dimensional catalysts for oxygen reduction reaction. *Prog. Nat. Sci. Mater. Int.* **2020**, *30*, 787–795. [CrossRef]
18. Zhao, S.; Wang, D.W.; Amal, R.; Dai, L. Carbon-based metal-free catalysts for key reactions involved in energy conversion and storage. *Adv. Mater.* **2019**, *31*, 1801526. [CrossRef]
19. Choi, H.J.; Jung, S.M.; Seo, J.M.; Chang, D.W.; Dai, L.M.; Baek, J.B. Graphene for energy conversion and storage in fuel cells and supercapacitors. *Nano Energy* **2012**, *1*, 534–551. [CrossRef]
20. Zhang, J.; Zhao, F.; Zhang, Z.; Chen, N.; Qu, L. Dimension-tailored functional graphene structures for energy conversion and storage. *Nanoscale* **2013**, *5*, 3112–3126. [CrossRef] [PubMed]
21. Wang, D.-W.; Su, D. Heterogeneous nanocarbon materials for oxygen reduction reaction. *Energy Environ. Sci.* **2014**, *7*, 576–591. [CrossRef]
22. Geng, D.S.; Ding, N.; Hor, T.S.A.; Liu, Z.L.; Sun, X.L.; Zong, Y. Potential of metal-free "graphene alloy" as electrocatalysts for oxygen reduction reaction. *J. Mater. Chem. A* **2015**, *3*, 1795–1810. [CrossRef]
23. Jiang, Y.; Guo, F.; Liu, Y.; Xu, Z.; Gao, C. Three-dimensional printing of graphene-based materials for energy storage and conversion. *SusMat* **2021**, *1*, 304–323. [CrossRef]
24. Mazanek, V.; Luxa, J.; Matejkova, S.; Kucera, J.; Sedmidubsky, D.; Pumera, M.; Sofer, Z. Ultrapure graphene is a poor electrocatalyst: Definitive proof of the key role of metallic impurities in graphene-based electrocatalysis. *ACS Nano* **2019**, *13*, 1574–1582. [CrossRef] [PubMed]
25. Wang, B.; Liu, B.; Dai, L. Non-N-doped carbons as metal-free electrocatalysts. *Adv. Sustain. Syst.* **2021**, *5*, 2000134. [CrossRef]
26. Kumar, R.; Sahoo, S.; Joanni, E.; Singh, R.K.; Maegawa, K.; Tan, W.K.; Kawaguchi, G.; Kar, K.K.; Matsuda, A. Heteroatom doped graphene engineering for energy storage and conversion. *Mater. Today* **2020**, *39*, 47–65. [CrossRef]

27. Tong, X.; Cherif, M.; Zhang, G.X.; Zhan, X.X.; Ma, J.G.; Almesrati, A.; Vidal, F.; Song, Y.J.; Claverie, J.P.; Sun, S.H. N, P-codoped graphene dots supported on N-doped 3D graphene as metal-free catalysts for oxygen reduction. *ACS Appl. Mater. Inter.* **2021**, *13*, 30512–30523. [CrossRef] [PubMed]
28. Xue, Y.; Wu, B.; Bao, Q.; Liu, Y. Controllable synthesis of doped graphene and its applications. *Small* **2014**, *10*, 2975–2991. [CrossRef] [PubMed]
29. Wood, K.N.; O'Hayre, R.; Pylypenko, S. Recent progress on nitrogen/carbon structures designed for use in energy and sustainability applications. *Energy Environ. Sci.* **2014**, *7*, 1212–1249. [CrossRef]
30. Poh, H.L.; Pumera, M. P-element-doped graphene: Heteroatoms for electrochemical enhancement. *Chemelectrochem* **2015**, *2*, 190–199. [CrossRef]
31. Dai, L.; Xue, Y.; Qu, L.; Choi, H.J.; Baek, J.B. Metal-free catalysts for oxygen reduction reaction. *Chem. Rev.* **2015**, *115*, 4823–4892. [CrossRef]
32. Langer, R.; Błoński, P.; Hofer, C.; Lazar, P.; Mustonen, K.; Meyer, J.C.; Susi, T.; Otyepka, M. Tailoring electronic and magnetic properties of graphene by phosphorus doping. *ACS Appl. Mater. Inter.* **2020**, *12*, 34074–34085. [CrossRef] [PubMed]
33. Yang, N.; Li, L.; Li, J.; Ding, W.; Wei, Z. Modulating the oxygen reduction activity of heteroatom-doped carbon catalysts via the triple effect: Charge, spin density and ligand effect. *Chem. Sci.* **2018**, *9*, 5795–5804. [CrossRef]
34. Qian, J.; Shen, C.; Yan, J.; Xi, F.; Dong, X.; Liu, J. Tailoring the electronic properties of graphene quantum dots by P doping and their enhanced performance in metal-free composite photocatalyst. *J. Phys. Chem. C* **2018**, *122*, 349–358. [CrossRef]
35. Wang, L.; Sofer, Z.; Zboril, R.; Cepe, K.; Pumera, M. Phosphorus and halogen co-doped graphene materials and their electrochemistry. *Chem.-Eur. J* **2016**, *22*, 15444–15450. [CrossRef]
36. Lin, L.H.; Fu, L.; Zhang, K.Y.; Chen, J.; Zhang, W.L.; Tang, S.L.; Du, Y.W.; Tang, N.J. P-superdoped graphene: Synthesis and magnetic properties. *ACS Appl. Mater. Inter.* **2019**, *11*, 39062–39067. [CrossRef]
37. Sun, L.; Yuan, G.; Gao, L.; Yang, J.; Chhowalla, M.; Gharahcheshmeh, M.H.; Gleason, K.K.; Choi, Y.S.; Hong, B.H.; Liu, Z. Chemical vapour deposition. *Nat. Rev. Methods Primers* **2021**, *1*, 5. [CrossRef]
38. Bulusheva, L.G.; Arkhipov, V.E.; Popov, K.M.; Sysoev, V.I.; Makarova, A.A.; Okotrub, A.V. Electronic structure of nitrogen- and phosphorus-doped graphenes grown by chemical vapor deposition method. *Materials* **2020**, *13*, 1173. [CrossRef]
39. Ovezmyradov, M.; Magedov, I.V.; Frolova, L.V.; Chandler, G.; Garcia, J.; Bethke, D.; Shaner, E.A.; Kalugin, N.G. Chemical vapor deposition of phosphorous- and boron-doped graphene using phenyl-containing molecules. *J. Nanosci. Nanotechnol.* **2015**, *15*, 4883–4886. [CrossRef]
40. Larrude, D.G.; Garcia-Basabe, Y.; Freire Junior, F.L.; Rocco, M.L.M. Electronic structure and ultrafast charge transfer dynamics of phosphorous doped graphene layers on a copper substrate: A combined spectroscopic study. *RSC Adv.* **2015**, *5*, 74189–74197. [CrossRef]
41. Some, S.; Kim, J.; Lee, K.; Kulkarni, A.; Yoon, Y.; Lee, S.; Kim, T.; Lee, H. Highly air-stable phosphorus-doped n-type graphene field-effect transistors. *Adv. Mater.* **2012**, *24*, 5481–5486. [CrossRef]
42. He, S.-M.; Huang, C.-C.; Liou, J.-W.; Woon, W.-Y.; Su, C.-Y. Spectroscopic and electrical characterizations of low-damage phosphorous-doped graphene via ion implantation. *ACS Appl. Mater. Inter.* **2019**, *11*, 47289–47298. [CrossRef] [PubMed]
43. Shin, D.W.; Kim, T.S.; Yoo, J.B. Phosphorus doped graphene by inductively coupled plasma and triphenylphosphine treatments. *Mater. Res. Bull.* **2016**, *82*, 71–75. [CrossRef]
44. Rao, Y.F.; Yuan, M.; Luo, F.; Wang, Z.P.; Li, H.; Yu, J.B.; Chen, X.P. One-step laser fabrication of phosphorus-doped porous graphene electrodes for high-performance flexible microsupercapacitor. *Carbon* **2021**, *180*, 56–66. [CrossRef]
45. Liu, R.; Zhao, J.; Huang, Z.; Zhang, L.; Zou, M.; Shi, B.; Zhao, S. Nitrogen and phosphorus co-doped graphene quantum dots as a nano-sensor for highly sensitive and selective imaging detection of nitrite in live cell. *Sens. Actuators B Chem.* **2017**, *240*, 604–612. [CrossRef]
46. Guo, Z.; Wu, H.; Li, M.; Tang, T.; Wen, J.; Li, X. Phosphorus-doped graphene quantum dots loaded on TiO_2 for enhanced photodegradation. *Appl. Surf. Sci.* **2020**, *526*, 146724. [CrossRef]
47. Wang, W.J.; Xu, S.F.; Li, N.; Huang, Z.Y.; Su, B.Y.; Chen, X.M. Sulfur and phosphorus co-doped graphene quantum dots for fluorescent monitoring of nitrite in pickles. *Spectrochim. Acta A* **2019**, *221*, 117211. [CrossRef]
48. Compton, O.C.; Nguyen, S.T. Graphene oxide, highly reduced graphene oxide, and graphene: Versatile building blocks for carbon-based materials. *Small* **2010**, *6*, 711–723. [CrossRef] [PubMed]
49. Gao, W.; Alemany, L.B.; Ci, L.; Ajayan, P.M. New insights into the structure and reduction of graphite oxide. *Nat. Chem.* **2009**, *1*, 403–408. [CrossRef] [PubMed]
50. Li, J.-C.; Hou, P.-X.; Liu, C. Heteroatom-doped carbon nanotube and graphene-based electrocatalysts for oxygen reduction reaction. *Small* **2017**, *13*, 1702002. [CrossRef]
51. Feng, L.; Qin, Z.; Huang, Y.; Peng, K.; Wang, F.; Yan, Y.; Chen, Y. Boron-, sulfur-, and phosphorus-doped graphene for environmental applications. *Sci. Total Environ.* **2020**, *698*, 134239. [CrossRef] [PubMed]
52. Kaushal, S.; Kaur, M.; Kaur, N.; Kumari, V.; Singh, P.P. Heteroatom-doped graphene as sensing materials: A mini review. *RSC Adv.* **2020**, *10*, 28608–28629. [CrossRef]
53. Some, S.; Shackery, I.; Kim, S.J.; Jun, S.C. Phosphorus-doped graphene oxide layer as a highly efficient flame retardant. *Chem.-Eur. J.* **2015**, *21*, 15480–15485. [CrossRef] [PubMed]

54. Zhang, C.; Mahmood, N.; Yin, H.; Liu, F.; Hou, Y. Synthesis of phosphorus-doped graphene and its multifunctional applications for oxygen reduction reaction and lithium ion batteries. *Adv. Mater.* **2013**, *25*, 4932–4937. [CrossRef] [PubMed]
55. Li, R.; Wei, Z.; Gou, X.; Xu, W. Phosphorus-doped graphene nanosheets as efficient metal-free oxygen reduction electrocatalysts. *RSC Adv.* **2013**, *3*, 9978–9984. [CrossRef]
56. An, M.C.; Du, C.Y.; Du, L.; Sun, Y.R.; Wang, Y.J.; Chen, C.; Han, G.K.; Yin, G.P.; Gao, Y.Z. Phosphorus-doped graphene support to enhance electrocatalysis of methanol oxidation reaction on platinum nanoparticles. *Chem. Phys. Lett.* **2017**, *687*, 1–8. [CrossRef]
57. Chu, K.; Wang, F.; Tian, Y.; Wei, Z. Phosphorus doped and defects engineered graphene for improved electrochemical sensing: Synergistic effect of dopants and defects. *Electrochim. Acta* **2017**, *231*, 557–564. [CrossRef]
58. Tian, Y.; Wei, Z.; Zhang, K.H.; Peng, S.; Zhang, X.; Liu, W.M.; Chu, K. Three-dimensional phosphorus-doped graphene as an efficient metal-free electrocatalyst for electrochemical sensing. *Sens. Actuators B-Chem.* **2017**, *241*, 584–591. [CrossRef]
59. Zhang, X.; Wang, K.P.; Zhang, L.N.; Zhang, Y.C.; Shen, L. Phosphorus-doped graphene-based electrochemical sensor for sensitive detection of acetaminophen. *Anal. Chim. Acta* **2018**, *1036*, 26–32. [CrossRef] [PubMed]
60. Fan, X.; Xu, H.; Zuo, S.S.; Liang, Z.P.; Yang, S.H.; Chen, Y. Preparation and supercapacitive properties of phosphorus-doped reduced graphene oxide hydrogel. *Electrochim. Acta* **2020**, *330*, 135207. [CrossRef]
61. Nie, G.S.; Deng, H.C.; Huang, J.; Wang, C.Y. Phytic acid assisted formation of phosphorus-doped graphene aerogel as electrode material for high-performance supercapacitor. *Int. J. Electrochem. Sci.* **2020**, *15*, 12578–12586. [CrossRef]
62. Bi, Z.H.; Huo, L.; Kong, Q.Q.; Li, F.; Chen, J.P.; Ahmad, A.; Wei, X.X.; Xie, L.J.; Chen, C.M. Structural evolution of phosphorus species on graphene with a stabilized electrochemical interface. *ACS Appl. Mater. Inter.* **2019**, *11*, 11421–11430. [CrossRef] [PubMed]
63. Shumba, M.; Nyokong, T. Development of nanocomposites of phosphorus-nitrogen co-doped graphene oxide nanosheets and nanosized cobalt phthalocyanines for electrocatalysis. *Electrochim. Acta* **2016**, *213*, 529–539. [CrossRef]
64. Wen, Y.Y.; Rufford, T.E.; Hulicova-Jurcakova, D.; Wang, L.Z. Nitrogen and phosphorous co-doped graphene monolith for supercapacitors. *Chemsuschem* **2016**, *9*, 513–520. [CrossRef]
65. Yu, X.; Kang, Y.; Park, H.S. Sulfur and phosphorus co-doping of hierarchically porous graphene aerogels for enhancing supercapacitor performance. *Carbon* **2016**, *101*, 49–56. [CrossRef]
66. Qu, K.G.; Zheng, Y.; Zhang, X.X.; Davey, K.; Dai, S.; Qiao, S.Z. Promotion of electrocatalytic hydrogen evolution reaction on nitrogen-doped carbon nanosheets with secondary heteroatoms. *ACS Nano* **2017**, *11*, 7293–7300. [CrossRef] [PubMed]
67. Wang, C.N.; Luo, S.Y.; Yang, Y.Y.; Ren, D.S.; Yu, X. Defect-rich graphene architecture induced by nitrogen and phosphorus dual doping for high-performance supercapacitors. *Energy Technol.-Ger.* **2020**, *8*, 1900685. [CrossRef]
68. Wang, K.; Li, Z. Synthesis of nitrogen and phosphorus dual-doped graphene oxide as high-performance anode material for lithium-ion batteries. *J. Nanosci. Nanotechol.* **2020**, *20*, 7673–7679. [CrossRef] [PubMed]
69. Thirumal, V.; Pandurangan, A.; Jayavel, R.; Venkatesh, K.S.; Palani, N.S.; Ragavan, R.; Ilangovan, R. Single pot electrochemical synthesis of functionalized and phosphorus doped graphene nanosheets for supercapacitor applications. *J. Mater. Sci. Mater. Electron.* **2015**, *26*, 6319–6328. [CrossRef]
70. Momodu, D.; Madito, M.J.; Singh, A.; Sharif, F.; Karan, K.; Trifkovic, M.; Bryant, S.; Roberts, E.P.L. Mixed-acid intercalation for synthesis of a high conductivity electrochemically exfoliated graphene. *Carbon* **2021**, *171*, 130–141. [CrossRef]
71. Song, J.; Yu, Z.; Gordin, M.L.; Hu, S.; Yi, R.; Tang, D.; Walter, T.; Regula, M.; Choi, D.; Li, X.; et al. Chemically bonded phosphorus/graphene hybrid as a high performance anode for sodium-ion batteries. *Nano Lett.* **2014**, *14*, 6329–6335. [CrossRef] [PubMed]
72. Gong, K.; Du, F.; Xia, Z.; Durstock, M.; Dai, L. Nitrogen-doped carbon nanotube arrays with high electrocatalytic activity for oxygen reduction. *Science* **2009**, *323*, 760–764. [CrossRef]
73. Jiao, Y.; Zheng, Y.; Jaroniec, M.; Qiao, S.Z. Design of electrocatalysts for oxygen- and hydrogen-involving energy conversion reactions. *Chem. Soc. Rev.* **2015**, *44*, 2060–2086. [CrossRef] [PubMed]
74. Zhao, D.; Zhuang, Z.; Cao, X.; Zhang, C.; Peng, Q.; Chen, C.; Li, Y. Atomic site electrocatalysts for water splitting, oxygen reduction and selective oxidation. *Chem. Soc. Rev.* **2020**, *49*, 2215–2264. [CrossRef]
75. Zhang, X.L.; Lu, Z.S.; Fu, Z.M.; Tang, Y.A.; Ma, D.W.; Yang, Z.X. The mechanisms of oxygen reduction reaction on phosphorus doped graphene: A first-principles study. *J Power Sources* **2015**, *276*, 222–229. [CrossRef]
76. Bai, X.W.; Zhao, E.J.; Li, K.; Wang, Y.; Jiao, M.G.; He, F.; Sun, X.X.; Sun, H.; Wu, Z.J. Theoretical insights on the reaction pathways for oxygen reduction reaction on phosphorus doped graphene. *Carbon* **2016**, *105*, 214–223. [CrossRef]
77. Lei, W.; Deng, Y.P.; Li, G.R.; Cano, Z.P.; Wang, X.L.; Luo, D.; Liu, Y.S.; Wang, D.L.; Chen, Z.W. Two-dimensional phosphorus-doped carbon nanosheets with tunable porosity for oxygen reactions in zinc-air batteries. *ACS Catal.* **2018**, *8*, 2464–2472. [CrossRef]
78. Balaji, S.S.; Ganesh, P.A.; Moorthy, M.; Sathish, M. Efficient electrocatalytic activity for oxygen reduction reaction by phosphorus-doped graphene using supercritical fluid processing. *Mater. Sci.* **2020**, *43*, 151. [CrossRef]
79. Poon, K.C.; Wan, W.Y.; Su, H.B.; Sato, H. One-minute synthesis via electroless reduction of amorphous phosphorus-doped graphene for oxygen reduction reaction. *ACS Appl. Energy Mater.* **2021**, *4*, 5388–5391. [CrossRef]
80. Yang, N.; Zheng, X.; Li, L.; Li, J.; Wei, Z. Influence of phosphorus configuration on electronic structure and oxygen reduction reactions of phosphorus-doped graphene. *J. Phys. Chem. C* **2017**, *121*, 19321–19328. [CrossRef]
81. Dong, L.; Hu, C.; Huang, X.; Chen, N.; Qu, L. One-pot synthesis of nitrogen and phosphorus co-doped graphene and its use as high-performance electrocatalyst for oxygen reduction reaction. *Chem.—Asian J.* **2015**, *10*, 2609–2614. [CrossRef] [PubMed]

82. Li, R.; Wei, Z.; Gou, X. Nitrogen and phosphorus dual-doped graphene/carbon nanosheets as bifunctional electrocatalysts for oxygen reduction and evolution. *ACS Catal.* **2015**, *5*, 4133–4142. [CrossRef]
83. Jang, D.; Lee, S.; Kim, S.; Choi, K.; Park, S.; Oh, J.; Park, S. Production of P, N co-doped graphene-based materials by a solution process and their electrocatalytic performance for oxygen reduction reaction. *ChemNanoMat* **2018**, *4*, 118–123. [CrossRef]
84. Zhou, L.J.; Zhang, C.Y.; Cai, X.Y.; Qian, Y.; Jiang, H.F.; Li, B.S.; Lai, L.F.; Shen, Z.X.; Huang, W. N, P co-doped hierarchical porous graphene as a metal-free bifunctional air cathode for Zn-air batteries. *Chemelectrochem* **2018**, *5*, 1811–1816. [CrossRef]
85. Ge, L.P.; Wang, D.; Yang, P.X.; Xu, H.; Xiao, L.H.; Zhang, G.X.; Lu, X.Y.; Duan, Z.Z.; Meng, F.; Zhang, J.Q.; et al. Graphite N-C-P dominated three-dimensional nitrogen and phosphorus co-doped holey graphene foams as high-efficiency electrocatalysts for Zn-air batteries. *Nanoscale* **2019**, *11*, 17010–17017. [CrossRef] [PubMed]
86. Zhang, X.R.; Zhang, X.; Xiang, X.; Pan, C.; Meng, Q.H.; Hao, C.; Tian, Z.Q.; Shen, P.K.; Jiang, S.P. Nitrogen and phosphate co-doped graphene as efficient bifunctional electrocatalysts by precursor modulation strategy for oxygen reduction and evolution reactions. *Chemelectrochem* **2021**, *8*, 3262–3272. [CrossRef]
87. Qiao, X.C.; Liao, S.J.; You, C.H.; Chen, R. Phosphorus and nitrogen dual doped and simultaneously reduced graphene oxide with high surface area as efficient metal-free electrocatalyst for oxygen reduction. *Catalysts* **2015**, *5*, 981–991. [CrossRef]
88. Chai, G.L.; Qiu, K.P.; Qiao, M.; Titirici, M.M.; Shang, C.X.; Guo, Z.X. Active sites engineering leads to exceptional ORR and OER bifunctionality in P, N co-doped graphene frameworks. *Energy Environ. Sci.* **2017**, *10*, 1186–1195. [CrossRef]
89. Yang, Z.R.; Wu, J.; Zheng, X.J.; Wang, Z.J.; Yang, R.Z. Enhanced catalytic activity for the oxygen reduction reaction with co-doping of phosphorus and iron in carbon. *J. Power Sources* **2015**, *277*, 161–168. [CrossRef]
90. Razmjooei, F.; Singh, K.P.; Bae, E.J.; Yu, J.S. A new class of electroactive Fe- and P-functionalized graphene for oxygen reduction. *J. Mater. Chem. A* **2015**, *3*, 11031–11039. [CrossRef]
91. Qiao, X.C.; Peng, H.L.; You, C.H.; Liu, F.F.; Zheng, R.P.; Xu, D.W.; Li, X.H.; Liao, S.J. Nitrogen, phosphorus and iron doped carbon nanospheres with high surface area and hierarchical porous structure for oxygen reduction. *J. Power Sources* **2015**, *288*, 253–260. [CrossRef]
92. Zhang, J.T.; Dai, L.M. Nitrogen, phosphorus, and fluorine tri-doped graphene as a multifunctional catalyst for self-powered electrochemical water splitting. *Angew. Chem. Int. Ed.* **2016**, *55*, 13296–13300. [CrossRef] [PubMed]
93. Dou, S.; Shen, A.L.; Ma, Z.L.; Wu, J.H.; Tao, L.; Wang, S.Y. N-, P- and S-tridoped graphene as metal-free electrocatalyst for oxygen reduction reaction. *J. Electroanal. Chem.* **2015**, *753*, 21–27. [CrossRef]
94. Wang, Y.S.; Zhang, B.W.; Xu, M.H.; He, X.Q. Tunable ternary (P, S, N)-doped graphene as an efficient electrocatalyst for oxygen reduction reaction in an alkaline medium. *RSC Adv.* **2015**, *5*, 86746–86753. [CrossRef]
95. Chen, W.; Sin, M.; Wei, P.J.; Zhang, Q.L.; Liu, J.G. Synergistic enhancement of electrocatalytic activity toward oxygen reduction reaction in alkaline electrolytes with pentabasic (Fe, B, N, S, P)-doped reduced graphene oxide. *Chin. J. Chem.* **2016**, *34*, 878–886. [CrossRef]
96. Lin, H.L.; Chu, L.; Wang, X.J.; Yao, Z.Q.; Liu, F.; Ai, Y.N.; Zhuang, X.D.; Han, S. Boron, nitrogen, and phosphorous ternary doped graphene aerogel with hierarchically porous structures as highly efficient electrocatalysts for oxygen reduction reaction. *New J. Chem.* **2016**, *40*, 6022–6029. [CrossRef]
97. Dong, F.; Cai, Y.X.; Liu, C.; Liu, J.Y.; Qiao, J.L. Heteroatom (B, N and P) doped porous graphene foams for efficient oxygen reduction reaction electrocatalysis. *Int. J. Hydrog. Energy* **2018**, *43*, 12661–12670. [CrossRef]
98. Liu, J.; Zhu, Y.Y.; Du, F.L.; Jiang, L.H. Iron/nitrogen/phosphorus co-doped three-dimensional porous carbon as a highly efficient electrocatalyst for oxygen reduction reaction. *J. Electrochem. Soc.* **2019**, *166*, F935–F941. [CrossRef]
99. Yang, R.; Xie, J.F.; Liu, Q.; Huang, Y.Y.; Lv, J.Q.; Ghausi, M.A.; Wang, X.Y.; Peng, Z.; Wu, M.X.; Wang, Y.B. A trifunctional ni-n/p-o-codoped graphene electrocatalyst enables dual-model rechargeable Zn-CO$_2$/ Zn-O$_2$ batteries. *J. Mater. Chem. A* **2019**, *7*, 2575–2580. [CrossRef]
100. Siahkalroudi, Z.M.; Aghabarari, B.; Vaezi, M.; Rodriguez-Castellon, E.; Martinez-Huerta, M.V. Effect of secondary heteroatom (S, P) in N-doped reduced graphene oxide catalysts to oxygen reduction reaction. *Mol. Catal.* **2021**, *502*, 111372. [CrossRef]
101. Yao, S.X.; Lyu, D.D.; Wei, M.; Chu, B.X.; Huang, Y.L.; Pan, C.; Zhang, X.R.; Tian, Z.Q.; Shen, P.K. N, S, P co-doped graphene-like carbon nanosheets developed via in situ engineering strategy of carbon p(z)-orbitals for highly efficient oxygen redox reaction. *Flatchem* **2021**, *27*, 100250. [CrossRef]
102. Nguyen, D.C.; Tran, D.T.; Doan, T.L.L.; Kim, N.H.; Lee, J.H. Constructing MoPx@MnPy heteronanoparticle-supported mesoporous N, P-codoped graphene for boosting oxygen reduction and oxygen evolution reaction. *Chem. Mater.* **2019**, *31*, 2892–2904. [CrossRef]
103. Guo, W.H.; Ma, X.X.; Zhang, X.L.; Zhang, Y.Q.; Yu, D.L.; He, X.Q. Spinel CoMn$_2$O$_4$ nanoparticles supported on a nitrogen and phosphorus dual doped graphene aerogel as efficient electrocatalysts for the oxygen reduction reaction. *RSC Adv.* **2016**, *6*, 96436–96444. [CrossRef]
104. Hu, X.L.; Dai, X.H.; He, X.Q. A N,P-co-doped 3D graphene/cobalt-embedded electrocatalyst for the oxygen reduction reaction. *New J. Chem.* **2017**, *41*, 15236–15243. [CrossRef]
105. Jiang, H.; Li, C.; Shen, H.B.; Liu, Y.S.; Li, W.Z.; Li, J.E. Supramolecular gel-assisted synthesis Co2P particles anchored in multielement co-doped graphene as efficient bifunctional electrocatalysts for oxygen reduction and evolution. *Electrochim. Acta* **2017**, *231*, 344–353. [CrossRef]

106. Xuan, L.L.; Liu, X.J.; Wang, X. Cobalt phosphate nanoparticles embedded nitrogen and phosphorus-codoped graphene aerogels as effective electrocatalysts for oxygen reduction. *Front. Mater.* **2019**, *6*, 22. [CrossRef]
107. Wang, R.; Dong, X.Y.; Du, J.; Zhao, J.Y.; Zang, S.Q. MOF-derived bifunctional Cu3P nanoparticles coated by a N, P-codoped carbon shell for hydrogen evolution and oxygen reduction. *Adv. Mater.* **2018**, *30*, 1703711. [CrossRef]
108. Wu, D.Y.; Zhu, C.; Shi, Y.T.; Jing, H.Y.; Hu, J.W.; Song, X.D.; Si, D.H.; Liang, S.X.; Hao, C. Biomass-derived multilayer-graphene-encapsulated cobalt nanoparticles as efficient electrocatalyst for versatile renewable energy applications. *ACS Sustain. Chem. Eng.* **2019**, *7*, 1137–1145. [CrossRef]
109. Hao, X.Q.; Jiang, Z.Q.; Zhang, B.A.; Tian, X.N.; Song, C.S.; Wang, L.K.; Maiyalagan, T.; Hao, X.G.; Jiang, Z.J. N-doped carbon nanotubes derived from graphene oxide with embedment of FeCo nanoparticles as bifunctional air electrode for rechargeable liquid and flexible all-solid-state zinc-air batteries. *Adv. Sci.* **2021**, *8*, 2004572. [CrossRef]
110. Ni, B.X.; Chen, R.; Wu, L.M.; Sun, P.C.; Chen, T.H. Encapsulated FeP nanoparticles with in-situ formed P-doped graphene layers: Boosting activity in oxygen reduction reaction. *Sci. China-Mater.* **2021**, *64*, 1159–1172. [CrossRef]
111. Choi, C.H.; Chung, M.W.; Kwon, H.C.; Park, S.H.; Woo, S.I. B, N- and P, N-doped graphene as highly active catalysts for oxygen reduction reactions in acidic media. *J. Mater. Chem. A* **2013**, *1*, 3694–3699. [CrossRef]
112. Gracia-Espino, E. Behind the synergistic effect observed on phosphorus nitrogen codoped graphene during the oxygen reduction reaction. *J. Phys. Chem. C* **2016**, *120*, 27849–27857. [CrossRef]
113. Han, C.L.; Chen, Z.Q. The mechanism study of oxygen reduction reaction (ORR) on non-equivalent P, N co-doped graphene. *Appl. Surf. Sci.* **2020**, *511*, 145382. [CrossRef]
114. Zheng, X.J.; Yang, Z.R.; Wu, J.; Jin, C.; Tian, J.H.; Yang, R.Z. Phosphorus and cobalt co-doped reduced graphene oxide bifunctional electrocatalyst for oxygen reduction and evolution reactions. *RSC Adv.* **2016**, *6*, 64155–64164. [CrossRef]
115. Hellgren, N.; Berlind, T.; Gueorguiev, G.K.; Johansson, M.P.; Stafström, S.; Hultman, L. Fullerene-like bcn thin films: A computational and experimental study. *Mater. Sci. Eng. B* **2004**, *113*, 242–247. [CrossRef]
116. Broitman, E.; Gueorguiev, G.K.; Furlan, A.; Son, N.T.; Gellman, A.J.; Stafström, S.; Hultman, L. Water adsorption on fullerene-like carbon nitride overcoats. *Thin Solid Film.* **2008**, *517*, 1106–1110. [CrossRef]
117. Gueorguiev, G.K.; Goyenola, C.; Schmidt, S.; Hultman, L. Cfx: A first-principles study of structural patterns arising during synthetic growth. *Chem. Phys. Lett.* **2011**, *516*, 62–67. [CrossRef]
118. Nan, H.Y.; Ni, Z.H.; Wang, J.; Zafar, Z.; Shi, Z.X.; Wang, Y.Y. The thermal stability of graphene in air investigated by raman spectroscopy. *J. Raman Spectrosc.* **2013**, *44*, 1018–1021. [CrossRef]

Review

Manipulation on Two-Dimensional Amorphous Nanomaterials for Enhanced Electrochemical Energy Storage and Conversion

Juzhe Liu [1,2,†], Rui Hao [1,2,†], Binbin Jia [1,†], Hewei Zhao [1,*] and Lin Guo [1,*]

1. Beijing Advanced Innovation Center for Biomedical Engineering, Key Laboratory of Bio-Inspired Smart Interfacial Science and Technology, School of Chemistry, Beihang University, Beijing 100191, China; liujuzhe@buaa.edu.cn (J.L.); haorui628@buaa.edu.cn (R.H.); jiabinbin@buaa.edu.cn (B.J.)
2. School of Physics, Beihang University, Beijing 100191, China
* Correspondence: zhaohewei@buaa.edu.cn (H.Z.); guolin@buaa.edu.cn (L.G.)
† These authors contributed equally to this work.

Abstract: Low-carbon society is calling for advanced electrochemical energy storage and conversion systems and techniques, in which functional electrode materials are a core factor. As a new member of the material family, two-dimensional amorphous nanomaterials (2D ANMs) are booming gradually and show promising application prospects in electrochemical fields for extended specific surface area, abundant active sites, tunable electron states, and faster ion transport capacity. Specifically, their flexible structures provide significant adjustment room that allows readily and desirable modification. Recent advances have witnessed omnifarious manipulation means on 2D ANMs for enhanced electrochemical performance. Here, this review is devoted to collecting and summarizing the manipulation strategies of 2D ANMs in terms of component interaction and geometric configuration design, expecting to promote the controllable development of such a new class of nanomaterial. Our view covers the 2D ANMs applied in electrochemical fields, including battery, supercapacitor, and electrocatalysis, meanwhile we also clarify the relationship between manipulation manner and beneficial effect on electrochemical properties. Finally, we conclude the review with our personal insights and provide an outlook for more effective manipulation ways on functional and practical 2D ANMs.

Keywords: manipulation; two-dimension amorphous; component interaction; geometric configuration; electrochemistry

1. Introduction

With the intensification of the global energy crisis, electrochemical energy storage and transformation has become one of the most concerned research hotspots in the world. Therefore, it is necessary to develop efficient, clean, and sustainable energy technologies, such as supercapacitor, battery, and electrocatalysis. As the core parts of these systems, electrode materials have experienced vigorous development and achieved multi-size, multi-dimensional, and multi-component precise regulation to adapt the diverse and complex energy storage and transformation processes [1–4].

Electrochemical performance is closely related to the structure of electrode materials. The two-dimensionalization of electrode materials can increase electrochemically active surface area (ECSA) and facilitate ion diffusion for enhanced electrochemical performance, which has drawn extensive attention [5–8]. Different from conventional material control strategies mainly concentrated upon composition, morphology, and dimension, crystal phase control demonstrates some superiority, especially for enhancing performance. Many materials have more than one phase, which is mainly determined by chemical bonds and thermodynamic parameters. By precisely controlling various structural parameters, it is possible to obtain non-thermodynamically stable phase structure with disordered atomic arrangement over a long range and only short-range order over a few atoms, that is amorphous structure. The materials with amorphous structure are isotropic, lack grain boundaries, and endowed with inherent abundant defects, which have come into people's attention and worked as advanced electrode materials [9]. For instance, Lei et al. found amorphous titanium dioxide to be an efficient electrode material for sodium ion batteries with impressive charge storage capacity and cycle life [10]. Therefore, it is challenging but meaningful to combine the merits of two-dimensional and amorphous structures for developing well-performed, two-dimensional amorphous nanomaterials (2D ANMs). Compared with conventional materials, 2D ANMs generally exhibit distinctive features: i) ultra-high specific surface area and plentiful defects, which can provide more exposed active sites; ii) favorable diffusion paths and distances, which are conductive to the insertion/extraction of reactants and products; iii) strong in-plane covalent bond and lacking of grain boundary enhancing mechanical properties for extended volume or shape change; iv) flexible morphology and composition providing an additional degree of freedom for further modification; v) unprecedented electron state induced by confinement of electrons in 2D scale, which may facilitate electron transfer and electrode reactions.

Up to now, many 2D ANMs have been synthesized and applied in energy storage and transformation processes, including carbon materials [11], black phosphorus [12], metal compounds [13–16], etc. For example, Guo et al. developed amorphous cobalt-vanadium hydr(oxy)oxide nanosheets as an efficient electrocatalytic material for oxygen evolution reaction (OER) superior to the crystalline counterparts [17]; Wu and co-workers prepared 2D amorphous Cr_2O_3/graphene nanosheets by rapidly heating hydrous chlorides, which exhibited ultrahigh reversible capacity and outstanding cycle life in Li-ion battery outperforming crystalline nanoparticles [18]. However, the development of 2D ANMs is relatively lagging for their immature synthetic methods and drawbacks in application, such as low conductivity and instability. On the face of it, some strategies have been proposed to manipulate 2D ANMs based on their flexible structures and compositions for enhanced electrochemical energy storage and conversion [19,20]. This review aims at illuminating the modes and roles of manipulating 2D ANMs in electrochemical fields (supercapacitor, battery, and electrocatalysis) in terms of geometric configuration design and component interaction. First, we pay attention to manipulation strategy of 2D ANMs in recent years. Following this, we emphatically discuss their application and electrochemical mechanism. Finally, we conclude our personal insights and provide outlook for the development of 2D manipulative amorphous nanomaterials. We hope that the integration of 2D manipulative amorphous nanomaterials in electrochemistry will offer great opportunities to address the challenges driven by the increasing global electrochemical energy storage and transformation processes.

2. Manipulation Strategy of 2D ANMs
2.1. Synthesis Methods

J. Kotakoski et al. firstly used electron irradiation to create 2D amorphous carbon material in 2011, which opened new possibilities for preparing 2D ANMs [21]. For some intrinsic bulk materials that are amorphous nanomaterials, the desired 2D ANMs can be obtained by exfoliation. For example, dimethylformamide (DMF) is an ideal solvent to exfoliate bulk MoS_2 relative to other solvents due to its low surface tension (~40 mJ m^{-2}) [22]. In view of this fact, 2D amorphous MoS_3 nanosheets can be successfully obtained by the exfoliation of the bulk amorphous MoS_3 material in DMF solvent under ultrasonic irradiation (Figure 1a) [23]. Other types of 2D ANMs have been obtained by many reliable methods including thermal decomposition [24], electrodeposition [25,26], template method [27–29], phase transformation [30,31], and element doping [17,32]. Synthesis of amorphous noble metal nanostructures is always a great challenge for their strong and isotropic nature of metallic bonds. In view of this, Li et al. proposed a simple method for preparing amorphous noble metal nanosheets by directly annealing metal acetylacetonate with alkali salt (Figure 1b) [33]. The synthesis temperature was between the melting point of metal acetylacetone and the melting point of alkali salt. When alkali salt was removed by deionized water, high yield amorphous noble metal nanomaterials can be successfully obtained, including monometal nanosheets, bimetal nanosheets, and trimetal nanosheets. Guo et al. utilized sacrificial template strategy to yield a library of ten distinct 2D ultrathin amorphous metal hydroxide nanosheets [28]. The key point of the synthesis is based on the balance between the etching rate of the Cu_2O template and deposition rate of the metal hydroxide. As shown in Figure 1c, Cu_2O was first employed as a sacrificial template to promote the 2D planar growth of metal hydroxides into a nanosheet structure. Then, $S_2O_3^{2-}$ can react with Cu_2O to produce OH^- ions. Finally, after the concentrations of OH^- ions increased to the precipitation threshold, metal ions could combine with OH^- to form 2D amorphous sheet structure. In general, most of them are based on the classical 2D crystalline nanomaterials synthetic theory by introducing some mechanisms of inhibiting crystallization. The common inhibition factors involve shorting reaction time, reducing reaction temperature, destroying crystal structure, etc.

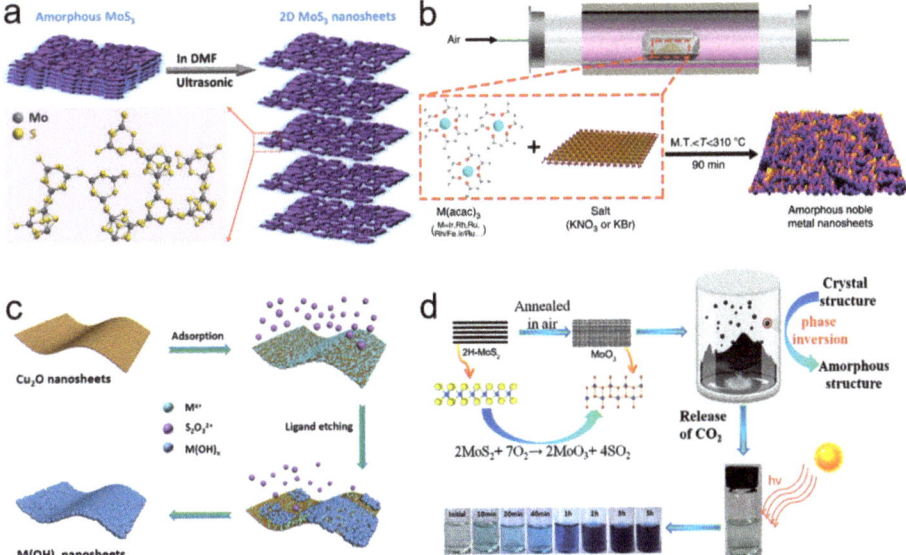

Figure 1. (a) Schematic illustration of exfoliation 2D amorphousMoS$_3$ nanosheets and their chemical structure. Reprinted with permission from Ref. [23]. Copyright Royal Society of Chemistry, 2019. (b) Schematic illustration of the general synthetic process for amorphous noble metal nanosheets. Reprinted with permission from Ref. [33]. (c) The schematic illustration of the synthesis of amorphous metal hydroxide nanosheets. Reprinted with permission from Ref. [28]. Copyright Royal Society of Chemistry, 2019. (d) Schematic illustration of the formation mechanism for amorphous MoO$_3$ nanosheets. Reprinted with permission from Ref. [31]. Copyright Wiley-VCH, 2017.

It needs to be clarified that some target amorphous products are difficult to prepare by one synthesis method and other methods should be involved. Xu et al. combined the oxidation of MoS$_2$ and supercritical CO$_2$ treatment strategy to prepare amorphous molybdenum oxide (MoO$_3$) nanosheets [31]. As shown in Figure 1d, single-layer or few layers of crystalline MoS$_2$ were firstly exfoliated. Then, oxygen atoms replaced sulfur atoms to destroy the regular atomic arrangement of MoS$_2$ during the annealing process. Finally, the stable amorphous MoO$_3$ was obtained by the adsorption of CO$_2$.

2.2. Manipulation Modes

2D amorphous material has flexible structure and composition that allows dexterous manipulation. As mentioned above, various strategies have been proposed to manipulate 2D ANMs for enhanced electrochemical performance. We generalize and conceptualize these strategies to be two major categories of geometric configuration design and component interaction. According to our understanding, the geometric configuration design mainly includes spatial structure design at micro/nano scale and coordination environment design at atomic scale. The component interaction mainly includes elemental interaction and heterophase compositing. Here, the relevant enhanced effects and implementation approaches will be introduced.

2.2.1. Geometric Configuration Design

Spatial Structure Design

Spatial structure design at micro/nano scale can be deemed as the manipulation on the shape, size, packing form, and porous structure of 2D ANMs, which can be controlled by template design and reactive conditions [34]. Specifically, endowing 2D ANMs with porous structure should be an advisable way for enhanced electrochemical property. In electrocatalysis process, porous nanostructure can provide large surface area and abundant active sites, ensure effective penetration of electrolyte ions and escape of products, and alleviate stacking problem of nanosheets. As a typical case for creating pores on 2D ANMs, Guo group proposed a universal strategy combining confined method and ion exchange strategy to synthesize a series of 2D porous amorphous metal oxide nanosheets, such as Fe_2O_3, Cr_2O_3, ZrO_2, SnO_2, and Al_2O_3 [35]. The schematic illustration for synthesis of ultrathin amorphous metal oxide nanosheets was demonstrated in Figure 2a. Firstly, lamellar oleate was introduced as a host matrix to restrict the Cu_2O template. Secondly, the target metal ion was replaced by Cu^+ ions and introduced into 2D space through ion exchange strategy to form corresponding amorphous $M(OH)_x$-oleate complex precursor. Finally, porous structure and disorder atom arrangement were achieved for metal oxide product by removing oleate and hydrone in heat treatment.

Coordination Environment Design

Tuning coordination environment at atomic scale can change the state of active sites, which afford improved electrochemical efficiency. Defect design is the most commonly used way and atomic-scale defects can be classified as anion vacancy, cation vacancy, associated vacancy, pits, distortions, and disorder [36]. Creating defects is generally deemed to be conductive to the mobility and adsorption of reactants and optimize reactive energy paths. In contrast to crystalline materials, precise design, and identification on defects are relatively difficult for amorphous ones with disorder atomic structure assembling massive and various defects, especially for 2D ANMs. Nonetheless, some efforts have been devoted to defect manipulation on 2D ANMs. Selective component removal or addition should be an effective way. Typically, Hou et al. developed amorphous MoS_x monolayer nanosheets with abundant Mo defects using the space-confined strategy [37]. The synthesis details are shown in Figure 2b. The precursor of layered double hydroxide with MoS_4^{2-} (LDH-MoS_4^{2-}) was first synthesized via dispersing a layered double oxide (LDO) in an aqueous solution of $(NH_4)_2MoS_4$. Afterwards, the obtained precursor was calcined in a N_2 atmosphere to form amorphous MoS_x monolayer nanosheets in the interlayer space of LDO. Finally, amorphous MoS_x monolayer nanosheets were successfully obtained by washing in a nonoxidative HCl solution to dissolve LDO (the host layers). In this process, the generation of Mo defects can be adjusted by calcination temperature, which affects the S/Mo atomic ratio.

2.2.2. Component Interaction

Elemental Interaction

Doping or coupling other elements may be a feasible method to enhance the electrochemical performance of 2D ANMs due to the multielement synergy effect. Commonly adopted strategies are direct coupling and post-doping. Wei et al. fabricated Fe-doped amorphous $VOPO_4$ in solvothermal environment by one-pot two-phase colloidal method (Figure 2c) [38]. The oil phase consisted of oleylamine (OM) and octadecene (ODE) dissolved with Fe and V precursor, which is mixed with water phase containing sodium dihydrogen phosphate, and then was sealed in an autoclave and heated to get the final product. Figure 2d demonstrated a typical post-doping way that crystalline CoMo ultrathin hydroxide was firstly constructed by coprecipitation reaction and then amorphous Fe-doped CoMo ultrathin hydroxide nanosheets was obtained by ion exchange process [39].

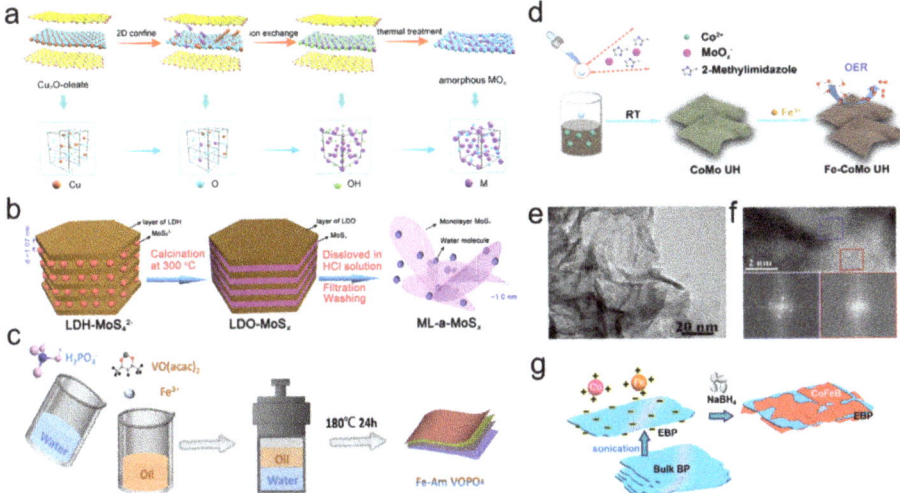

Figure 2. (**a**) Schematic illustration for synthesis of amorphous metal oxide ultrathin nanosheets. Reprinted with permission from Ref. [35]. Copyright American Chemical Society, 2020. (**b**) Schematic of the preparation process for amorphous MoS$_x$ monolayer nanosheets with abundant Mo defects. Reprinted with permission from Ref. [37]. Copyright Elsevier, 2020. (**c**) Schematic representation of the one-pot preparation processes of Fe-doped amorphous VOPO$_4$. Reprinted with permission from Ref. [38]. Copyright Elsevier, 2021. (**d**) Schematic illustration of amorphous Fe-doped CoMo ultrathin hydroxide nanosheets. Reprinted with permission from Ref. [39]. Copyright Royal Society of Chemistry, 2021. (**e**) HRTEM image of CoV-Fe hydroxide nanosheets and (**f**) corresponding FFT patterns of selected regions marked by blue and red squares, respectively. Reprinted with permission from Ref. [40]. Copyright Wiley-VCH, 2020. (**g**) Schematic illustrations of the synthesis of EBP/CoFeB nanosheets. Reprinted with permission from Ref. [41]. Copyright American Chemical Society, 2021.

Heterophase Compositing

Compositing is a common means to combine a different phase with 2D ANMs. It can integrate advantages and realize optimized design on interfacial structure, holistic architecture and physical property, embody in enhanced conductivity, modulated electron structure and active sites, improved stability, etc. Recently, the introduction of crystal phase into the amorphous phase to form a crystalline/amorphous hybrid dual-phase structure has attracted much attention, due to the unique properties produced by the phase boundary. The flexible amorphous structure has abundant active centers, which can enhance the electrochemical activity, while the crystalline structure possesses a highly symmetrical nonflexible structure, which can enhance the stability of the material. Yan et al. prepared hybrid dual-phase materials by doping Fe in CoV hydroxide nanosheets composed of a large number of crystalline and amorphous domain mixtures (Figure 2e,f) [40].

The unique interfaces of the catalyst promote the exposure of the active center, adjust the local coordination environment and electronic structure, and reduce the thermodynamic barrier during the OER catalytic reaction. Carbon materials are desirable candidates to form compositing structure with 2D ANMs. Wen et al. reported an exfoliated black phosphorus/CoFeB nanosheet (EBP/CoFeB) implemented by three steps under a N$_2$ atmosphere (Figure 2g) [41]. First, EBP was obtained from bulk phosphorus by liquid stripping;

then metal ions (Co^{2+}, Fe^{2+}) were adsorbed on the surface of EBP through electrostatic interaction; finally, CoFeB nanosheets were grown on EBP through chemical reduction initiated by $NaBH_4$. Both of them provide good demonstration on manipulating 2D ANMs by compositing way. Besides, many amorphous nanosheets deposited on various conductive substrates have been successfully synthesized to enhance the electrochemical performance, such as nickel foam [42,43], graphite [44,45], graphene [27], and TiO_2 mesh [46].

3. Manipulating 2D ANMs for Batteries and Supercapacitors

Well-manipulated 2D ANMs are attractive and show considerable application potential for diverse electrochemical systems, benefiting from their unique properties including abundant pores for ion storage capacitance, larger interlayer distance for ion de/intercalation, enhanced conductivity and elemental interaction by compositing. In this section, we mainly concentrate on introducing the manipulative strategies on 2D ANMs in renewable energy technologies including rechargeable battery (Lithium-ion battery (LIB), Sodium-ion battery (SIB), Potassium-ion battery (KIB)) and Supercapacitor (SC).

3.1. Rechargeable Battery

Geometric configuration is an effective way to operate 2D ANM in order to overcome the obstacles of volume expansion and faded capacity for electrode materials in long cycles [47]. Specifically, by introducing heterostructure, the electrochemical cycle-life and rate performance should be apparently improved. Guo group reported a breathable 2D MnO_2 artificial leaf with atomic thickness (b-MnO_2 ALAT) and proposed the manipulation approach of 2D ANMs by modifying the pore structure and compositing crystalline skeleton on the amorphous substrate (Figure 3a–c) [48]. This obtained ultrathin leaf-like structure comprises of amorphous microporous mesophyll-like nanosheet as substrate and vein-like crystalline skeleton as support (Figure 3d). As shown in Figure 3e, when used as the anode material for LIBs, it delivered high capacity of 520 mAh g^{-1} and extremely stable cycle life over 2500 cycles at 1.0 A g^{-1}, overcoming the disadvantages of pure MnO_2 with irreversible capacity loss and poor cycling behavior. The outstanding electrochemical performance was elaborated as follows (Figure 3f): First, 2D nanostructure possessed large surface area, which can accommodate the volume changes associated with electrochemical reactions; Second, porous amorphous structure guaranteed the effective wetting and penetration of electrolyte, offered continuous charge transport pathway, and buffered volume changes and shortened ion diffusion distance. Third, the vein-like crystalline support could perfectly solve the closely stacking problem of 2D ultrathin nanosheets since it could leave a small inter space between overlapped nanosheets and effectively increase the number of lithium-storage sites and ion diffusion rate.

Based on the same strategy, Xu et al. explored a novel non-van der Waals (non-vdW) heterostructure of 2D amorphous MoO_{3-x} (aMoO_{3-x}) nanosheet on Ti_3C_2-MXene (Figure 3g,h), which displayed superior electrochemical properties than counterparts. Density functional theory (DFT) calculations (Figure 3i,j) suggested that the amorphous non-vdW heterostructure can strongly stabilize aMoO_{3-x} nanosheet contributing to the improved stability and conductivity as well as facile Li ion diffusion. The restacked 2D heterostructures provide additional 2D Li^+ diffusion pathways (Figure 3k), where a significant amount of Li^+ can be stored on the surface defects and surface vacant sites of aMoO_{3-x}. When applied as the anode for LIB, it shows excellent rate capability (Figure 3l) and high reversible capacity (Figure 3m), which is more outstanding than those of self-assembled aMoO_{3-x}/MXene vdW heterostructure and bulk aMoO_{3-x} [49]. Huang et al. integrated 2D porous amorphous Si nanoflakes with ultralong multiwalled carbon nanotubes (MWCNTs) as a freestanding film electrode with high volumetric capacity and energy density. The interconnected network can prevent adjacent amorphous nanoflakes from restacking and the 2D porous structure provides large electrode/electrolyte contact area, both of which can enhance fast Li^+ transportation and suppress the volume change [50]. Wang et al.

successfully composited amorphous MoS₂ with different carbon-based nanomaterials as LIB anode for increased conductivity and energy density [51].

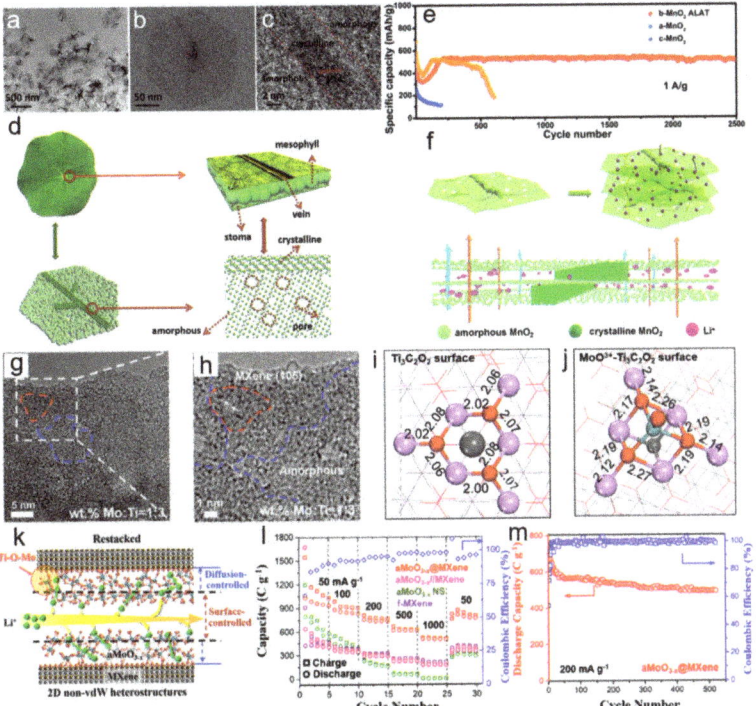

Figure 3. (**a**,**b**) TEM images of b-MnO₂ ALAT; (**c**) HRTEM image of b-MnO₂ ALAT; (**d**) Illustration of the breathable artificial leaves structure; (**e**) Cycling performance of b-MnO₂ ALAT at 1 A g^{-1}; (**f**) Schematic illustration of advantageous features of b-MnO₂ ALAT for energy storage. Reprinted with permission from Ref. [48]. Copyright Wiley-VCH, 2019. (**g**,**h**) HRTEM images of 2D heterostructures of aMoO$_{3-x}$@MXene non-vdW heterostructures; (**i**,**j**) Ti-O bond lengths in MXenes before and after the adsorption of MoO^{3+}, respectively; (**k**) Illustration of facile capacitor-like interlayer diffusion and diffusion-controlled interlayer diffusion; (**l**) Cycling performance at different rates for aMoO$_{3-x}$ NS, self-assembled aMoO$_{3-x}$//MXene vdW heterostructures, and aMoO$_{3-x}$@MXene non-vdW heterostructures; (**m**) Cycling performance for aMoO$_{3-x}$@MXene non-vdW heterostructures at 200 mA g^{-1}. Reprinted with permission from Ref. [49]. Copyright Elsevier, 2021.

Component interaction is another satisfactory approach to further optimize the relationship of 2D amorphous structure and electrochemical properties. SIB and KIB are considered to replace LIB and become the protagonist of the next generation of energy storage. However, the ion de-intercalation and volume expansion problems caused by the large ion radius of sodium and potassium are still serious obstacles in the actual application process. Hence, adaptable 2D ANMs should be ideal candidates for SIB and KIB, benefiting from synergistic effects including improved conductivity, enlarged interlayer, reformed wettability and introduced vacancies. Wang et al. successfully explored a nanohybrid of amorphous vanadium oxide on V₂C MXene (a-VO$_x$/V₂C) (Figure 4a). The coexistence of a-VO$_x$ and V₂C is demonstrated in Figure 4b,c. Electron paramagnetic resonance revealed that the a-VO$_x$/V₂C electrode with disordered V-O framework generated more oxygen vacancies than c-VO$_x$ (Figure 4d), which would favor fast Na$^+$ insertion/extraction. When used as the anode for SIBs, it possesses more excellent cycling performance compared to the c-VO$_x$ (Figure 4e) [52]. Yu and co-workers prepared 2D amorphous MoS$_3$-on-reduced graphene oxide (MoS$_3$-on-rGO) (Figure 4f), which exhibited a superior compatibility in KIBs. The contact angle tests (Figure 4g,h) showed that the amorphous MoS$_3$-on-rGO electrode was endowed with a superior wetting property to the carbonate electrolyte owning to higher surface free energy of amorphous MoS$_3$ and unique 3D interconnected porous structure, contributing to an optimized ion diffusion kinetics during the electrochemical process. When applied as the anode for KIBs, it exhibits high specific capacity (541 mAh g^{-1} at a current density of 0.2 A g^{-1}) and excellent long-term cycling stability, which is significantly superior to the corresponding crystal sample (Figure 4i,j) [53]. Ji et al. took full advantage of element interaction to obtain phosphorus-doped amorphous carbon nanosheet (P-CNS) through thermal treatment, which achieved high performance as anode material for SIB. The long cycle stability exhibited a high specific capacity of 149 mAh g^{-1} for 5000 cycles at 5 A g^{-1} [54]. Amorphous FeO$_x$ nanosheets with loose packing characteristic was developed by Hong and co-workers, which showed high electrochemical performance that specific capacity can be maintained at 263.4 mAh g^{-1} as an anode material for SIBs [55]. Bao et al. obtained a promising cathode material of SIB through coating amorphous FePO$_4$ nanosheets on carbon nanosphere, which displayed high initial discharge capacity of 126.4 mAh g^{-1} at a current density of 20 mA g^{-1} and superior cycling performance [56].

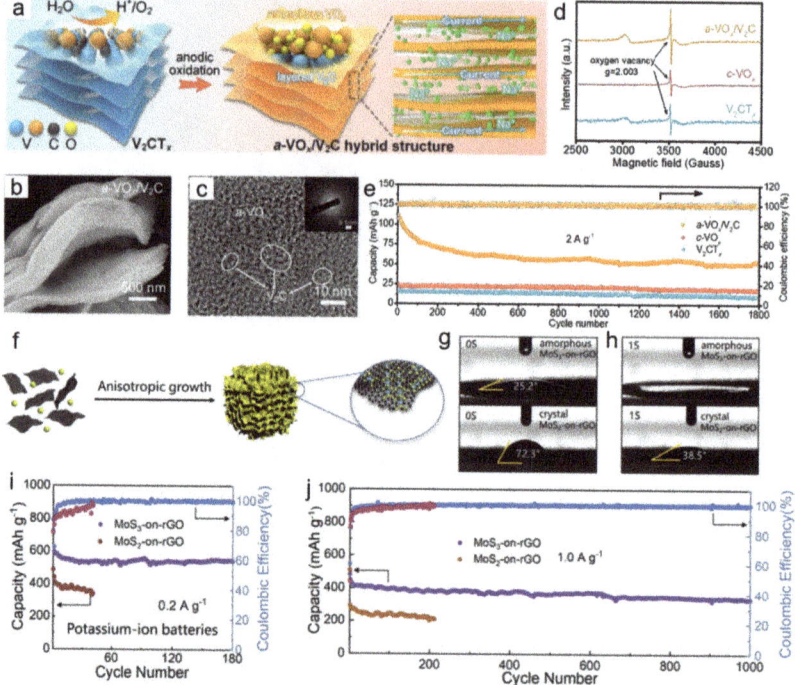

Figure 4. (**a**) Schematic illustration showing the synthesis and structure; (**b**) SEM image; (**c**) HRTEM image, the inset is the SAED pattern; (**d**) EPR spectra; (**e**) Cycling performances of a-VO$_x$/V$_2$C. Reprinted with permission from Ref. [52]. Copyright Wiley-VCH, 2021. (**f**) Schematic illustration of growth behaviors of amorphous MoS$_3$-on-rGO; (**g**,**h**) Contact angle images of electrolyte on the electrode surface of amorphous MoS$_3$-on-rGO and crystal MoS$_2$-on-rGO. (**i**) Cyclic performance of KIB at 0.2 A g^{-1}. (**j**) Long cycle performance of KIB at 1.0 A g^{-1}. Reprinted with permission from Ref. [53]. Copyright Wiley-VCH, 2021.

3.2. Supercapacitor

Supercapacitors (SCs) are being increasingly used to complement or partially replace batteries in various energy storage application owning to its high power density and exceptionally long lifetime. In the electrochemical reactions of SCs, the materials need to allow favorable diffusion of the electrolyte ions to access the active materials and cope with the strain and stress during charging/discharging process. Hence, manipulating 2D ANMs is now realized as an effective method to increase the electrochemical performance. The design of hole structure is considered to be one of the effective means to improve the performance of SCs. Qiu et al. developed a general approach for the synthesis of 2D porous carbon nanosheets from bio-sources-derived carbon precursors by an integrated procedure of intercalation, pyrolysis, and activation (Figure 5a,b). The as-prepared nanosheets possess optimized porous structures, which can shorten the ion transport distance during the charging/discharging process. When used as the electrode material in SCs, it shows a significantly improved rate performance with a high specific capacitance of 246 F g^{-1} and capacitance retention of 82% at 100 A g^{-1}, being nearly twice than that of carbon particulates directly obtained from gelatin (Figure 5c) [57]. In addition, the synergistic interaction between different elements is also very effective in improving the performance of 2D ANMs in SC. Chen et al. synthesized Ni–OH nanosheets via a one-pot hydrothermal method. Then, a cation exchange reaction was conducted to exchange amorphous Ni(OH)$_2$ with metal ions (Co^{2+}, Mn^{2+}, Cu^{2+} and Zn^{2+}) to obtain a series of bimetal nanosheets (Figure 5d). Due to the higher activity from the combined contribution of Ni and Co, the NiCo–OH nanosheets show a superior specific capacity, rate performance, and cycling stability compared to that of pure Ni(OH)$_2$ nanosheets (Figure 5e) [58]. Zhang et al. explored amorphous Co–Ni pyrophosphates nanosheets through controllably adjusting the ratios of Co and Ni. The optimized amorphous Ni–Co pyrophosphate showed much higher specific capacitance than monometallic Ni and Co pyrophosphates and exhibits excellent cycling ability [59]. Chen et al. proposed hydrothermal synthesis strategy to prepare amorphous NiCoMn–OH nanosheets, which was used as positive electrode materials for SC. The strong synergy between the transition metal ions in amorphous NiCoMn–OH is deemed to significantly promote the electrochemical activity, rate capability, and cycling stability. It is worth mentioning that the robust synthesis method was also used to fabricate the NiCoMn–OH porous network on conductive Ni foam (Figure 5f) and showed a specific capacity close to its theoretical value, indicating a complete utilization of the electroactive material [60]. Similarly, Zhu et al. fabricated ultrathin amorphous Co–Fe–B nanosheets on the 3D nickel foam substrate (Figure 5g) and the obtained sample showed an excellent specific capacitance (981 F g^{-1} at the 1 A g^{-1}) and superior rate performance (Figure 5h) [61].

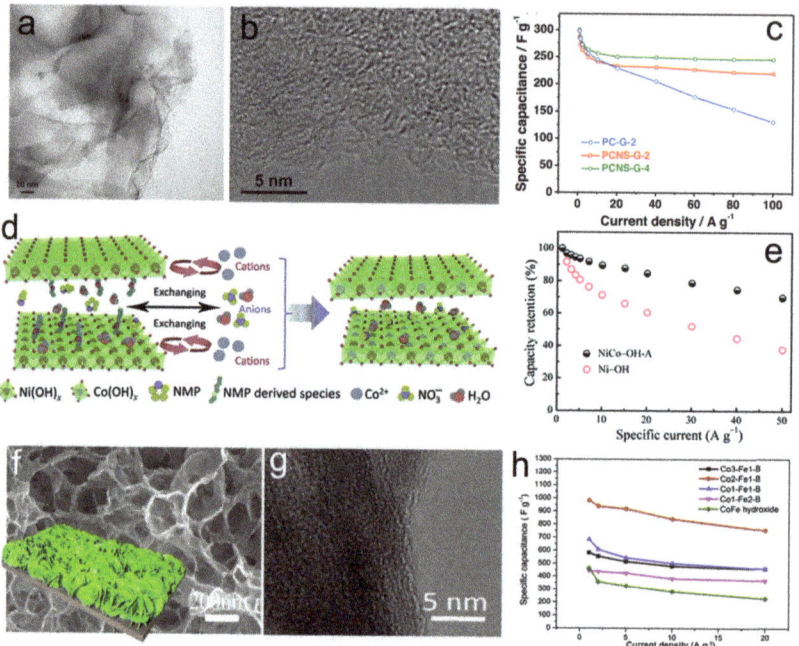

Figure 5. (**a**) TEM image of 2D porous carbon nanosheets; (**b**) HRTEM image of 2D porous carbon nanosheets; (**c**) Specific capacitances of 2D porous carbon nanosheets at different current densities. Reprinted with permission from Ref. [57]. Copyright Wiley-VCH, 2015. (**d**) Schematic diagram of the mechanism for ion exchange reactions; (**e**) Rate performance of amorphous Ni–Co hydroxide nanosheets. Reprinted with permission from Ref. [58]. Copyright Royal Society of Chemistry, 2018. (**f**) FESEM image and schematic illustration of the microstructure of NiCoMn–OH on Ni foams. Reprinted with permission from Ref. [60]. Copyright Elsevier, 2019. (**g**) HRTEM of the $Co_{0.2}Ni_{0.8}$ pyrophosphate nanosheets. (**h**) The cycling performance at the current density of 1.5 A g^{-1} for the $Co_{0.2}Ni_{0.8}$ pyrophosphate and $Co_{0.2}Ni_{0.8}NH_4PO_4 \cdot H_2O$ precursor. Reprinted with permission from Ref. [61]. Copyright Elsevier, 2019.

4. Manipulating 2D ANMs for Electrocatalysis

To build a clean future, massive efforts are underway to achieve high-efficiency and high-selectivity electrocatalysis systems for utilizing renewable energy and producing higher-value chemicals. Most typical electrocatalysis processes include nitrogen reduction reaction (NRR), carbon dioxide reduction reaction (CRR), oxygen reduction reaction (ORR) and water splitting which involves anodic oxygen evolution reaction (OER) and cathodic hydrogen evolution reaction (HER), etc. Catalysts play core role in diminishing energy loss and optimizing kinetics in these processes. Designing catalytic materials into 2D amorphous structure should be a wise and promising way since it can realize superiority combination of extended exposed area, abundant active sites, tunable electron states, and faster ion transport capacity. Accordingly, a batch of 2D ANMs have been developed to satisfy the urgent need. Here, we provide collective knowledge of manipulating 2D ANMs for electrocatalysis based on catalytic processes.

4.1. Water Splitting

Electrochemical water splitting is generally deemed as one of the most convenient and promising strategies to transform intermittent energy (e.g., solar and wind) to produce hydrogen. HER and OER are two- and four-electron processes, respectively, and thus OER is more kinetically sluggish and suffers from higher overpotentials. Traditional catalysts such as Pt or IrO$_2$, etc. with high activity are scarce and high cost. Some well-manipulated 2D ANMs have been demonstrated to possess satisfactory performance and compelling potential to substitute traditional noble metal catalysts.

Tuning composition or introducing hetero atoms for element interaction should be an effective way to develop efficient 2D amorphous catalytic materials. Based on this modification strategy, Guo and co-workers developed a simple, yet robust one-step coprecipitation method to fabricate ultrathin amorphous cobalt-vanadium hydr(oxy)oxide nanosheets (CoV-UAH) with a thickness of ~0.7 nm (Figure 6a) [17]. The involvement of V is proved to give rise to the formation of ultrathin amorphous structure, which allows facile transformation to the desirable active phase consisting of V-doped cobalt oxyhydroxide species. First-principle simulations suggest that V doping can optimize reaction free energies of neighboring Co sites, leading to a theoretical low overpotential (Figure 6b). Thus, the talented material possesses large electrochemically active surface areas, low charge-transfer resistance, and impressive performance for the OER with a low overpotential of 0.250 V at a current density of 10 mA cm^{-2} and Tafel slope of 44 mV dec^{-1} superior to counterparts and commercial IrO$_2$ (Figure 6c). In their other work, amorphous delafossite analogue nanosheets were fabricated by in situ electrochemical self-reconstruction, which features special structure of Ag intercalated into bimetallic cobalt-iron (oxyhydr)oxide layers (Figure 6d,e) [62]. The introduction of Ag modulates can regulate integral electron state and optimize electrocatalysis energetics, leading to superior OER activity and stability. Haik and co-workers doped Ga into amorphous cobalt boride nanosheets on Au support for smoothing growth of nanosheets and modifying surface electronic structure, thereby achieving a well-performed electrode with 230 mV overpotential to attain a current density of 10 mA cm^{-2} [63]. In another case, Co ion-intercalation can tune the structure of amorphous cobalt manganese oxide nanosheets at the atomic-scale to expose more active sites and allow easy penetration of electrolyte ions [64]. In HER field, highly active amorphous CoMoS4 nanosheets were constructed by coupling Mo to cobalt-based nanosheets, which showed favorable free energy change for H* adsorption and remarkable activity [65]. Some other metals can be benign dopants, such as Fe [38,39], Mo [66], Ag [62], etc. [67]. Surely, coupling nonmetal with 2D ANMs should also bring up beneficial synergistic effect. It is found that phosphating of metal (hydr)oxides (e.g., CoFe hydroxide, FeMnO$_x$) nanosheets can result in amorphization so as to obtain highly active 2D amorphous catalyst for water splitting including both HER and OER with optimized catalytic sites and faster electronic transport [68,69]. Amorphous CoBP ternary alloy nanosheets were demonstrated to a well-performed HER catalyst [70]. Its remarkable activity can be attributed to synergistic effect of elements P and B, which accelerates dissociation of H$_2$O, weakens surface H absorption, and suppresses Co oxidation.

In addition to element interaction, compositing hetero phases or substrates with 2D amorphous nanomaterials should also be a promising manner to realize multi-advantage integration. Chen et al. constructed a 2D heterostructure of EBP/CoFeB, which exhibits excellent OER activity with an ultralow overpotential of 227 mV at 10 mA cm^{-2} and excellent stability (Figure 6f) [41]. This nanohybrid structure not only optimizes the reactive intermediate absorption, but also combines the high conductivity of black phosphorus. Combining amorphous phase with crystalline phase also seems to be a wise way to promote catalytic properties of catalysts. Huang and co-workers rationally designed channel-rich RuCu nanosheets composed of crystallized Ru and amorphous Cu for overall water splitting in pH-universal electrolytes [71]. The amorphous/crystalline compositing structure is endowed with highly active electron transfer and optimized electronic structures. Furthermore, Hu and co-workers developed amorphous NiO nanosheets coupled with crystalline Ni and MoO$_3$ nanoparticles, which exhibited two heterostructures of Ni–NiO and MoO$_3$–NiO (Figure 6g) [72]. The deliberately manipulated structure dramatically diminishes the energetic barrier and works as catalytically active centers, synergistically improving the overall water splitting. The similar strategy was also adopted to develop crystalline platinum oxide-decorated amorphous cobalt oxide hydroxide nanosheets and amorphous

RuCu nanosheets grown on crystalline Cu nanotubes as HER catalysts [46,73]. Besides, enhanced catalytic effect by compositing hetero phases can be achieved by assembling amorphous nanosheets with nanodots [74], carbon nanofibers [75], metal oxide [76–78], etc. [79,80]. It should be mentioned that anchoring amorphous nanosheets on high conductivity substrate should be an effective way to afford electrolysis at large current density. Zhao and co-worker electrodeposited amorphous mesoporous nickel-iron composite nanosheets directly onto macro-porous nickel foam as OER electrode, which can deliver current densities of 500 and 1000 mA cm^{-2} at overpotentials of 240 and 270 mV, respectively [42]. Zhang et al. anchored amorphous MoS$_2$ nanosheet arrays on carbon cloth to form a three-dimensional nanostructure with abundant exposed edge sites [81]. The composite exhibited satisfactory catalytic activity and durability for the HER in acidic solutions. In addition, Ni foil [82], graphite foil [44], and Ti plate [83] were also proved to be good substrates.

Defect design and porousness manipulation should be regarded as micro-design on geometric configuration, which can be easily carried out on 2D amorphous structure. Shao et al. found sulfur defects could modulate the electron state and Gibbs free energies of amorphous Mo–FeS nanosheets, leading to preferable OER kinetics (Figure 6h) [84]. Mo defects could be created on monolayer amorphous molybdenum sulfide and identified as catalytically active sites for HER [37]. In addition, porousness and channel design also show positive effect for electrocatalysis of water splitting [71]. Amorphous cobalt phosphate nanosheets, amorphous CoS$_x$(OH)$_y$ nanosheets, and amorphous NiCoFe phosphate nanosheets with well-designed porous structure were developed as efficient OER catalysts [85–87]. The porous characteristic can provide large free space, increased distribution of the active centers, and facile movement of reactants and products, thus enhanced catalysis performance can be achieved. Except for micro-design, the spatial arrangement of amorphous nanosheet should be noteworthy. Yang et al. realized in situ vertical growth of amorphous FePO$_4$ nanosheets (Figure 6i) on Ni foam which demonstrated excellent catalytical activity and stability in overall water splitting [88]. This type of geometric configuration is favorable to electron transport, electrolyte diffusion, and structural stability in catalysis processes.

4.2. Electrochemical Reduction Reactions

Some 2D ANMs were also manipulated to catalyze electrochemical reduction reactions, such as ORR, CRR, and NRR. As to positive effect induced by component interaction for electrochemical reduction reactions, similar ways of tuning composition or introducing hetero atoms and compositing hetero phases or substrates are also advisable. Wang et al. tailored amorphous NiFeB nanosheets for enhancing the electrocatalytic NRR kinetics by adjusting the ratio of Ni/Fe/B (Figure 7a) [89]. The collective merits of amorphous structure and optimized element ratio gives rise to synergistic effect in creating active sites and facilitating the N$_2$ adsorption capacity, thereby leading to superior NRR activity with a high NH$_3$ formation rate of 3.24 µg h^{-1} cm^{-2}. Sun and co-workers developed biomass-derived oxygen-doped amorphous carbon nanosheets with high electrochemical selectivity and activity for NRR [90]. Similarly, highly efficient ORR catalysts with abundant active sites and high conductivity were fabricated by doping N or P into amorphous carbon sheets [91,92]. As mentioned above, amorphous nanosheets possess flexible morphology and structure, which can provide an ideal platform for supporting or embedding nanoparticles. In a work by Lu et al., a unique hybrid catalyst of Pd hydride nanocubes encapsulated within amorphous Ni–B nanosheets (PdH$_x$@Ni–B) was synthesized and demonstrated an impressive ORR activity (1.05 mg$_{Pd}$$^{-1}$ at 0.90 V versus reversible hydrogen electrode) and stability (10,000 potential cycles) (Figure 7b) [93]. Gao et al. constructed a composite catalyst of Au nanocrystals@amorphous MnO$_2$ nanosheets with CO faradic efficiency (FE) of 90.5% for CRR at −0.7 V versus reversible hydrogen electrode [94]. The core/shell nanostructure brings about the interaction between Au and amorphous MnO$_2$ nanosheets, which contributed to its performance (Figure 7c). Specially, Yuan et al. reported the mass-production of amorphous SnO$_x$ nanoflakes modified by BiO$_x$ species from nanoparticles to single atoms, which exhibited an FE of HCOOH over 90% in CRR [20].

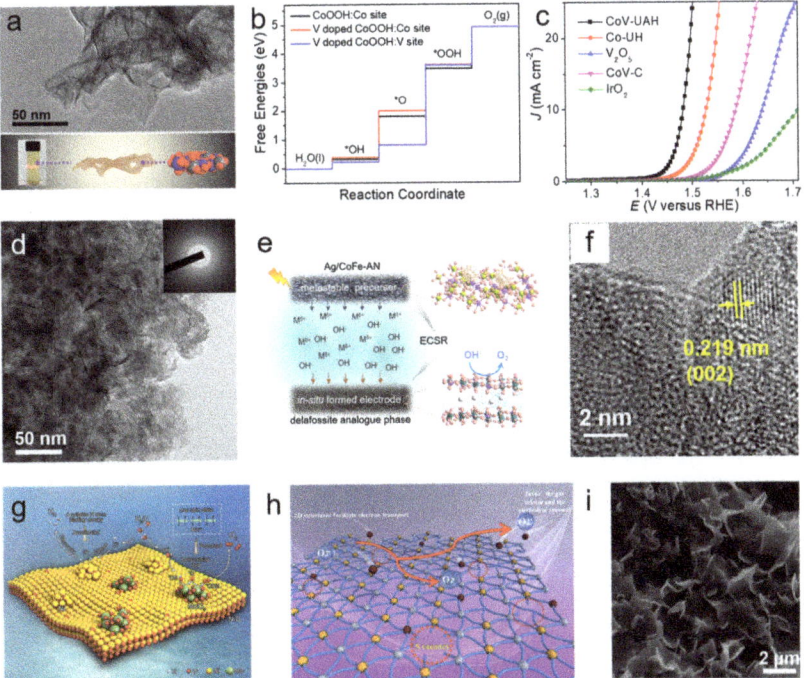

Figure 6. (a) The TEM image of CoV-UAH; (b) The free-energy landscape for V doped cobalt oxyhydroxide; (c) Linear sweep voltammetry curves for CoV-UAH and counterparts. Reprinted with permission from Ref. [17]. Copyright Royal Society of Chemistry, 2018. (d) The TEM image of amorphous delafossite analogue nanosheets, and inset is selected area electron diffraction (SAED); (e) Self-reconstruction process. Reprinted with permission from Ref. [62]. (f) HRTEM image of EBP/CoFeB sample. Reprinted with permission from Ref. [41]. Copyright American Chemical Society, 2021. (g) The enhanced mechanism of hybrid nanocatalysts for overall water splitting. Reprinted with permission from Ref. [72]. Copyright Wiley-VCH, 2020. (h) Activity mechanism for Mo–FeS nanosheets with sulfur defects. Reprinted with permission from Ref. [84]. Copyright American Chemical Society, 2020. (i) High magnification SEM image of amorphous FePO$_4$ nanosheets on Ni foam. Reprinted with permission from Ref. [88]. Copyright Wiley-VCH, 2017.

4.3. Electrochemical Reduction Reactions

Some 2D ANMs were also manipulated to catalyze electrochemical reduction reactions, such as ORR, CRR, and NRR. As to positive effect induced by component interaction for electrochemical reduction reactions, similar ways of tuning composition or introducing hetero atoms and compositing hetero phases or substrates are also advisable. Wang et al. tailored amorphous NiFeB nanosheets for enhancing the electrocatalytic NRR kinetics by adjusting the ratio of Ni/Fe/B (Figure 7a) [89]. The collective merits of amorphous structure and optimized element ratio gives rise to synergistic effect in creating active sites and facilitating the N$_2$ adsorption capacity, thereby leading to superior NRR activity with a high NH$_3$ formation rate of 3.24 μg h^{-1} cm^{-2}. Sun and co-workers developed biomass-derived oxygen-doped amorphous carbon nanosheets with high electrochemical selectivity and activity for NRR [90]. Similarly, highly efficient ORR catalysts with abundant active sites and high conductivity were fabricated by doping N or P into amorphous carbon sheets [91,92]. As mentioned above, amorphous nanosheets possess flexible morphology and structure, which can provide an ideal platform for supporting or embedding nanoparticles. In a work by Lu et al., a unique hybrid catalyst of Pd hydride nanocubes encapsulated within amorphous Ni–B nanosheets (PdH_x@Ni–B) was synthesized and demonstrated an impressive ORR activity (1.05 mg_{Pd}^{-1} at 0.90 V versus reversible hydrogen electrode) and stability (10,000 potential cycles) (Figure 7b) [93]. Gao et al. constructed a composite catalyst of Au nanocrystals@amorphous MnO_2 nanosheets with CO faradic efficiency (FE) of 90.5% for CRR at −0.7 V versus reversible hydrogen electrode [94]. The core/shell nanostructure brings about the interaction between Au and amorphous MnO_2 nanosheets, which contributed to its performance (Figure 7c). Specially, Yuan et al. reported the mass-production of amorphous SnO_x nanoflakes modified by BiO_x species from nanoparticles to single atoms, which exhibited an FE of HCOOH over 90% in CRR [20].

For electrochemical reduction reactions, the optimization of geometric configuration is also recommendable. As typical cases, amorphous MoO_{3-x} monolayers with oxygen vacancies and amorphous MoS_3 nanosheets with sulfur vacancies can work as efficient NRR catalysts [95,96]. The vacancy defects are able to modulate electron state of catalysts and reduce energetics barriers for facilitating NRR process and simultaneously suppressing HER (Figure 7d). Additionally, porous design is achieved on amorphized FeB_2 nanosheets for boosted NRR activity with an NH_3 yield of 39.8 µg h^{-1} mg^{-1} (Figure 7e) [97]. The porous amorphous structure could upraise the d-band center of a-FeB_2 and strengthen the absorption of key *N_2H intermediate, thereby reducing reaction barrier (Figure 7f).

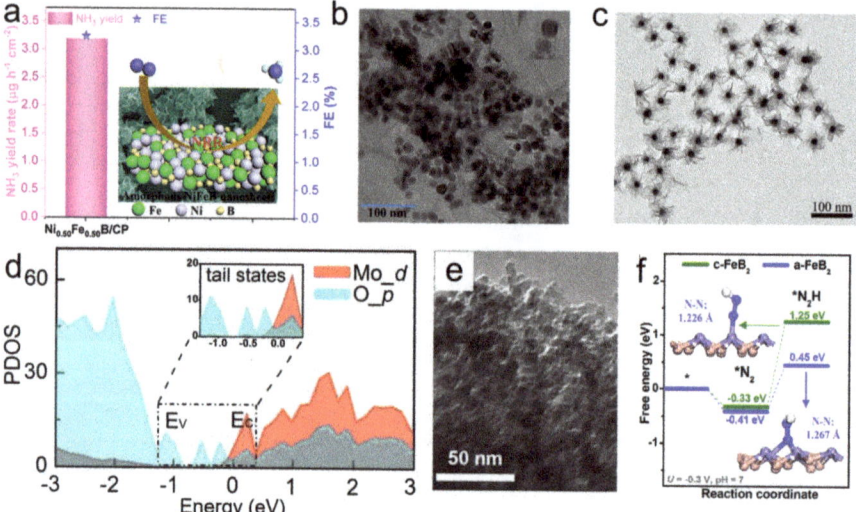

Figure 7. (a) Schematic diagram of NRR on amorphous NiFeB nanosheets. Reprinted with permission from Ref. [89]. Copyright American Chemical Society, 2020. (b) TEM image of PdH_x@Ni–B. Reprinted with permission from Ref. [93]. Copyright Wiley-VCH, 2017. (c) TEM image of Au nanocrystals@amorphous MnO_2 nanosheets. Reprinted with permission from Ref. [94]. Copyright American Chemical Society, 2021. (d) Anderson tail states of amorphous MoO_{3-x}. Reprinted with permission from Ref. [95]. Copyright Wiley-VCH, 2019. (e) TEM image of porous FeB_2 nanosheets. (f) Free energy diagrams of *N_2 and *N_2H adsorption on crystalline FeB_2 and amorphous FeB_2. Reprinted with permission from Ref. [97]. Copyright Elsevier, 2021.

5. Conclusions and Outlook

The interests in 2D ANMs are growing continuously along with the extensive study of amorphous material science. These materials are promising candidates for facilitating the key processes of electrochemical energy storage and conversion systems due to their unique advantages of large specific surface area, excellent "in-plane" charge-carrier transport, abundant defects, etc. However, some issues still exist for their electrochemical application: (i) synthesis systems and mechanisms are lacking and ambiguous, which limit their categories and quantity production; (ii) most of the synthesized 2D ANMs are metal oxide with poor conductivity; (iii) the dispersity and structural stability of 2D ANMs are unsatisfactory due to the high surface energy; (iv) intrinsic activity still deserves improvement. Hence, it is indeed necessary to explore more effective and reliable methods to optimize this family of materials for preferable electrochemical application.

In this review, we summarized effective strategies to manipulate on 2D ANMs and their applications in battery, supercapacitor, and electrocatalysis. We conceptualized these strategies to be geometric configuration design and component interaction, concretely embodying in spatial structure and coordination environment design as well as elemental interaction and heterophase compositing. For geometric configuration, the introduction of pores or defects within nanosheets can provide more active sites, superior electrolyte diffusion and ion transport kinetics. As to component interaction, heteroatom doping can change the band structure and electronic properties, while heterophase compositing enable advantage integration to achieve improved conductivity and stability. Thanks to the flexible structure of 2D ANMs, these elaborate manipulations can be realized by deliberate synthetic routes. These manipulated 2D ANMs with optimized structures and properties demonstrated enhanced electrochemical performance, while discriminatory manipulation ways are related with different applications.

2D ANMs are intriguing, and manipulating them for purposive application is promising. Despite the visible progress that has been witnessed, there are still many issues to be addressed: (i) 2D amorphous structure is mysterious, which retards our deeper cognition; (ii) immature synthesis methods; (iii) controllable manipulation means are still lacking, especially at atomic scale; (iv) in-depth understanding to the roles of well-built structure in electrochemical processes is insufficient. To meet these challenges, more advanced characterization techniques are needed to clarify the nature of 2D amorphous structures, formation mechanisms, and functional rules. Meanwhile, some experience can be selectively drawn from crystalline systems. As such, the discovery, manipulation, and application of 2D ANMs following the success of amorphous materials have opened up a new pave for sustainable energy applications. It is believed that the development of new 2D ANMs and their derived materials will further not only play a role in improving the performance of sustainable energy devices and contribute to resolving the current environmental and energy crises, but also stimulate the advances in amorphous science field.

Author Contributions: L.G. designed the manuscript; J.L., R.H., B.J. and H.Z. wrote the manuscript and prepared the figures. H.Z. revised the manuscript. All authors have read and agreed to the published version of the manuscript.

Funding: This work is supported by the National Natural Science Foundation of China (51532001, 51802010, 52073008, 51772011, U1910208); China Postdoctoral Science Foundation (2019TQ0020, 2019TQ0013, 2020TQ0023, 2019M660398, 2020M670088 and 2020M680295).

Institutional Review Board Statement: Not applicable.

Informed Consent Statement: Not applicable.

Data Availability Statement: Not applicable.

Conflicts of Interest: The authors declare no conflict of interest.

References

1. McCreery, R.L. Advanced carbon electrode materials for molecular electrochemistry. *Chem. Rev.* **2008**, *108*, 2646–2687. [CrossRef] [PubMed]
2. Augustyn, V.; Simonbc, P.; Dunn, B. Pseudocapacitive oxide materials for high-rate electrochemical energy storage. *Energy Environ. Sci.* **2014**, *7*, 1597–1614. [CrossRef]
3. Suen, N.-T.; Hung, S.-F.; Quan, Q.; Zhang, N.; Xu, Y.-J.; Chen, H.M. Electrocatalysis for the oxygen evolution reaction: Recent development and future perspectives. *Chem. Soc. Rev.* **2017**, *46*, 337–365. [CrossRef] [PubMed]

4. Luo, M.; Guo, S. Strain-controlled electrocatalysis on multimetallic nanomaterials. *Nat. Rev. Mater.* **2017**, *2*, 17059. [CrossRef]
5. Anasori, B.; Lukatskaya, M.R.; Gogotsi, Y. 2D metal carbides and nitrides (MXenes) for energy storage. *Nat. Rev. Mater.* **2017**, *2*, 16098. [CrossRef]
6. Tan, C.; Cao, X.; Wu, X.-J.; He, Q.; Yang, J.; Zhang, Z.; Chen, J.; Zhao, W.; Han, S.; Nam, G.-H.; et al. Recent advances in ultrathin two-dimensional nanomaterials. *Chem. Rev.* **2017**, *117*, 6225–6331. [CrossRef]
7. Chia, X.; Pumera, M. Characteristics and performance of two-dimensional materials for electrocatalysis. *Nat. Catal.* **2018**, *1*, 909–921. [CrossRef]
8. Pomerantseva, E.; Gogotsi, Y. Two-dimensional heterostructures for energy storage. *Nat. Energy* **2017**, *2*, 17089. [CrossRef]
9. Li, Q.; Xu, Y.; Zheng, S.; Guo, X.; Xue, H.; Pang, H. Recent progress in some amorphous materials for supercapacitors. *Small* **2018**, *14*, 1800426–1800444. [CrossRef]
10. Zhou, M.; Xu, Y.; Wang, C.; Li, Q.; Xiang, J.; Liang, L.; Wu, M.; Zhao, H.; Lei, Y. Amorphous TiO_2 inverse opal anode for high-rate sodium ion batteries. *Nano Energy* **2017**, *31*, 514–524. [CrossRef]
11. Toh, C.; Zhang, H.; Lin, J.; Mayorov, A.S.; Wang, Y.-P.; Orofeo, C.M.; Ferry, D.B.; Andersen, H.; Kakenov, N.; Guo, Z.; et al. Synthesis and properties of free-standing monolayer amorphous carbon. *Nature* **2020**, *557*, 199–203. [CrossRef]
12. Bellus, M.Z.; Yang, Z.; Hao, J.; Lau, S.P.; Zhao, H. Amorphous two-dimensional black phosphorus with exceptional photocarrier transport properties. *2D Mater.* **2017**, *4*, 025063. [CrossRef]
13. Zhang, L.; Liu, P.F.; Li, Y.H.; Wang, C.W.; Zu, M.Y.; Fu, H.Q.; Yang, X.H.; Yang, H.G. Accelerating neutral hydrogen evolution with tungsten modulated amorphous metal hydroxides. *ACS Catal.* **2018**, *8*, 5200–5205. [CrossRef]
14. Radhakrishnan, T.; Aparna, M.P.; Chatanathodi, R.; Sandhyarani, N. Amorphous rhenium disulfide nanosheets: A methanol-tolerant transition metal dichalcogenide catalyst for oxygen reduction reaction. *ACS Appl. Nano Mater.* **2019**, *2*, 4480–4488. [CrossRef]
15. Li, H.; Gao, Y.; Wang, C.; Yang, G. A simple electrochemical route to access amorphous mixed-metal hydroxides for supercapacitor electrode materials. *Adv. Energy Mater.* **2014**, *5*, 1401767. [CrossRef]
16. Chodankar, N.R.; Dubal, D.P.; Ji, S.-H.; Kim, D.-H. Self-assembled nickel pyrophosphate-decorated amorphous bimetal hydroxides 2D-on-2D nanostructure for high-energy solid-state asymmetric supercapacitor. *Small* **2019**, *15*, 1901145. [CrossRef]
17. Liu, J.; Ji, Y.; Nai, J.; Niu, X.; Luo, Y.; Guo, L.; Yang, S. Ultrathin amorphous cobalt-vanadium hydr (oxy) oxide catalysts for the oxygen evolution reaction. *Energy Environ. Sci.* **2018**, *11*, 1736–1741. [CrossRef]
18. Zhao, C.; Zhang, H.; Si, W.; Wu, H. Mass production of two-dimensional oxides by rapid heating of hydrous chlorides. *Nat. Commun.* **2016**, *7*, 1–8. [CrossRef]
19. Xu, H.; Fei, B.; Cai, G.; Ha, Y.; Liu, J.; Jia, H.; Zhang, J.; Liu, M.; Wu, R. Boronization-induced ultrathin 2D nanosheets with abundant crystalline-amorphous phase boundary supported on nickel foam toward efficient water splitting. *Adv. Energy Mater.* **2020**, *10*, 1902714. [CrossRef]
20. Yuan, T.; Hu, Z.; Zhao, Y.; Fang, J.; Lv, J.; Zhang, Q.; Zhuang, Z.; Gu, L.; Hu, S. Two-dimensional amorphous SnO_x from liquid metal: Mass production, phase transfer, and electrocatalytic CO_2 reduction toward formic acid. *Nano Lett.* **2020**, *20*, 2916–2922. [CrossRef]
21. Kotakoski, J.; Krasheninnikov, A.V.; Kaiser, U.; Meyer, J.C. From point defects in graphene to two-dimensional amorphous carbon. *Phys. Rev. Lett.* **2011**, *106*, 105505. [CrossRef]
22. Gan, X.; Zhao, H.; Wong, K.; Lei, D.; Zhang, Y.; Quan, X. Covalent functionalization of MoS_2 nanosheets synthesized by liquid phase exfoliation to construct electrochemical sensors for Cd (II). *Talanta* **2017**, *182*, 38–48. [CrossRef]
23. Fu, W.; Yang, S.; Yang, H.; Guo, B.; Huang, Z. 2D amorphous MoS_3 nanosheets with porous network structures for scavenging toxic metal ions from synthetic acid mine drainage. *J. Mater. Chem. A* **2019**, *7*, 18799–18806. [CrossRef]
24. Selvaraj, A.R.; Muthusamy, A.; Inho-Cho; Kim, H.-J.; Senthil, K.; Prabakara, K. Ultrahigh surface area biomass derived 3D hierarchical porous carbon nanosheet electrodes for high energy density supercapacitors. *Carbon* **2021**, *174*, 463–474. [CrossRef]
25. Liu, W.; Liu, H.; Dang, L.; Zhang, H.; Wu, X.; Yang, B.; Li, Z.; Zhang, X.; Lei, L.; Jin, S. Amorphous cobalt-iron hydroxide nanosheet electrocatalyst for efficient electrochemical and photo-electrochemical oxygen evolution. *Adv. Funct. Mater.* **2017**, *27*, 1603904. [CrossRef]
26. Yu, L.; Zhou, H.; Sun, J.; Mishra, I.K.; Luo, D.; Yu, F.; Yu, Y.; Chen, S.; Ren, Z. Amorphous NiFe layered double hydroxide nanosheets decorated on 3D nickel phosphide nanoarrays: A hierarchical core-shell electrocatalyst for efficient oxygen evolution. *J. Mater. Chem. A* **2018**, *6*, 13619–13623. [CrossRef]
27. Zhao, H.; Zhu, Y.; Li, F.; Hao, R.; Wang, S.; Guo, L. A generalized strategy for the synthesis of large-size ultrathin two-dimensional metal oxide nanosheets. *Angew. Chem. Int. Ed.* **2017**, *129*, 8892–8896. [CrossRef]
28. Jia, B.; Hao, R.; Huang, Z.; Hu, P.; Li, L.; Zhang, Y.; Guo, L. Creating ultrathin amorphous metal hydroxide and oxide nanosheet libraries. *J. Mater. Chem. A* **2019**, *7*, 4383–4388. [CrossRef]
29. Zhao, H.; Yue, Y.; Zhang, Y.; Li, L.; Guo, L. Ternary artificial nacre reinforced by ultrathin amorphous alumina with exceptional mechanical properties. *Adv. Mater.* **2016**, *28*, 2037–2042. [CrossRef]
30. Jiang, Y.; Song, Y.; Pan, Z.; Meng, Y.; Jiang, L.; Wu, Z.; Yang, P.; Gu, Q.; Sun, D.; Hu, L. Rapid amorphization in metastable $CoSeO_3 \cdot H_2O$ nanosheets for ultrafast lithiation kinetics. *ACS Nano* **2018**, *12*, 5011–5020. [CrossRef]
31. Liu, W.; Xu, Q.; Cui, W.; Zhu, C.; Qi, Y. CO_2-assisted fabrication of two-dimensional amorphous molybdenum oxide nanosheets for enhanced plasmon resonances. *Angew. Chem. Int. Ed.* **2017**, *56*, 1600–1604. [CrossRef] [PubMed]

32. Lin, Z.; Du, C.; Yan, B.; Wang, C.; Yang, G. Two-dimensional amorphous NiO as a plasmonic photocatalyst for solar H_2 evolution. *Nat. Commun.* **2018**, *9*, 1–11. [CrossRef] [PubMed]
33. Wu, G.; Zheng, X.; Cui, P.; Jiang, H.; Wan, X.; Qu, Y.; Chen, W.; Lin, Y.; Li, H.; Han, X.; et al. A general synthesis approach for amorphous noble metal nanosheets. *Nat. Commun.* **2019**, *10*, 1–11. [CrossRef] [PubMed]
34. Qin, C.; Fan, A.; Ren, D.; Luan, C.; Yang, J.; Liu, Y.; Zhang, X.; Dai, X.; Wang, M. Amorphous NiMS (M: Co, Fe or Mn) holey nanosheets derived from crystal phase transition for enhanced oxygen evolution in water splitting. *Electrochim. Acta* **2019**, *323*, 134756. [CrossRef]
35. Jia, B.; Yang, J.; Hao, R.; Li, L.; Guo, L. Confined synthesis of ultrathin amorphous metal-oxide nanosheets. *ACS Mater. Lett.* **2020**, *2*, 610–615. [CrossRef]
36. Thomas, S.; Jung, H.; Kim, S.; Jun, B.; Lee, C.; Lee, S. Two-dimensional haeckelite h567: A promising high capacity and fast Li diffusion anode material for lithium-ion batteries. *Carbon* **2019**, *148*, 344–353. [CrossRef]
37. Wang, D.; Li, H.; Du, N.; Hou, W. Amorphous molybdenum sulfide monolayer nanosheets for highly efficient electrocatalytic hydrogen evolution. *Chem. Eng. J.* **2020**, *398*, 125685. [CrossRef]
38. Wei, Q.; Tan, X.; Zhang, J.; Yang, L.; Cao, L.; Dong, B. Fe doped amorphous single layered vanadyl phosphate nanosheets as highly efficient electrocatalyst for water oxidation. *J. Colloid Interface Sci.* **2021**, *586*, 505–513. [CrossRef]
39. Zeng, L.; Cao, B.; Wang, X.; Liu, H.; Shang, J.; Lang, J.; Cao, X.; Gu, H. Ultrathin amorphous iron-doped cobalt-molybdenum hydroxide nanosheets for advanced oxygen evolution reactions. *Nanoscale* **2021**, *13*, 3153–3160. [CrossRef]
40. Kuang, M.; Zhang, J.; Liu, D.; Tan, H.; Dinh, K.N.; Yang, L.; Ren, H.; Huang, W.; Fang, W.; Yao, J.; et al. Amorphous/crystalline heterostructured cobalt-vanadium-iron (oxy) hydroxides for highly efficient oxygen evolution reaction. *Adv. Energy Mater.* **2020**, *10*, 2002215. [CrossRef]
41. Chen, H.; Chen, J.; Ning, P.; Chen, X.; Liang, J.; Yao, X.; Chen, D.; Qin, L.; Huang, Y.; Wen, Z. 2D heterostructure of amorphous CoFeB coating black phosphorus nanosheets with optimal oxygen intermediate absorption for improved electrocatalytic water oxidation. *ACS Nano* **2021**, *15*, 12418–12428. [CrossRef]
42. Lu, X.; Zhao, C. Electrodeposition of hierarchically structured three-dimensional nickel-iron electrodes for efficient oxygen evolution at high current densities. *Nat. Commun.* **2015**, *6*, 6616. [CrossRef]
43. Yoon, S.; Yun, J.-Y.; Lim, J.-H.; Yoo, B. Enhanced electrocatalytic properties of electrodeposited amorphous cobalt-nickel hydroxide nanosheets on nickel foam by the formation of nickel nanocones for the oxygen evolution reaction. *J. Alloys Compd.* **2017**, *693*, 964–969. [CrossRef]
44. Ye, Y.-J.; Zhang, N.; Liu, X.-X. Amorphous NiFe(oxy)hydroxide nanosheet integrated partially exfoliated graphite foil for high efficiency oxygen evolution reaction. *J. Mater. Chem. A* **2017**, *5*, 24208–24216. [CrossRef]
45. Gao, Y.Q.; Li, H.B.; Yang, G.W. Amorphous Co $(OH)_2$ nanosheet electrocatalyst and the physical mechanism for its high activity and long-term cycle stability. *J. Appl. Phys.* **2016**, *119*, 034902. [CrossRef]
46. Wang, Z.; Ren, X.; Shi, X.; Asiri, A.M.; Wang, L.; Li, X.; Sun, X.; Zhang, Q.; Wang, H. A platinum oxide decorated amorphous cobalt oxide hydroxide nanosheet array towards alkaline hydrogen evolution. *J. Mater. Chem. A* **2018**, *6*, 3864–3868. [CrossRef]
47. Mahmood, N.; Tang, T.; Hou, Y. Nanostructured anode materials for lithium ion batteries: Progress, challenge and perspective. *Adv. Energy Mater.* **2016**, *6*, 1600374. [CrossRef]
48. Jia, B.; Chen, W.; Luo, J.; Yang, Z.; Li, L.; Guo, L. Construction of MnO_2 artificial leaf with atomic thickness as highly stable battery anodes. *Adv. Mater.* **2019**, *32*, 1906582. [CrossRef]
49. Yan, P.; Ji, L.; Liu, X.; Guan, Q.; Guo, J.; Shen, Y.; Zhang, H.; Wei, W.; Cui, X.; Xu, Q. 2D amorphous-MoO_{3-x}@Ti_3C_2-MXene non-van der Waals heterostructures as anode materials for lithium-ion batteries. *Nano Energy* **2021**, *86*, 106139. [CrossRef]
50. Wang, Z.; Li, Y.; Huang, S.; Liu, L.; Wang, Y.; Jin, J.; Kong, D.; Zhang, L.; Schmidtbe, O.G. PVD customized 2D porous amorphous silicon nanoflakes percolated with carbon nanotubes for high areal capacity lithium ion batteries. *J. Mater. Chem. A* **2020**, *8*, 4836–4843. [CrossRef]
51. Liu, X.; Zhang, X.; Ma, S.; Tong, S.; Han, X.; Wang, H. Flexible amorphous MoS_2 nanoflakes/N-doped carbon microtubes/reduced graphite oxide composite paper as binder free anode for full cell lithium ion batteries. *Electrochim. Acta* **2020**, *333*, 135568. [CrossRef]
52. Zhang, W.; Peng, J.; Hua, W.; Liu, Y.; Wang, J.; Liang, Y.; Lai, W.; Jiang, Y.; Huang, Y.; Zhang, W.; et al. Architecting amorphous vanadium oxide/MXene nanohybrid via tunable anodic oxidation for high-performance sodium-ion batteries. *Adv. Energy Mater.* **2021**, *11*, 2100757. [CrossRef]
53. Ma, M.; Zhang, S.; Wang, L.; Yao, Y.; Shao, R.; Shen, L.; Yu, L.; Dai, J.; Jiang, Y.; Cheng, X.; et al. Harnessing the Volume Expansion of MoS_3 Anode by Structure Engineering to Achieve High Performance Beyond Lithium-Based Rechargeable Batteries. *Adv. Mater.* **2021**, *33*, 2106232. [CrossRef]
54. Hou, H.; Shao, L.; Zhang, Y.; Zou, G.; Chen, J.; Ji, X. Large-area carbon nanosheets doped with phosphorus: A high-performance anode material for sodium-ion batteries. *Adv. Sci.* **2017**, *4*, 1600243. [CrossRef] [PubMed]
55. Sun, R.; Gao, J.; Wu, G.; Liu, P.; Guo, W.; Zhou, H.; Ge, J.; Hu, Y.; Xue, Z.; Li, H.; et al. Amorphous metal oxide nanosheets featuring reversible structure transformations as sodium-ion battery anodes. *Cell Rep.* **2020**, *1*, 100118. [CrossRef]
56. Zhang, Z.; Han, Y.; Xu, J.; Ma, J.; Zhou, X.; Bao, J. Construction of amorphous $FePO_4$ nanosheets with enhanced sodium storage properties. *ACS Appl. Energy Mater.* **2018**, *1*, 4395–4402. [CrossRef]

57. Fa, X.; Yu, C.; Yang, J.; Ling, Z.; Hu, C.; Zhang, M.; Qiu, J. A Layered-nanospace-confinement strategy for the synthesis of two-dimensional porous carbon nanosheets for high-rate performance supercapacitors. *Adv. Energy Mater.* **2015**, *5*, 1401761.
58. Huang, C.; Song, X.; Qin, Y.; Xu, B.; Chen, H.C. Cation exchange reaction derived amorphous bimetal hydroxides as advanced battery materials for hybrid supercapacitors. *J. Mater. Chem. A* **2018**, *6*, 21047–21055. [CrossRef]
59. Chen, C.; Zhang, N.; He, Y.; Liang, B.; Ma, R.; Liu, X. Controllable fabrication of amorphous Co-Ni pyrophosphates for tuning electrochemical performance in supercapacitors. *ACS Appl. Mater. Interfaces* **2016**, *8*, 23114–23121. [CrossRef]
60. Chen, H.; Qin, Y.; Cao, H.; Song, X.; Huang, C.; Feng, H.; Zhao, X.S. Synthesis of amorphous nickel-cobalt-manganese hydroxides for supercapacitor-battery hybrid energy storage system. *Energy Storage Mater.* **2019**, *17*, 194–203. [CrossRef]
61. Meng, Q.; Xu, W.; Zhu, S.; Liang, Y.; Cui, Z.; Yang, X.; Inoue, A. Low-cost fabrication of amorphous cobalt-iron-boron nanosheets for high-performance asymmetric supercapacitors. *Electrochim. Acta* **2019**, *296*, 198–205. [CrossRef]
62. Liu, J.; Hu, Q.; Wang, Y.; Yang, Z.; Fan, X.; Liu, L.-M.; Guo, L. Achieving delafossite analog by in situ electrochemical self-reconstruction as an oxygen-evolving catalyst. *Proc. Natl. Acad. Sci. USA* **2020**, *117*, 21906–21913. [CrossRef]
63. Haq, T.u.; Mansour, S.A.; Munir, A.; Haik, Y. Gold-supported gadolinium doped CoB amorphous sheet: A new benchmark electrocatalyst for water oxidation with high turnover frequency. *Adv. Funct. Mater.* **2020**, *30*, 1910309. [CrossRef]
64. Zhao, C.; Yu, C.; Huang, H.; Han, X.; Liu, Z.; Qiu, J. Co ion-intercalation amorphous and ultrathin microstructure for high-rate oxygen evolution. *Energy Storage Mater.* **2018**, *10*, 291–296. [CrossRef]
65. Ren, X.; Wu, D.; Ge, R.; Sun, X.; Ma, H.; Yan, T.; Zhang, Y.; Du, B.; Wei, Q.; Chen, L. Self-supported $CoMoS_4$ nanosheet array as an efficient catalyst for hydrogen evolution reaction at neutral pH. *Nano Res.* **2018**, *11*, 2024–2033. [CrossRef]
66. Zhang, H.; Jiang, H.; Hu, Y.; Saha, P.; Li, C. Mo-triggered amorphous Ni_3S_2 nanosheets as efficient and durable electrocatalysts for water splitting. *Mater. Chem. Front.* **2018**, *2*, 1462–1466. [CrossRef]
67. Chen, L.; Ren, X.; Teng, W.; Shi, P. Amorphous Nickel-Cobalt-Borate nanosheet arrays for efficient and durable water oxidation electrocatalysis under near-neutral conditions. *Chem. Eur. J.* **2017**, *23*, 9741–9745. [CrossRef]
68. Zhang, W.; Li, Y.; Zhou, L.; Zheng, Q.; Xie, F.; Lam, K.H.; Lin, D. Ultrathin amorphous CoFeP nanosheets derived from CoFe LDHs by partial phosphating as excellent bifunctional catalysts for overall water splitting. *Electrochim. Acta* **2019**, *323*, 134595. [CrossRef]
69. Shen, L.; Zhang, Q.; Luo, J.; Fu, H.C.; Chen, X.H.; Wu, L.L.; Luo, H.Q.; Li, N.B. Heteroatoms adjusting amorphous FeMn-based nanosheets via a facile electrodeposition method for full water splitting. *ACS Sustain. Chem. Eng.* **2021**, *9*, 5963–5971. [CrossRef]
70. Sun, H.; Xu, X.; Yan, Z.; Chen, X.; Jiao, L.; Cheng, F.; Chen, J. Superhydrophilic amorphous Co-B-P nanosheet electrocatalysts with Pt-like activity and durability for the hydrogen evolution reaction. *J. Mater. Chem. A* **2018**, *6*, 22062–22069. [CrossRef]
71. Yao, Q.; Huang, B.; Zhang, N.; Sun, M.; Shao, Q.; Huang, X. Channel-rich RuCu nanosheets for pH-Universal overall water splitting electrocatalysis. *Angew. Chem. Int. Ed.* **2019**, *58*, 13983–13988. [CrossRef]
72. Li, X.; Wang, Y.; Wang, J.; Da, Y.; Zhang, J.; Li, L.; Zhong, C.; Deng, Y.; Han, X.; Hu, W. Sequential electrodeposition of bifunctional catalytically active structures in MoO_3/Ni-NiO composite electrocatalysts for selective hydrogen and oxygen evolution. *Adv. Mater.* **2020**, *32*, 2003414. [CrossRef]
73. Cao, D.; Wang, J.; Xu, H.; Cheng, D. Growth of highly active amorphous RuCu nanosheets on Cu nanotubes for the hydrogen evolution reaction in wide pH values. *Small* **2020**, *16*, 2000924. [CrossRef]
74. Zhao, H.; Yang, Y.; Dai, X.; Qiao, H.; Yong, J.; Luan, X.; Yu, L.; Luan, C.; Wang, Y.; Zhang, X. NiCo-DH nanodots anchored on amorphous NiCo-Sulfide sheets as efficient electrocatalysts for oxygen evolution reaction. *Electrochim. Acta* **2019**, *295*, 1085–1092. [CrossRef]
75. Sukanya, R.; Chen, S.-M. Amorphous cobalt boride nanosheets anchored surface-functionalized carbon nanofiber: An bifunctional and efficient catalyst for electrochemical sensing and oxygen evolution reaction. *J. Colloid Interface Sci.* **2020**, *580*, 318–331. [CrossRef]
76. Liu, Q.; Cao, F.; Wu, F.; Lu, H.; Li, L. Ultrathin Amorphous $Ni(OH)_2$ Nanosheets on Ultrathin α-Fe_2O_3 Films for Improved Photoelectrochemical Water Oxidation. *Adv. Mater. Interfaces* **2016**, *3*, 1600256. [CrossRef]
77. Fang, M.; Han, D.; Xu, W.-B.; Shen, Y.; Lu, Y.; Cao, P.; Han, S.; Xu, W.; Zhu, D.; Liu, W.; et al. Surface-guided formation of amorphous mixed-metal oxyhydroxides on ultrathin MnO_2 nanosheet arrays for efficient electrocatalytic oxygen evolution. *Adv. Energy Mater.* **2020**, *10*, 2001059. [CrossRef]
78. Niu, Y.; Li, W.; Wu, X.; Feng, B.; Yu, Y.; Hu, W.; Li, C.M. Amorphous nickel sulfide nanosheets with embedded vanadium oxide nanocrystals on nickel foam for efficient electrochemical water oxidation. *J. Mater. Chem. A* **2019**, *7*, 10534–10542. [CrossRef]
79. Masa, J.; Sinev, I.; Mistry, H.; Ventosa, E.; de la Mata, M.; Arbiol, J.; Muhler, M.; Cuenya, B.R.; Schuhmann, W. Ultrathin high surface area nickel boride (Ni_xB) nanosheets as highly efficient electrocatalyst for oxygen evolution. *Adv. Energy Mater.* **2017**, *7*, 1700381. [CrossRef]
80. Zou, X.; Liu, Y.; Li, G.-D.; Wu, Y.; Liu, D.-P.; Li, W.; Li, H.-W.; Wang, D.; Zhang, Y.; Zou, X. Ultrafast formation of amorphous bimetallic hydroxide films on 3D conductive sulfide nanoarrays for large-current-density oxygen evolution electrocatalysis. *Adv. Mater.* **2017**, *29*, 1700404. [CrossRef]
81. Zhang, X.; Zhang, Y.; Yu, B.-B.; Yin, X.-L.; Jiang, W.-J.; Jiang, Y.; Hu, J.-S.; Wan, L.-J. Physical vapor deposition of amorphous MoS_2 nanosheet arrays on carbon cloth for highly reproducible large-area electrocatalysts for the hydrogen evolution reaction. *J. Mater. Chem. A* **2015**, *3*, 19277–19281. [CrossRef]

82. Ahn, B.-W.; Kim, T.-Y.; Kim, S.-H.; Song, Y.-I.; Suh, S.-J. Amorphous MoS$_2$ nanosheets grown on copper@nickel-phosphorous dendritic structures for hydrogen evolution reaction. *Appl. Surf. Sci.* **2018**, *432*, 183–189. [CrossRef]
83. Zhang, J.; Hu, Y.; Liu, D.; Yu, Y.; Zhang, B. Enhancing oxygen evolution reaction at high current densities on amorphous-like Ni-Fe-S ultrathin nanosheets via oxygen incorporation and electrochemical tuning. *Adv. Sci.* **2017**, *4*, 1600343. [CrossRef] [PubMed]
84. Shao, Z.; Meng, H.; Sun, J.; Guo, N.; Xue, H.; Huang, K.; He, F.; Li, F.; Wang, Q. Engineering of amorphous structures and sulfur defects into ultrathin FeS nanosheets to achieve superior electrocatalytic alkaline oxygen evolution. *ACS Appl. Mater. Interfaces* **2020**, *12*, 51846–51853. [CrossRef]
85. Guo, X.; Xu, Y.; Cheng, Y.; Zhang, Y.; Pang, H. Amorphous cobalt phosphate porous nanosheets derived from two-dimensionalcobalt phosphonate organic frameworks for high performance of oxygen evolution reaction. *Appl. Mater. Today* **2020**, *18*, 100517. [CrossRef]
86. Zeng, Y.; Chen, L.; Chen, R.; Wang, Y.; Xie, C.; Tao, L.; Huang, L.; Wang, S. One-step, room temperature generation of porous and amorphous cobalt hydroxysulfides from layered double hydroxides for superior oxygen evolution reactions. *J. Mater. Chem. A* **2018**, *6*, 24311–24316. [CrossRef]
87. Sial, M.A.Z.G.; Lin, H.; Wang, X. Microporous 2D NiCoFe phosphate nanosheets supported on Ni foam for efficient overall water splitting in alkaline media. *Nanoscale* **2018**, *10*, 12975–12980. [CrossRef]
88. Yang, L.; Guo, Z.; Huang, J.; Xi, Y.; Gao, R.; Su, G.; Wang, W.; Cao, L.; Dong, B. Vertical growth of 2D amorphous FePO$_4$ nanosheet on Ni foam: Outer and inner structural design for superior water splitting. *Adv. Mater.* **2017**, *29*, 1704574. [CrossRef]
89. Wang, Y.; Tian, Y.; Zhang, J.; Yu, C.; Cai, R.; Wang, J.; Zhang, Y.; Wu, J.; Wu, Y. Tuning morphology and electronic structure of amorphous NiFeB nanosheets for enhanced electrocatalytic N$_2$ reduction. *ACS Appl. Energy Mater.* **2020**, *3*, 9516–9522. [CrossRef]
90. Huang, H.; Xia, L.; Cao, R.; Niu, Z.; Chen, H.; Liu, Q.; Li, T.; Shi, X.; Asiri, A.M.; Sun, X. A Biomass-derived carbon-based electrocatalyst for efficient N$_2$ fixation to NH$_3$ under ambient conditions. *Chem. Eur. J.* **2019**, *25*, 1914–1917. [CrossRef]
91. Li, Q.; Kong, D.; Zhao, X.; Cai, Y.; Ma, Z.; Huang, Y.; Wang, H. Short-range amorphous carbon nanosheets for oxygen reduction electrocatalysis. *Nanoscale* **2020**, *2*, 5769–5776. [CrossRef]
92. Poon, K.C.; Wan, W.Y.; Su, H.; Sato, H. One-minute synthesis via electroless reduction of amorphous phosphorus-doped graphene for oxygen reduction reaction. *ACS Appl. Energy Mater.* **2021**, *4*, 5388–5391. [CrossRef]
93. Lu, Y.; Wang, J.; Peng, Y.; Fisher, A.; Wang, X. Highly efficient and durable Pd hydride nanocubes embedded in 2D amorphous NiB nanosheets for oxygen reduction reaction. *Adv. Energy Mater.* **2017**, *7*, 1700919. [CrossRef]
94. Zhang, J.; Sun, W.; Ding, L.; Wu, Z.; Gao, F. Au Nanocrystals@defective amorphous MnO$_2$ nanosheets Core/Shell nanostructure with effective CO$_2$ adsorption and activation toward CO$_2$ electroreduction to CO. *ACS Sustain. Chem. Eng.* **2021**, *9*, 5230–5239. [CrossRef]
95. Liu, W.; Li, C.; Xu, Q.; Yan, P.; Niu, C.; Shen, Y.; Yuan, P.; Jia, Y. Anderson localization in 2D amorphous MoO$_{3-x}$ monolayers for electrochemical ammonia synthesis. *ChemCatChem* **2019**, *11*, 5412–5416. [CrossRef]
96. Chu, K.; Nan, H.; Li, Q.; Guo, Y.; Tian, Y.; Liu, W. Amorphous MoS$_3$ enriched with sulfur vacancies for efficient electrocatalytic nitrogen reduction. *J. Energy Chem.* **2021**, *53*, 132–138. [CrossRef]
97. Chu, K.; Gu, W.; Li, Q.; Liu, Y.; Tian, Y.; Liu, W. Amorphization activated FeB$_2$ porous nanosheets enable efficient electrocatalytic N$_2$ fixation. *J. Energy Chem.* **2021**, *53*, 82–89. [CrossRef]

Communication

Three-Dimensionally Ordered Macro/Mesoporous Nb₂O₅/Nb₄N₅ Heterostructure as Sulfur Host for High-Performance Lithium/Sulfur Batteries

Haoxian Chen [1,†], Jiayi Wang [2,†], Yan Zhao [3,*], Qindan Zeng [1], Guofu Zhou [1] and Mingliang Jin [1,*]

1. National Center for International Research on Green Optoelectronics, South China Academy of Advanced Optoelectronics, South China Normal University, Guangzhou 510006, China; haoxian.chen@ecs-scnu.org (H.C.); qin-dan.zeng@ecs-scnu.org (Q.Z.); guofu.zhou@m.scnu.edu.cn (G.Z.)
2. School of Information and Optoelectronic Science and Engineering, South China Normal University, Guangzhou 510006, China; jiayi.wang@zq-scnu.org
3. School of Materials Science and Engineering, Hebei University of Technology, Tianjin 300130, China
* Correspondence: yanzhao1984@hebut.edu.cn (Y.Z.); mingliang.jin@zq-scnu.org (M.J.)
† H. Chen and J. Wang contributed equally to this work.

Abstract: The severe shuttle effect of soluble polysulfides hinders the development of lithium–sulfur batteries. Herein, we develop a three-dimensionally ordered macro/mesoporous (3DOM) Nb_2O_5/Nb_4N_5 heterostructure, which combines the strong adsorption of Nb_2O_5 and remarkable catalysis effect of Nb_4N_5 by the promotion "adsorption-transformation" mechanism in sulfur reaction. Furthermore, the high electrocatalytic activity of Nb_4N_5 facilitates ion/mass transfer during the charge/discharge process. As a result, cells with the S-Nb_2O_5/Nb_4N_5 electrode delivered outstanding cycling stability and higher discharge capacity than its counterparts. Our work demonstrates a new routine for the multifunctional sulfur host design, which offers great potential for commercial high-performance lithium–sulfur batteries.

Keywords: Nb_2O_5; Nb_4N_5; heterostructure; lithium-sulfur batteries

1. Introduction

Electronic devices play a vital role in modern society, setting high standards for corresponding energy storage systems [1,2]. Lithium–sulfur batteries (LSBs) are demonstrated as one of the most promising candidates owing to their high theoretical energy density, low cost, and environmental friendliness [3–5]. However, the solid-electrolyte interphase (SEI) has been found to have poor mechanical strength and Li-ion conductivity. The formation of unstable SEI causes safety issues and faster decay of capacity in the anode side for LSB [6]. Artificial SEI fabricated by fluorinated electrolyte and ultrathin bilayer SEI are thus applied to protect the electrodes and suppress Li dendrite growth [7,8]. Besides, the development of LSBs is hindered by low conductivity of sulfur and its discharge product, repeated volume change, and severe lithium polysulfides (LiPS) shuttle effect [9–12].

To enhance the performance of LSBs, several kinds of sulfur hosts have developed by researchers, including carbon materials, conductive polymer, and metallic chalcogenides, among others [13–17]. A sulfur host can fasten charge transfer in LSB, capture LiPS, and catalyze each step of conversion of this chemical species. Qiao et al. combined iron phosphide (FeP) with reduced graphene oxide (rGO) to construct a sulfiphilic composite [18]. The catalytic properties of FeP and the electron transport properties of rGO are integrated by the synergistic effect, which results in high coulombic efficiency and capacity of the cell loaded with this kind of sulfiphilic host. Recently, polar metal oxides were found to deliver great potential to serve as a sulfur host relying on polar-polar interactions with the LiPS [19–21]. As an oxygen-rich material, anions of metal oxides work as active sites to absorb the LiPS. Among the family of oxides, Nb_2O_5 shows high LiPS anchor ability

due to the strong metal-sulfur bond [22,23]. However, as an insulator with a wide band gap energy, the Nb_2O_5 sulfur host is still limited by poor electronic conductivity. While, transition metal nitride is reported to possess high conductivity and catalytic activity, and have been widely used as a sulfur host for LSBs [24–27]. When different catalysts are applied in LSB, them can catalyze various processes in the conversion of LiPS. Wang et al. found that FeP is able to catalyze the liquid-liquid-solid process, while Fe_3O_4 can promote the solid-liquid conversion. When the batteries were assembled with these two catalysts, the cycle stability and capacity retention of the battery was improved simultaneously [28]. Hence, it is a feasible way to combine the Nb_2O_5 and transition metal nitride to achieve a high-performance sulfur host.

Herein, we develop a three-dimensionally ordered macro/mesoporous (3DOM) Nb_2O_5/Nb_4N_5 heterostructure, which combines the strong adsorption of Nb_2O_5 and remarkable catalysis effect of Nb_4N_5, promoting the "adsorption-transformation" mechanism in the lithium-sulfur battery. Furthermore, the high electrocatalytic activity of Nb_4N_5 can provide a fast ion transfer routine during the cycling process and the ordered porous structure not only provides sufficient space for sulfur loading, but also improves the electrolyte infiltration. Therefore, the S-Nb_2O_5/Nb_4N_5 electrode delivers satisfying cycling stability and remarkable discharge capacity.

2. Materials and Methods

2.1. Materials Preparation

The polymethyl methacrylate (PMMA) template was prepared according to our reported methods [29]. In the typical procedure of the synthesis of 3DOM Nb_2O_5/Nb_4N_5, 20 mL ethanol was mixed with 1.35 g of niobium pentachloride ($NbCl_5$, Aladdin, Shanghai, China) under magnetic stirring. When a clear solution was formed, the prepared PMMA template was immersed in the precursor solution for 12 h. Subsequently, the precursor solution was removed from the PMMA template through vacuum filtration. The obtained sample was put into a porcelain boat and dried in air at 60 °C. Subsequent calcination at 600 °C in air for 3 h was employed to remove the PMMA template. The obtained 3DOM Nb_2O_5 was heated under NH_3 to prepare 3DOM Nb_2O_5/Nb_4N_5.

2.2. Characterization

X-ray diffraction (XRD, D8 Focus Bruker, Karlsruhe, Germany), scanning electron microscopy/ energy dispersive spectroscopy (FE–SEM/EDS, ZEISS Ultra 55, Oberkochen, Germany) and transmission electron microscopy (TEM, JEOL 2100, Tokyo, Japan) were employed to observe the phase and morphology of 3DOM Nb_2O_5/Nb_4N_5. The element value and bonding state were explored by the X–ray photoelectron spectra (XPS, Thermo Scientific ESCALAB 250Xi, Waltham, MA, USA). N_2 adsorption–desorption isotherms and pore distribution were tested using V-Sorb 2800P. Thermogravimetric analysis (TGA, PerkinElmer TGA-8000, Waltham, MA, USA) was used to determine the sulfur content of the samples.

2.3. Cell Assembling and Testing

All reagents for assembling and testing are purchased from Aladdin (Shanghai, China) without further purification. At a mass ratio of 75:25 (wt.%), Nb_2O_5/Nb_4N_5 and sulfur powder were ground together and melting-diffusion routine was conducted to obtain S-Nb_2O_5/Nb_4N_5. N-hydroxy-2-pyrrolidone (NMP) was used as a solvent, and S-Nb_2O_5/Nb_4N_5 and conductive carbon black with polyvinylidene fluoride powder (PVDF) were mixed to produce a black slurry, in a mass ratio of 8:1:1. Al foil, serving as current collector, was coated with the as-prepared slurry and dried at 60 °C overnight. CR-2032 coin-type cells were applied to study the prepared electrode. To assemble the cells, cathode was made with a diameter of 12 mm. In this kind of cell, Li foil is an anode while Celgard 2400 works as a separator. 1,3-dioxolane (DOL) and 1,2-dimethoxyethane (DME) ($v/v = 1/1$) were mixed with 1 M LiTFSI and 1% $LiNO_3$ additive to serve as an electrolyte of the cell. Cyclic voltametric test and electrochemical impedance spectroscopy were carried out on a CHI660E (CH Instruments, Inc., Austin, TX, USA) electrochemical workstation. Charge-discharge surveys were conducted on a Neware battery tester (Shenzhen, China) from 1.7 V to 2.8 V.

LiPS adsorption test: The polysulfides were produced by adding sulfur and lithium sulfides in DME at a specific molar ratio. All tested samples was mixed with diluted Li_2S_6 solution inside the glove box. The photographs of absorption result were collected after stirring for 2 h and aging for one-sixth of a day.

Symmetric cells: 0.5 mg cm^{-2} of active materials were dropped onto the circular disks (12 mm in diameter) of carbon cloths. The amount of 0.2 M Li_2S_6 solution used in symmetric cells was 30 µL. CV measurements were obtained on a CHI660E (CH Instruments, Inc., Austin, TX, USA) electrochemistry workstation from −0.8 to 0.8 V in 1 mV s^{-1}.

Li_2S nucleation test: 0.25 M Li_2S_8 electrolyte was prepared for the test. Nb_2O_5/Nb_4N_5 heterostructure materials/carbon cloth and the lithium metal were used as electrodes. During the test, all batteries were discharged at 2.06 V with a steady current of 0.112 mA, and then maintained at 2.05 V until the current was decreased to 10^{-5} A.

Linear sweep voltammetry (LSV) test: To investigate the oxidization behavior of Li_2S, LSV measurements were performed in methanol with 0.1 M Li_2S. Typically, to construct a three-electrode system, an Ag/AgCl electrode and platinum wire are used as the reference electrode and counter electrode, respectively. Moreover, the glass carbon electrode covered with prepared materials was used as the working electrode. The prepared materials were dispersed in NMP and added onto a glass carbon electrode to fabricate a working electrode. The tests were conducted by scanning from −0.4 to −0.2 V in 5 mV s^{-1}.

3. Results and Discussion

As observed in Figure 1, $NbCl_5$ first penetrated the closed-packing PMMA template by capillarity forces. Subsequently, the PMMA was removed by heating at 600 °C, thus by the formation of 3DOM Nb_2O_5. The construction of a heterostructure relied on the Nb_2O_5 nitridation treatment through NH_3 erosion at a high temperature. When the conversion of the LiPSs species occurs, satisfactory pores and voids were provided by the unique 3DOM structure, for storing and immobilizing sulfur. Working as a polar material, the Nb_2O_5 part in the heterojunction can strongly and chemically adsorb polysulfide. Soon after adsorption, the niobium nitride (Nb_4N_5) in the heterojunction enhances the catalysis of polysulfide and solid Li_2S nucleation is seen.

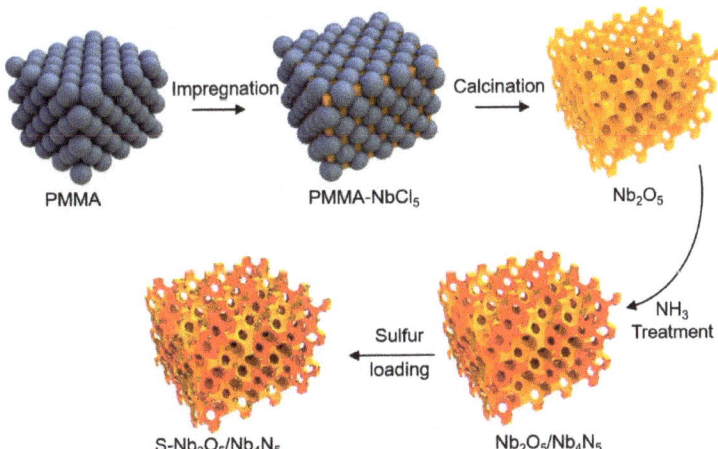

Figure 1. Schematic diagram of preparation of 3DOM S-Nb_2O_5/Nb_4N_5.

SEM images of Nb_2O_5/Nb_4N_5 are shown in Figure 2a,b. Figure 2a shows that the composites have three-dimensionally ordered porous structure and the pores are uniformly distributed and interconnected. The pore diameter is around 150–200 nm (Figure 2b) and sulfur can be absorbed intensely by these nanopores during galvanostatic charge–discharge cycling. Figure S1 shows the energy dispersive spectroscopy (EDS) result. Through the analysis of this mapping, the composites have 24% Nb_2O_5, while the other 76% is Nb_4N_5. The TEM result further confirms the 3DOM structure of Nb_2O_5/Nb_4N_5 in Figure 2c, and implies that the unique structure is preserved during the formation of the Nb_2O_5/Nb_4N_5 heterostructure. More importantly, the pore diameter is 160 nm, which is consistent with the SEM image. HRTEM image of Nb_2O_5/Nb_4N_5 (Figure 2d) demonstrates a distinctly different crystalline structure, which is attributed to Nb_2O_5 and Nb_4N_5. For further verification, fast Fourier transform (FFT) and inverse FFT patterns are collected, as shown in Figure 2e,f. Clear diffraction spots can be seen in both areas, demonstrating the excellent crystallinity of Nb_2O_5 and Nb_4N_5. The crystal plane spacings are measured to be 0.39 nm and 0.25 nm, which is consistent with typical (001) plane of Nb_2O_5 and (211) plane of Nb_4N_5. Furthermore, the scanning TEM image of Nb_2O_5/Nb_4N_5 is shown in Figure 2g–j, confirming uniform element distribution. The above results prove the successful preparation of Nb_2O_5/Nb_4N_5 heterostructure material.

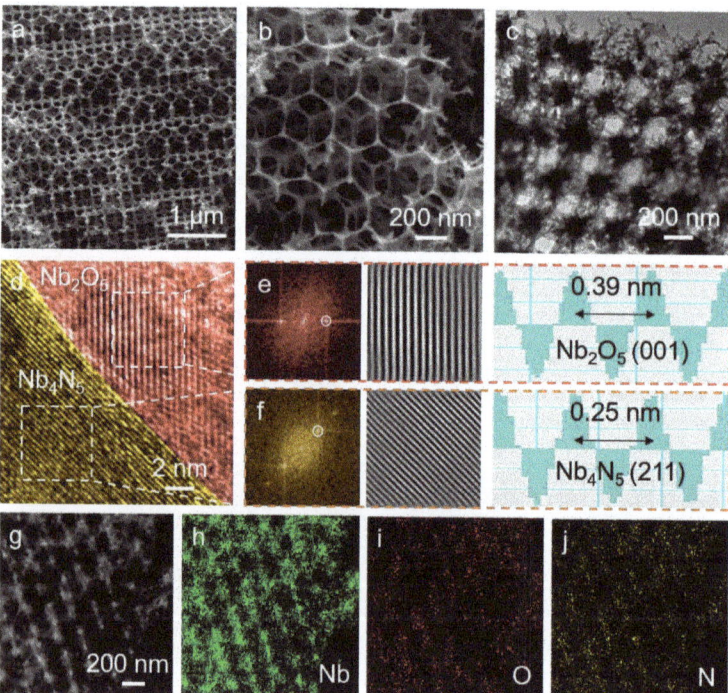

Figure 2. (**a**,**b**) SEM images of 3DOM Nb_2O_5/Nb_4N_5; (**c**) TEM and (**d**) HRTEM image of 3DOM Nb_2O_5/Nb_4N_5; (**e**,**f**) FFT patterns, inverse FFT patterns, and lattice spacing images of the selected area; (**g–j**) STEM image and the corresponding element distribution of 3DOM Nb_2O_5/Nb_4N_5.

XRD patterns of Nb_4N_5, Nb_2O_5/Nb_4N_5, and Nb_2O_5 are shown in Figure 3a; all the peaks are consistent with Nb_4N_5 (PDF#51-1327) and Nb_2O_5 (PDF#30-0873), indicating the high purity of the synthesized products. The 3DOM structures are further probed through the N_2 adsorption/desorption isotherms (Figure 3b); a similar specific surface area is obtained by Nb_4N_5 (34.6 $m^2\ g^{-1}$), Nb_2O_5/Nb_4N_5 (38.9 $m^2\ g^{-1}$), and Nb_2O_5 (40.9 $m^2\ g^{-1}$), indicating that the pore structures are well preserved during the phase transformation process. The pore distributions in Figure 3c indicate the existence of abundant micropores and mesopores, which can adsorb LiPS by relying the physical effect. Furthermore, the pore volume in Table S3 indicate that Nb_2O_5/Nb_4N_5 has a large number of bigger pores, which benefit the storing of sulfur. The XPS test is conducted to explore the chemical bonding environment of Nb_2O_5/Nb_4N_5. Correction for specimen charging is applied to XPS analysis according to the C 1 s peak at 284 eV. Typical Nb-O (209.9 eV, 207.3 eV) and Nb-N (206.8 eV) bonds are obtained (Figure 3d), indicating the co-existence state of Nb_2O_5 and Nb_4N_5. The high-resolution N 1s spectrum exhibits one major peak at 396.4 eV, which can be owing to the existence of Nb-N bonds of Nb_4N_5 (Figure 3e) [30]. Furthermore, one other major peak appears at 530.6 eV, which can be ascribed to Nb-O bonds of Nb_2O_5, and a sub peak of O-containing surface group emerged at 531.5 eV (Figure 3f) [22]. These results further confirm the successful construction of the Nb_2O_5/Nb_4N_5 heterostructure, which is expected to possess high adsorption and catalysis ability for LiPS.

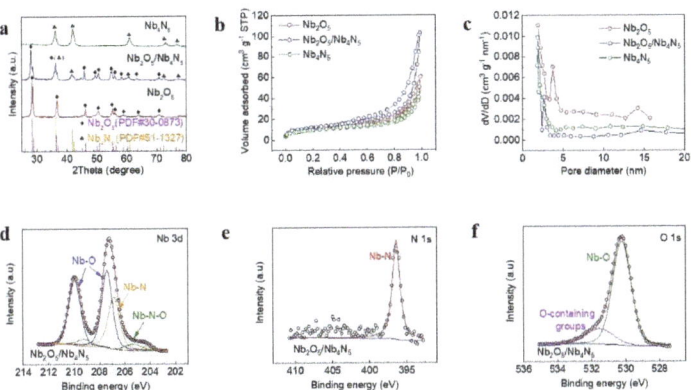

Figure 3. (a) XRD patterns; (b) N_2 adsorption/desorption isotherms and (c) pore distributions of Nb_4N_5, Nb_2O_5/Nb_4N_5 and Nb_2O_5. XPS spectra of Nb_2O_5/Nb_4N_5: (d) Nb 3d, (e) N 1s, and (f) O 1s.

The LiPS adsorption effect of Nb_2O_5/Nb_4N_5 is evaluated by the LiPS adsorption test using Li_2S_6 as a representative LiPS (Figure 4a). As is shown in this photograph, the glass bottles starting from left to right contained blank Li_2S_6 solution, 3DOM Nb_2O_5 with Li_2S_6 solution, 3DOM Nb_2O_5/Nb_4N_5 with Li_2S_6 solution, and 3DOM Nb_4N_5 with Li_2S_6 solution, respectively. The orange colour in the solution changed to a lighter brown after 3DOM Nb_4N_5 was added. Moreover, after 3DOM Nb_2O_5 is added into the solution, the color of the solution fades intensely and become much more transparent than that of the one with 3DOM Nb_4N_5, verifying that Li_2S_6 adsorption ability of Nb_2O_5 is much stronger than that of Nb_4N_5. At the same time, the solution with Nb_2O_5/Nb_4N_5 became completely colorless, suggesting the synergistic effect of three-dimensionally ordered porous Nb_2O_5 and the unique catalytic nature of Nb_4N_5 towards effective trapping of lithium polysulfides. The UV-vis curves comparison displays the vanishing of typical peaks related to S_6^{2-} and S_4^{2-}, demonstrating the strong adsorption of Nb_2O_5/Nb_4N_5. The highest current response delivered by the Nb_2O_5/Nb_4N_5 electrode can be observed in the LSV test (Figure 4b), indicating the enhanced Li_2S oxidation kinetics achieved by Nb_2O_5/Nb_4N_5. This result also implies reduction of the energy barrier of conversion of polysulfides by heterojunction, ensuring that the 3DOM Nb_2O_5/Nb_4N_5 electrodes promoted catalytic process of sulfur [31]. The TGA graph was performed as shown in Figure S2. The S content can reach 73% owing to the abundant hierarchical pore structure. To assess the enhanced electrochemical kinetics in depth, cyclic voltammetry (CV) characterization of the symmetric cells containing the 0.2 M Li_2S_6 electrolyte are performed with the scan rate of 1 mV s^{-1} (Figure 4c). The CV profile of 3DOM Nb_2O_5/Nb_4N_5 exhibits excellent reversibility, with two pairs of redox peaks (−0.03, 0.03 and −0.22, 0.22 V) probed. However, the CV of 3DOM Nb_2O_5 and Nb_4N_5 only obtains one pair of broadened redox reaction peaks, which is at −0.19 V and 0.19 V, separately. Moreover, the peak intensity of Nb_2O_5/Nb_4N_5 is higher than that of Nb_2O_5 and Nb_4N_5, indicating the limited transformation of polysulfides on the bare surface of Nb_2O_5/Nb_4N_5 heterojunction. The initial three CV cycles (Figure 4d) of the Nb_2O_5/Nb_4N_5 heterojunction are perfectly overlapped, suggesting excellent cycling stability. As is widely accepted by the scientific community, the transformation from Li_2S_4 to Li_2S contributes almost 75% of discharge capacity. As a result, the Li_2S deposition test is conducted, as shown in Figure 4e. It can be observed from Figure 4e that nucleation peak response of the heterostructure is earlier than that of Nb_2O_5 and Nb_4N_5, and nucleation capacity of 3DOM Nb_2O_5/Nb_4N_5 (283 mAh g^{-1}) is highest among the three samples. This may lead to a lower overpotential of the nucleation, electrocatalytic conversion of Li_2S, and adsorbent to polysulfide species [32].

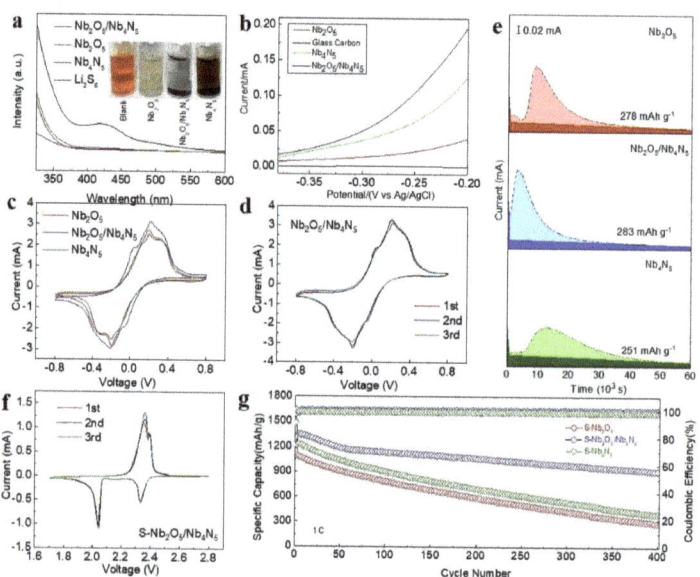

Figure 4. (**a**) LiPS adsorption test; (**b**) LSV test; (**c**,**d**) CV curves of symmetric cell; (**e**) Li_2S deposition test; (**f**) CV results of the cell with $S-Nb_2O_5/Nb_4N_5$ electrode; (**g**) long-term cycling tests at 1 C.

The electrochemical performance is tested by employing the $S-Nb_2O_5/Nb_4N_5$ electrode. The CV results are shown in Figure 4f—two distinct reduction peaks and one main oxide peak can be seen. The reduction peaks located at ~2.35 V and ~2.05 V represent the transformation from sulfur to Li_2S_4 and further into Li_2S. The oxide peak is produced by the regeneration of sulfur. The nearly overlapped curves indicate the excellent reversibility of the electrochemical reactions. Figure S3 show the Nyquist plots of the battery loaded with different samples in the frequency range 0.01–100 KHz. The $S-Nb_2O_5/Nb_4N_5$ cathode shows the smallest charge-transfer resistance (R_{ct}), denoting its fast kinetic process. The internal resistance (R_s) of all samples is similar and the R_{ct} of $S-Nb_2O_5/Nb_4N_5$ cathode is 46.79 Ω (Table S2). Furthermore, a long-term cycling test at 1 C was conducted (Figure 4g). The first 2 cycles at 0.2 C are applied for activation of the electrodes. A high discharge capacity of 1354 mAh g^{-1} at 1 C is obtained and a remarkable reversible capacity of 913 mAh g^{-1} can still be maintained after 400 cycles with a low capacity attenuation rate (0.08% per cycle), which is obviously improved compared with the $S-Nb_2O_5$ and $S-Nb_4N_5$ electrodes. Additionally, the voltage profiles at 1 C are provided in Figure S4; it can be seen that the $S-Nb_2O_5/Nb_4N_5$ electrode displays stable voltage plateau and negligible polarization behavior under prolonged cycling. Moreover, cycling test at 1 C with different mass loading is shown in Figure S5. The as-developed S-Nb_2O_5/Nb_4N_5 electrodes are capable of withstanding at 1 C at sulfur loading of 2 and 6 mg cm^{-2} (Figure S5), attributing the favorable mass/charge transfer and the catalyzed sulfur redox reactions in the Nb_2O_5/Nb_4N_5 matrix. On comparison of our work with other current works, it is seen that the Nb_2O_5/Nb_4N_5 electrode exhibits excellent electrochemical performance among recently published heterojunction materials for LSB (Table S4).

4. Conclusions

A 3DOM Nb_2O_5/Nb_4N_5 heterostructure was constructed through in-situ nitridation to serve as a multi-functional sulfur host. The porous structure with interconnected channels can accommodate sulfur as well as facilitate electrolyte infiltration. Strong LiPS immobilization of Nb_2O_5 and the remarkable catalysis effect of Nb_4N_5 are combined to realize the accelerated LiPS "adsorption-transformation" process. As a result, the LSBs with S-Nb_2O_5/Nb_4N_5 delivered enhanced kinetics and improved cycling stability and discharge capacity, indicating great capability of Nb_2O_5/Nb_4N_5 for high-performance LSBs.

Supplementary Materials: The following are available online at https://www.mdpi.com/article/10.3390/nano11061531/s1, Figure S1: EDS mapping of Nb2O5/Nb4N5, Figure S2: TGA profile of Nb2O5/Nb4N5, Figure S3: Nyquist plots of S-Nb4N5, S-Nb2O5/Nb4N5 and S-Nb2O5, Figure S4: Voltage profiles of S-Nb2O5/Nb4N5 at 1 C, Figure S5: Cycling tests at 1 C with different mass loading, Table S1: Conductivities of Nb4N5, Nb2O5/Nb4N5 and Nb2O5, Table S2: The resistance of Rs and Rct simulated from equivalent circuits, Table S3: Pore size distribution and/or pore volume of Nb4N5, Nb2O5/Nb4N5 and Nb2O5, Table S4: Comparison of electrochemical properties of our work with other works.

Author Contributions: Formal Analysis, H.C. and J.W.; investigation, H.C., Q.Z., and J.W.; writing—original draft, H.C. and J.W.; writing—review and editing, G.Z., M.J., and Y.Z.; supervision, M.J.; project administration, M.J. and Y.Z. All authors have read and agreed to the published version of the manuscript.

Funding: This work was funded by the Science and Technology Program of Guangzhou (No. 2019050001); Yunnan Expert Workstation (202005AF150028).

Data Availability Statement: Data is contained within this article and Supplementary material.

Conflicts of Interest: The authors declare no competing interest.

References

1. Pang, Q.; Shyamsunder, A.; Narayanan, B.; Kwok, C.Y.; Curtiss, L.A.; Nazar, L.F. Tuning the electrolyte network structure to invoke quasi-solid state sulfur conversion and suppress lithium dendrite formation in Li–S batteries. *Nat. Energy* **2018**, *3*, 783–791. [CrossRef]
2. Zhao, C.; Xu, G.-L.; Yu, Z.; Zhang, L.; Hwang, I.; Mo, Y.-X.; Ren, Y.; Cheng, L.; Sun, C.-J.; Ren, Y.; et al. A high-energy and long-cycling lithium–sulfur pouch cell via a macroporous catalytic cathode with double-end binding sites. *Nat. Nanotechnol.* **2021**, *16*, 166–173. [CrossRef]
3. Li, G.; Lu, F.; Dou, X.; Wang, X.; Luo, D.; Sun, H.; Yu, A.; Chen, Z. Polysulfide Regulation by the Zwitterionic Barrier toward Durable Lithium-Sulfur Batteries. *J. Am. Chem. Soc.* **2020**, *142*, 3583–3592. [CrossRef]
4. Wang, J.; Luo, D.; Li, J.; Zhang, Y.; Zhao, Y.; Zhou, G.; Shui, L.; Chen, Z.; Wang, X. "Soft on rigid" nanohybrid as the self-supporting multifunctional cathode electrocatalyst for high-performance lithium-polysulfide batteries. *Nano Energy* **2020**, *70*, 105293. [CrossRef]
5. Zhang, Y.; Li, G.; Wang, J.; Cui, G.; Wei, X.; Shui, L.; Kempa, K.; Zhou, G.; Wang, X.; Chen, Z. Hierarchical Defective Fe_{3-x}C@C Hollow Microsphere Enables Fast and Long-Lasting Lithium–Sulfur Batteries. *Adv. Funct. Mater.* **2020**, *30*, 2001165. [CrossRef]
6. Liu, Y.; Lin, D.; Liang, Z.; Zhao, J.; Yan, K.; Cui, Y. Lithium-coated polymeric matrix as a minimum volume-change and dendrite-free lithium metal anode. *Nat. Commun.* **2016**, *7*, 10992. [CrossRef] [PubMed]
7. Pathak, R.; Chen, K.; Gurung, A.; Reza, K.M.; Bahrami, B.; Wu, F.; Chaudhary, A.; Ghimire, N.; Zhou, B.; Zhang, W.H.; et al. Ultrathin Bilayer of Graphite/SiO2 as Solid Interface for Reviving Li Metal Anode. *Adv. Energy Mater.* **2019**, *9*, 1901486. [CrossRef]
8. Pathak, R.; Chen, K.; Gurung, A.; Reza, K.M.; Bahrami, B.; Pokharel, J.; Baniya, A.; He, W.; Wu, F.; Zhou, Y.; et al. Fluorinated hybrid solid-electrolyte-interphase for dendrite-free lithium deposition. *Nat. Commun.* **2020**, *11*, 93. [CrossRef] [PubMed]

9. Wang, X.; Luo, D.; Wang, J.; Sun, Z.; Cui, G.; Chen, Y.; Wang, T.; Zheng, L.; Zhao, Y.; Shui, L.; et al. Strain Engineering of a MXene/CNT Hierarchical Porous Hollow Microsphere Electrocatalyst for a High-Efficiency Lithium Polysulfide Conversion Process. *Angew. Chem. Int. Ed. Engl.* **2021**, *60*, 2371–2378. [CrossRef]
10. Wang, J.; Zhao, Y.; Li, G.; Luo, D.; Liu, J.; Zhang, Y.; Wang, X.; Shui, L.; Chen, Z. Aligned sulfur-deficient ZnS_{1-x} nanotube arrays as efficient catalyzer for high-performance lithium/sulfur batteries. *Nano Energy* **2021**, *84*, 105891. [CrossRef]
11. Zhu, Y.; Li, G.; Luo, D.; Wan, H.; Feng, M.; Yuan, D.; Hu, W.; Li, Z.; Gao, R.; Zhang, Z.; et al. Unsaturated coordination polymer frameworks as multifunctional sulfur reservoir for fast and durable lithium-sulfur batteries. *Nano Energy* **2021**, *79*, 105393. [CrossRef]
12. Moorthy, B.; Kwon, S.; Kim, J.H.; Ragupathy, P.; Lee, H.M.; Kim, D.K. Tin sulfide modified separator as an efficient polysulfide trapper for stable cycling performance in Li-S batteries. *Nanoscale Horiz.* **2019**, *4*, 214–222. [CrossRef] [PubMed]
13. Wang, D.; Luo, D.; Zhang, Y.; Zhao, Y.; Zhou, G.; Shui, L.; Chen, Z.; Wang, X. Deciphering interpenetrated interface of transition metal oxides/phosphates from atomic level for reliable Li/S electrocatalytic behavior. *Nano Energy* **2021**, *81*, 105602. [CrossRef]
14. Qiu, W.; Li, G.; Luo, D.; Zhang, Y.; Zhao, Y.; Zhou, G.; Shui, L.; Wang, X.; Chen, Z. Hierarchical Micro-Nanoclusters of Bimetallic Layered Hydroxide Polyhedrons as Advanced Sulfur Reservoir for High-Performance Lithium-Sulfur Batteries. *Adv. Sci.* **2021**, *8*, 2003400. [CrossRef]
15. Li, M.; Lu, J.; Shi, J.; Son, S.B.; Luo, D.; Bloom, I.; Chen, Z.; Amine, K. In Situ Localized Polysulfide Injector for the Activation of Bulk Lithium Sulfide. *J. Am. Chem. Soc.* **2021**, *143*, 2185–2189. [CrossRef]
16. Thangavel, R.; Kannan, A.G.; Ponraj, R.; Kaliyappan, K.; Yoon, W.S.; Kim, D.W.; Lee, Y.S. Cinnamon-Derived Hierarchically Porous Carbon as an Effective Lithium Polysulfide Reservoir in Lithium-Sulfur Batteries. *Nanomaterials* **2020**, *10*, 1220. [CrossRef]
17. Park, J.W.; Hwang, H.J.; Kang, H.J.; Bari, G.; Lee, T.G.; An, B.H.; Cho, S.Y.; Jun, Y.S. Hierarchical Porous, N-Containing Carbon Supports for High Loading Sulfur Cathodes. *Nanomaterials* **2021**, *11*, 408. [CrossRef] [PubMed]
18. Zhao, Z.; Pathak, R.; Wang, X.; Yang, Z.; Li, H.; Qiao, Q. Sulfiphilic FeP/rGO as a highly efficient sulfur host for propelling redox kinetics toward stable lithium-sulfur battery. *Electrochim. Acta* **2020**, *364*, 137117. [CrossRef]
19. Zhang, J.; Zhao, Y.; Zhang, Y.; Li, J.; Babaa, M.R.; Liu, N.; Bakenov, Z. Synthesis of microflower-like vacancy defective copper sulfide/reduced graphene oxide composites for highly efficient lithium-ion batteries. *Nanotechnology* **2020**, *31*, 095405. [CrossRef] [PubMed]
20. Liang, C.; Zhang, X.; Zhao, Y.; Tan, T.; Zhang, Y.; Bakenov, Z. Three-dimensionally ordered macro/mesoporous TiO_2 matrix to immobilize sulfur for high performance lithium/sulfur batteries. *Nanotechnology* **2018**, *29*, 415401. [CrossRef]
21. Zhang, Y.; Qiu, W.; Zhao, Y.; Wang, Y.; Bakenov, Z.; Wang, X. Ultra-fine zinc oxide nanocrystals decorated three-dimensional macroporous polypyrrole inverse opal as efficient sulfur hosts for lithium/sulfur batteries. *Chem. Eng. J.* **2019**, *375*, 122055. [CrossRef]
22. Zhou, J.; Liu, X.; Zhu, L.; Zhou, J.; Guan, Y.; Chen, L.; Niu, S.; Cai, J.; Sun, D.; Zhu, Y.; et al. Deciphering the Modulation Essence of p Bands in Co-Based Compounds on Li-S Chemistry. *Joule* **2018**, *2*, 2681–2693. [CrossRef]
23. Luo, D.; Zhang, Z.; Li, G.; Cheng, S.; Li, S.; Li, J.; Gao, R.; Li, M.; Sy, S.; Deng, Y.P.; et al. Revealing the Rapid Electrocatalytic Behavior of Ultrafine Amorphous Defective Nb_2O_{5-x} Nanocluster toward Superior Li-S Performance. *ACS Nano* **2020**, *14*, 4849–4860. [CrossRef] [PubMed]
24. Liu, Y.; Chen, M.; Su, Z.; Gao, Y.; Zhang, Y.; Long, D. Direct trapping and rapid conversing of polysulfides via a multifunctional Nb_2O_5-CNT catalytic layer for high performance lithium-sulfur batteries. *Carbon* **2021**, *172*, 260–271. [CrossRef]
25. Shang, C.; Li, G.; Wei, B.; Wang, J.; Gao, R.; Tian, Y.; Chen, Q.; Zhang, Y.; Shui, L.; Zhou, G.; et al. Dissolving Vanadium into Titanium Nitride Lattice Framework for Rational Polysulfide Regulation in Li–S Batteries. *Adv. Energy Mater.* **2020**, *11*, 2003020. [CrossRef]
26. Wang, X.; Li, G.; Li, M.; Liu, R.; Li, H.; Li, T.; Sun, M.; Deng, Y.; Feng, M.; Chen, Z. Reinforced polysulfide barrier by g-C_3N_4/CNT composite towards superior lithium-sulfur batteries. *J. Energy Chem.* **2021**, *53*, 234–240. [CrossRef]
27. Wei, B.; Shang, C.; Pan, X.; Chen, Z.; Shui, L.; Wang, X.; Zhou, G. Lotus Root-Like Nitrogen-Doped Carbon Nanofiber Structure Assembled with VN Catalysts as a Multifunctional Host for Superior Lithium-Sulfur Batteries. *Nanomaterials* **2019**, *9*, 1724. [CrossRef] [PubMed]
28. Zhao, Z.; Yi, Z.; Li, H.; Pathak, R.; Yang, Z.; Wang, X.; Qiao, Q. Synergetic effect of spatially separated dual co-catalyst for accelerating multiple conversion reaction in advanced lithium sulfur batteries. *Nano Energy* **2021**, *81*, 105621. [CrossRef]
29. Xing, Z.; Li, G.; Sy, S.; Chen, Z. Recessed deposition of TiN into N-doped carbon as a cathode host for superior Li-S batteries performance. *Nano Energy* **2018**, *54*, 1–9. [CrossRef]
30. Wang, J.; Li, G.; Luo, D.; Zhang, Y.; Zhao, Y.; Zhou, G.; Shui, L.; Wang, X.; Chen, Z. Engineering the Conductive Network of Metal Oxide-Based Sulfur Cathode toward Efficient and Longevous Lithium–Sulfur Batteries. *Adv. Energy Mater.* **2020**, *10*, 2002076. [CrossRef]
31. Ge, W.; Wang, L.; Li, C.; Wang, C.; Wang, D.; Qian, Y.; Xu, L. Conductive cobalt doped niobium nitride porous spheres as an efficient polysulfide convertor for advanced lithium-sulfur batteries. *J. Mater. Chem. A* **2020**, *8*, 6276–6282. [CrossRef]
32. Wu, Q.; Zhou, X.; Xu, J.; Cao, F.; Li, C. Adenine Derivative Host with Interlaced 2D Structure and Dual Lithiophilic-Sulfiphilic Sites to Enable High-Loading Li-S Batteries. *ACS Nano* **2019**, *13*, 9520–9532. [CrossRef] [PubMed]

Article

Synergistic Adsorption-Catalytic Sites TiN/Ta$_2$O$_5$ with Multidimensional Carbon Structure to Enable High-Performance Li-S Batteries

Chong Wang [1], Jian-Hao Lu [2], Zi-Long Wang [2], An-Bang Wang [2], Hao Zhang [2], Wei-Kun Wang [2,*], Zhao-Qing Jin [2,*] and Li-Zhen Fan [1,*]

[1] Beijing Advanced Innovation Center for Materials Genome Engineering, Institute of Advanced Materials and Technology, University of Science and Technology Beijing, Beijing 100083, China; wangchong18810@163.com

[2] Military Power Sources Research and Development Center, Research Institute of Chemical Defense, Beijing 100191, China; hahalujianhao@163.com (J.-H.L.); wangzilong0709@163.com (Z.-L.W.); wab_wang2000@163.com (A.-B.W.); dr.h.zhang@hotmail.com (H.Z.)

* Correspondence: wangweikun2002@163.com (W.-K.W.); jinzhaoqing1001@gmail.com (Z.-Q.J.); fanlizhen@ustb.edu.cn (L.-Z.F.)

Citation: Wang, C.; Lu, J.-H.; Wang, Z.-L.; Wang, A.-B.; Zhang, H.; Wang, W.-K.; Jin, Z.-Q.; Fan, L.-Z. Synergistic Adsorption-Catalytic Sites TiN/Ta$_2$O$_5$ with Multidimensional Carbon Structure to Enable High-Performance Li-S Batteries. *Nanomaterials* **2021**, *11*, 2882. https://doi.org/10.3390/nano11112882

Academic Editor: Byungwoo Park

Received: 17 September 2021
Accepted: 23 October 2021
Published: 28 October 2021

Publisher's Note: MDPI stays neutral with regard to jurisdictional claims in published maps and institutional affiliations.

Copyright: © 2021 by the authors. Licensee MDPI, Basel, Switzerland. This article is an open access article distributed under the terms and conditions of the Creative Commons Attribution (CC BY) license (https://creativecommons.org/licenses/by/4.0/).

Abstract: Lithium-sulfur (Li-S) batteries are deemed to be one of the most optimal solutions for the next generation of high-energy-density and low-cost energy storage systems. However, the low volumetric energy density and short cycle life are a bottleneck for their commercial application. To achieve high energy density for lithium-sulfur batteries, the concept of synergistic adsorptive–catalytic sites is proposed. Base on this concept, the TiN@C/S/Ta$_2$O$_5$ sulfur electrode with about 90 wt% sulfur content is prepared. TiN contributes its high intrinsic electron conductivity to improve the redox reaction of polysulfides, while Ta$_2$O$_5$ provides strong adsorption capability toward lithium polysulfides (LiPSs). Moreover, the multidimensional carbon structure facilitates the infiltration of electrolytes and the motion of ions and electrons throughout the framework. As a result, the coin Li-S cells with TiN@C/S/Ta$_2$O$_5$ cathode exhibit superior cycle stability with a decent capacity retention of 56.1% over 300 cycles and low capacity fading rate of 0.192% per cycle at 0.5 C. Furthermore, the pouch cells at sulfur loading of 5.3 mg cm^{-2} deliver a high areal capacity of 5.8 mAh cm^{-2} at low electrolyte/sulfur ratio (E/S, 3.3 μL mg^{-1}), implying a high sulfur utilization even under high sulfur loading and lean electrolyte operation.

Keywords: lithium-sulfur batteries; catalyst; TiN/Ta$_2$O$_5$; multidimensional carbon

1. Introduction

Lithium-sulfur (Li-S) batteries are ideal candidates to substitute conventional lithium-ion batteries [1]. However, despite their attractive merits including high theoretical energy density and abundant resources, the practical uses of Li-S batteries are still hampered by a series of deleterious defects facing widespread commercialization, such as the intrinsic poor electrical conductivity of sulfur and discharge products (Li$_2$S/Li$_2$S$_2$), the shuttling phenomenon originated from the dissolution of lithium polysulfide (LiPSs), and, particularly, the sluggish conversion of LiPSs to solid lithium sulfides, bringing about low sulfur utilization, fast capacity fading, and poor cyclability [2].

Accordingly, strategies have been developed to solve the above problems in the past few decades. It has been corroborated that the conversion of LiPSs on the interface between the electrolyte and host materials is decided by their moderate interactions and fast electron/ion exchange [3]. Catalysts such as transition metal-free polar materials [4,5], transition metal compounds [6–9], and metals [10] can not only capture LiPSs to decrease their dissolution and diffusion in the electrolyte but also boost the conversion between LiPSs and Li$_2$S$_2$/Li$_2$S. Among them, titanium nitride has been widely used as the catalytic material [11,12], due to its high electrical conductivity, which can propel the kinetics of the transformation of LiPSs. Goodenough and coworkers reported a mesoporous TiN with a high surface area, benefiting from its intrinsic electrical conductivity, fine porous framework, and appropriate adsorption ability of TiN; the TiN-S composite cathode exhibits high specific capacity and excellent rate capability [13]. However, TiN nanoparticles (TiN NPs) cannot effectively suppress the shuttle effects of LiPSs, because of their weak adsorption capacity toward LiPSs [14]. Kim et al. proposed an effective electrocatalyst Ta_2O_5 for LiPSs conversion in the Li-S system; the intrinsic chemical polarity of Ta_2O_5 could establish favorable interactions with the LiPSs [15]. However, Ta_2O_5 demonstrates insufficient electrical conductivity, because of its electron band structure [16], which does not provide desirable electron mobility and catalytic activity. Therefore, it is difficult to make full use of catalysis ability depending on the sole component, which is short in either adsorption or electrical conductivity. Especially with high sulfur loading for a thick cathode, the sluggish and incomplete conversion of LiPSs to solid lithium sulfides limits the full utilization of intermediates, leading to the shuttle effect and rapid capacity decay during cycling.

To attain high specific capacity under the condition of high sulfur content of the sulfur-based composite and thick sulfur cathode, we designed the TiN@C/S/Ta_2O_5 cathode for Li-S batteries. Both TiN and Ta_2O_5 have a synergy enhancement effect to promote the affinity with LiPSs and speed up the kinetics of sulfur conversion reaction. In addition, the multidimensional carbon structure, which is the mixture of Super P, CNT, and graphene, can offer high electrical conductivity and sustain the strain generated by the volumetric changes of the active materials during cycling [17]. Their characteristic superiorities endow our high fraction of sulfur cathode with good rate response capability and superior cyclability even under high sulfur loading and lean electrolyte/sulfur ration operation.

2. Experimental Section

2.1. Preparation of TiN@C/S/Ta_2O_5 Composite

The TiN@C/S/Ta_2O_5 composite was fabricated in a typical liquid-phase suspension process. A certain amount of Super P, CNT, graphene, and TiN NPs (the weight ratio of Super P: CNT: G: TiN = 2:2:1:10) was ball-milled to obtain the uniform slurry. Sulfur was synthesized based on the reaction between $Na_2S_2O_3$ and HCOOH [18]; the suspension of sulfur was injected into the mixed solution of TiN@C under vigorous stirring for more than 10 h. After that, the sediment was obtained by filtering, washed with distilled water several times to wipe off the soluble impurities, and dried under vacuum 60 °C for 24 h. Next, the TiN@C/S materials were uniformly dispersed in distilled water again, and an appropriate amount of Ta(OEt)$_5$ was added to the above suspension; the amorphous Ta_2O_5 was produced by the hydrolysis reactions between Ta(OEt)$_5$ and deionized (DI) water [19]. Subsequently, the mixture was stirred all night. Finally, the TiN@C/S/Ta_2O_5 (the weight ratio of Super P: CNT: G: TiN: Ta_2O_5 = 2:2:1:10:10) composite was collected by centrifugation, washed with deionized water several times, and dried at 60 °C for 24 h. The TiN@C/S (SuperP: CNT: G: TiN = 2:2:1:20) and Ta_2O_5@C/S (Super P: CNT: G: Ta_2O_5 = 2:2:1:20) composites were prepared with the procedure similar to TiN@C/S/Ta_2O_5 composite; the content of sulfur in both TiN@C/S and Ta_2O_5@C/S materials is also 90 wt%.

The sulfur composite cathodes were prepared by mixing the active material (TiN@C/S/Ta_2O_5), Super P, carbon nanotube, and a binder (LA133) in deionized water and isopropanol mixed solution with a weight ratio of 80:5:5:10. After stirring for 12 h, the cathode slurry was blade-cast onto Al foils, followed by drying at 60 °C for 24 h. Similarly, the TiN@C/S and Ta_2O_5@C/S composites were prepared.

2.2. Material Characterization Techniques

Thermogravimetric analysis (NETZSCH TG 209F3, NETZSCH Gerätebau GmbH, Selb, Germany) was carried out with a heating rate of 5 °C min^{-1} under an atmosphere of N_2. The metallic element content was measured by ICP-OES (Agilent 725ES & Agilent 5110, Agilent Technologies, Santa Clara, CA, USA). The morphology and structure of materials were recorded using scanning electron microscopy (Zeiss G300, Carl Zeiss Inc., Thornwood, New York, NY, USA) and transmission electron microscopy (JEOL Ltd., Tokyo, Japan). The elemental compositions and crystal structures of these samples were analyzed by X-ray photoelectron spectroscopy (XPS) (Thermo Scientific K-Alpha, Thermo Fisher Scientific, Waltham, MA, USA) and X-ray diffraction (Ultima IV, Rigaku Corporation, Tokyo, Japan).

2.3. Polysulfides Adsorption Experiment

Li_2S_6 solution with a concentration of 10 mmol L^{-1} was prepared by mixing lithium sulfide (Li_2S) and sulfur power with a molar ratio of 1:5; the mixture was added into 1,3-dioxolane (DOL)/1,2-dimethoxyethane (DME) (1:1, v/v) solution, followed by intense stirring for 24 h in an Ar atmosphere. A total of 20 mg of TiN, Ta_2O_5, and TiN/Ta_2O_5 were added into 30 mL of Li_2S_6 solution, respectively, and rested for 12 h. The supernatant and precipitates were researched by UV-vis spectrophotometry and XPS.

2.4. Assembly of the Symmetric Cell

The electrode powers (TiN, Ta_2O_5, and TiN/Ta_2O_5,), CNT, and polyvinylidene fluoride (PVDF) binder were dispersed into NMP with a weight ratio of 70:20:10 to form a homogeneous solution, and then it was coated onto the current collector. The symmetric cell used the electrodes as both cathode and anode, 30 μL electrolyte containing 0.5 mol L^{-1} of Li_2S_6, and 1.0 mol L^{-1} of LiTFSI in a 1:1 (v/v) mixture of 1,3-dioxolane (DOL), and 1,2-dimethoxyethane (DME) was added into each coin cell. Cyclic voltammetry (CV) and electrochemical impedance spectroscopy (EIS) measurements were collected on an electrochemical workstation (VersaSTAT3, ametek, Berwyn, PA, USA).

2.5. Electrochemical Measurement

Standard CR2025 coin cells were assembled using the prepared electrodes and polypropylene separator (Celgard 2400, Celgard, Charlotte, NC, USA), with lithium metal as the anode. Charge-discharge performances of both coin cells and pouch cells were tested between 1.8 and 2.6 V using a LAND CT2001A (Landian, Wuhan, China) multi-channel battery testing system at room temperature.

In this experiment, the electrolyte was 1M LiTFSI in DOl/DME (1:1 v/v) containing $LiNO_3$ as an additive (1 wt%), The E/S ratio in the coin cells with areal sulfur loading (1.5 mg cm^{-2}) was controlled to be 10 μL mg^{-1}. The pouch cells have average areal sulfur loading of 5.3 mg cm^{-2} and a decreased electrolyte/sulfur ratio of 3.3 μL mg^{-1}.

3. Results and Discussion

The fabrication process of TiN@C/S/Ta_2O_5 composite is shown in Figure 1. Firstly, Super P, CNT, graphene, and TiN were mixed and dispersed in deionized water to obtain the homogenous host materials. Sulfur was synthesized based on a disproportionated reaction. Then, the suspension of sulfur nanoparticles was added into the above host materials system. Through long-time stirring, all those materials were dispersed homogeneously without agglomeration, and the sulfur nanoparticles were evenly deposited in the TiN@C host. Finally, Ta(OEt)$_5$ was added to TiN@C/S slurry to shape the amorphous Ta_2O_5 by the hydrolysis reactions, which were well-dispersed on the external surface of TiN@C/S/Ta_2O_5.

Based on thermogravimetric analysis (TGA), the rationale design TiN@C/S/Ta_2O_5 composite displays a sulfur content of 90 wt% in the sulfur composite (Figure 2a). Therefore, it is difficult for the low content 5% of both TiN and Ta_2O_5 to respond in X-ray diffraction patterns (XRD) measurement (Ultima IV, Rigaku Corporation, Tokyo, Japan). The characteristic peaks of the TiN@C/S/Ta_2O_5 composites are following the standard sulfur PDF card S (JCPDS 08-0247) (Figure S1), while the diffraction peaks of TiN and Ta_2O_5 can scarcely be found. To prove the existence of TiN and Ta_2O_5, the TiN@C/S/Ta_2O_5 composites were washed with carbon disulfide to wipe off redundant sulfur. TiN diffraction peaks, which are in accordance with the standard PDF card of TiN, and amorphous tantalum oxide are discovered in the rinsed sample (Figure 2b). The mass ratio of Ta and Ti in TiN@C/S/Ta$_2$O$_5$ are 3.78 wt% and 3.52 wt% by ICP-OES testing. The surface chemical states of TiN@C/S/Ta$_2$O$_5$ were investigated by XPS under high vacuum, the display of strong peak intensity of Ta, and weak intensity Ti, further revealing the existence of an out-coated Ta$_2$O$_5$ layer in Figure 2c,d. The pore structures of the multidimensional carbon structure were studied using N$_2$ adsorption-desorption analysis. The BET-specified surface area of the multidimensional carbon structure was 352 m^2 g^{-1}. A dual distribution of micropore and mesopore was observed on the multidimensional carbon structure (Figure 2e,f); its electronic conductivity was 4.38×10^3 S m^{-1}.

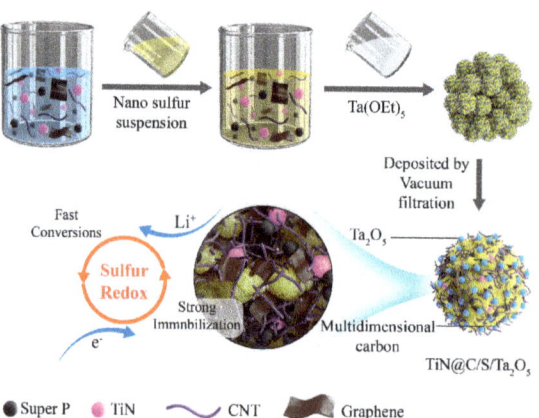

Figure 1. Schematic illustration of the synthesis route of the TiN@C/S/Ta$_2$O$_5$ composites.

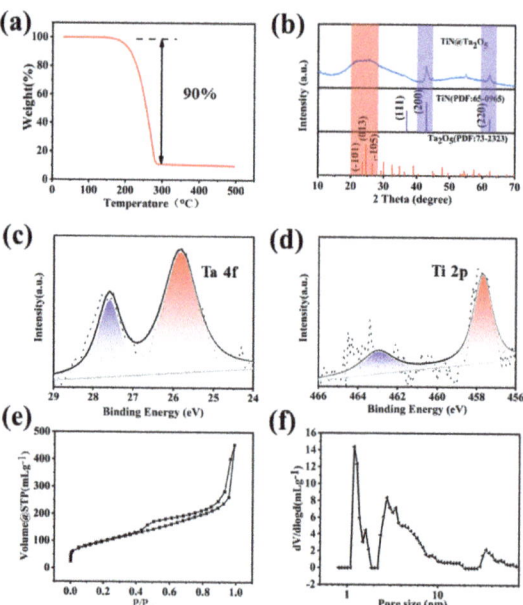

Figure 2. (a) TGA of TiN@C/S/Ta$_2$O$_5$ composites. (b) XRD patterns of TiN@C/Ta$_2$O$_5$ composites. (c,d) XPS spectra of TiN@C/S/Ta$_2$O$_5$ composites. (e) N$_2$ sorption isotherm and (f) pore size distribution based on QSDFT model of the multidimensional carbon structure.

The morphology of the TiN@C/S/Ta$_2$O$_5$ composites was investigated by scanning electron microscope (SEM); the TiN@C/S/Ta$_2$O$_5$ composite consists of a pile of clusters with an average size of about 0.2–1 μm (Figure 3a). The rough surface and hollow morphology guarantee the intimate contact between sulfur composites and electrolytes, leading to fast ion and electron transportation (Figure 3b).

Figure 3. (a,b) SEM image of TiN@C/S/Ta$_2$O$_5$ composite.

The end products of the TiN@C/S/Ta$_2$O$_5$ composite electrode were displayed by transition electron microscope (TEM) observation. The closely packed nano-clusters structure is further conformed (Figure 4a). As displayed in the high-magnification TEM image (Figure 4b), the lattice spacings at 0.212 nm correspond to the (200) plane of TiN. Meanwhile, as shown in the (EDS) elemental mapping images in Figure 4c,g, Ti, N, Ta, O, and S elements are observed; these results manifest the formation of TiN@C/S/Ta$_2$O$_5$ composite. In addition, the EDS elemental mapping images of one single bulk TiN@C/S/Ta$_2$O$_5$ show that more signals of Ta and O can be observed, and most of them are on the surface of the unit; this result confirms the formation of the external Ta$_2$O$_5$ coating layer. Moreover, the existence of a few elements Ti and N imply that most TiN may be implanted inside the TiN@C/S/Ta$_2$O$_5$ composite (Figure S2).

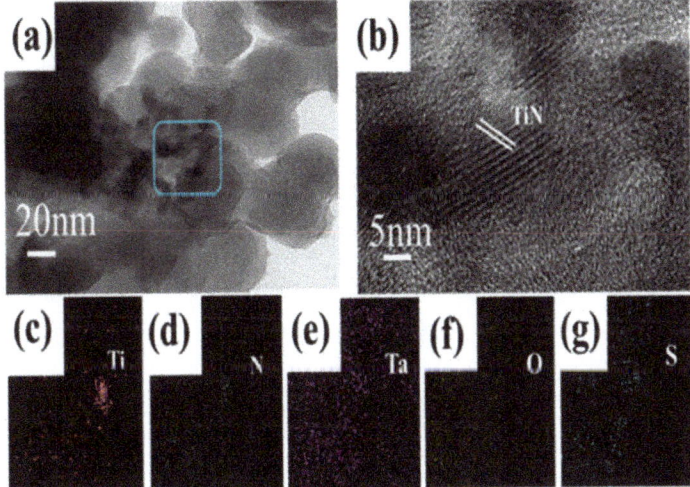

Figure 4. (a) TEM image of TiN@C/S/Ta$_2$O$_5$ material. (b) High-resolution TEM image of TiN@C/S/Ta$_2$O$_5$ material. (c–g) EDS elemental mapping images of TiN@C/S/Ta$_2$O$_5$ material with the selective regions shown in (a).

Polysulfides adsorption experiments were systematically implemented using the Li$_2$S$_6$ electrolyte for modeling polysulfide intermediates. In contrast, the Li$_2$S$_6$ solution turns almost transparent after being adsorbed by Ta$_2$O$_5$ and TiN/Ta$_2$O$_5$, while an obvious yellowish color still can be viewed for that with TiN. The absorbance was analyzed by UV-vis absorption spectra; it can be seen that the adsorption intensity with Ta$_2$O$_5$ and TiN/Ta$_2$O$_5$ solution decreases compared with others (Figure 5a). The X-ray photoelectron spectroscopy (XPS) technique was put into use to understand the chemical interaction of catalysts before and after the absorption of Li$_2$S$_6$; it can be observed that Ti 2p peaks have hardly any binding energy shift before and after the adsorption of Li$_2$S$_6$ (Figure 5b). However, the additional peak at 407.3 and 399.3 eV in the N 1 s spectrum of the Li$_2$S$_6$-treated TiN corresponds to Ti-N-S bonding (Figure 5c), which should be ascribed to the electronegativity and polar of nitrogen; it ensures a strongly interaction with LiPSs and means that the exposed N sites are utilized as the main active sites for absorbing Li$_2$S$_6$ [20]. After interacting with Li$_2$S$_6$, a large positive shift can be observed in both Ta 4f and O1s peaks in (Figure 5d,e); the shifts of these peak positions are considered to be produced by the strong binding interaction between Li$_2$S$_6$ and Ta$_2$O$_5$, confirming their strong LiPSs adsorption capability [15].

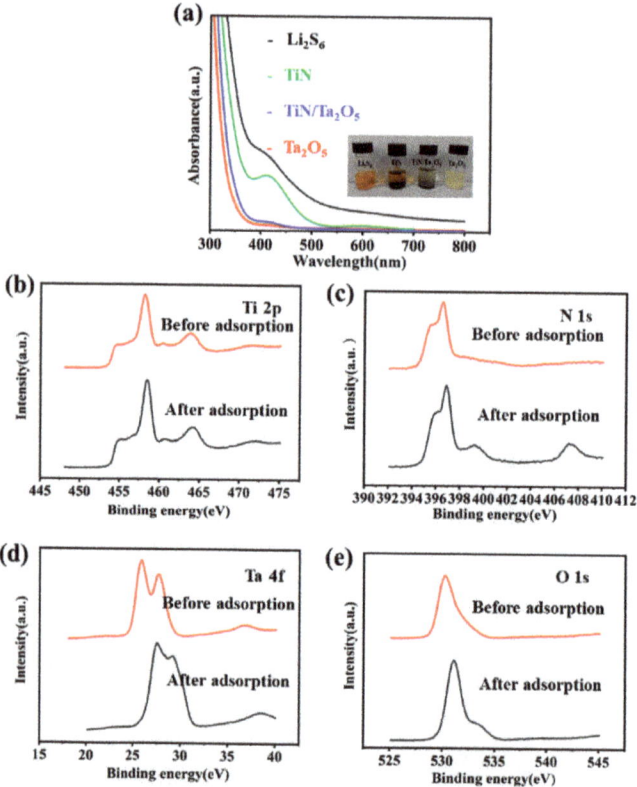

Figure 5. (a) UV-vis spectra of the Li$_2$S$_6$ solution with TiN, Ta$_2$O$_5$, TiN/Ta$_2$O$_5$, and bare Li$_2$S$_6$ solution. (b–e) XPS spectra of Ti 2p, N 1s, Ta 4f, and O1s before and after adsorbed Li$_2$S$_6$.

Symmetrical cells were assembled using TiN, Ta$_2$O$_5$, and TiN/Ta$_2$O$_5$ identical electrodes and Li$_2$S$_6$ electrolyte to investigate the LiPSs conversion dynamics [21]. The TiN/Ta$_2$O$_5$ compound presents the strongest redox current peaks among different samples (Figure 6a), which could be attributed to the integrated adsorption and catalytic ability of the TiN/Ta$_2$O$_5$ compound. Ta$_2$O$_5$ shows strong adsorption capability for polysulfides, but their intrinsically low electrical conductivity will impede the reaction kinetics of soluble LiPSs conversion into insoluble Li$_2$S/Li$_2$S$_2$. Similarly, although TiN exhibits good electrical conductivity, its weak affinities with lithium polysulfides cannot retain LiPSs to suppress the shuttling effect. Moreover, electrochemical impedance spectroscopy (EIS) analysis was employed to understand the interfacial charge transfer kinetics [22] (Figure 6b). TiN/Ta$_2$O$_5$ shows the smallest charge transfer resistance; the interface impedance of symmetric cells can explain the chemical affinity ability of the electrode materials to LiPSs and the ability of TiN/Ta$_2$O$_5$ to accept electrons when interacting with LiPSs, further reflecting its fast charge transfer and facile sulfur redox reactions at the TiN/Ta$_2$O$_5$ and polysulfides interface. All the above results reveal that the synergistic adsorptive-catalytic effect of TiN/Ta$_2$O$_5$ toward enhanced LiPSs conversions.

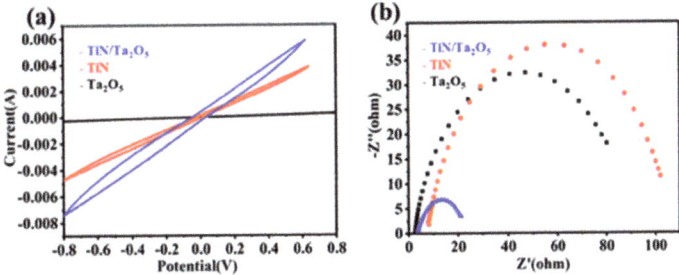

Figure 6. (a) CV curves of TiN, Ta$_2$O$_5$, and TiN/Ta$_2$O$_5$ symmetric cells with 0.1 M Li$_2$S$_6$ electrolyte. (b) EIS spectra of Li$_2$S$_6$ symmetrical cells.

To investigate the improved reaction kinetics of the TiN@C/S/Ta$_2$O$_5$ composites, the expedited polysulfides redox kinetics between solid-liquid-solid conversions was further indicated by the CV curves of the cells with TiN@C/S/Ta$_2$O$_5$, Ta$_2$O$_5$@C/S, and TiN@C/S electrodes at a scan rate of 0.1 mV s^{-1} [23]. All the assembled cells show the typical pair of redox peaks. Compared with Ta$_2$O$_5$@C/S and TiN@C/S, CV curves of the TiN@C/S/Ta$_2$O$_5$ electrode display visibly stronger peak current intensity and closer peak position (Figure 7a). Additionally, the CV curves overlap well upon several cycles (Figure S3). At the same time, the voltage plateaus of galvanostatic charge/discharge profiles are also correspondent to peaks in the CV curves. The TiN@C/S/Ta$_2$O$_5$ cathode exhibits the highest specific capacity and a relatively lowest overpotential at 0.5C during the first cycle [24] (Figure 7b), when adsorptive-catalytic sites, both TiN and Ta$_2$O$_5$, were implanted within the multidimensional carbon structure. On the one hand, the dispersed-distribution Ta$_2$O$_5$ coating suggests stronger LiPSs adsorption capability due to the polar Ta-O bonding; it is found that long-chain polysulfides Li$_2$S$_6$ and Li$_2$S$_8$ are easily deformed after adsorbing on the surface of strong polar active sites Ta$_2$O$_5$. In this way, the Ta$_2$O$_5$ facilitates the fragmentation reactions of long-chain polysulfides into shorter chains and accelerates the kinetics of polysulfides conversion reactions by reducing the activation energy [25]. On the other hand, its high electrical conductivity of TiN can improve the reduction reaction kinetics by promoting electron transport in the electrode, resulting in rapid conversion of LiPSs into Li$_2$S. The TiN@C/S/Ta$_2$O$_5$ electrode owns the dual advantages of excellent trapping capability of LiPSs (Ta$_2$O$_5$) and superior electronic conductivity (TiN) to achieve the adsorption-catalysis synergy. The rate properties and corresponding charge-discharge profiles of these electrodes are displayed (Figure 7c and Figure S4); the TiN@C/S/Ta$_2$O$_5$ delivers the best rate performance with the highest discharge capacity of 1112 mAh g^{-1} at 1C compared with Ta$_2$O$_5$@C/S (942 mAh g^{-1}) and TiN@C/S (993 mAh g^{-1}), the reversible capacity of 1216 mAh g^{-1} when the current returns to 0.1 C. These results furtherly confirm the improved catalytic activity and kinetics for LiPSs conversion, which is due to the elaborate design of TiN@C/S/Ta$_2$O$_5$ composite. The charge/discharge voltage profiles of

TiN@C/S/Ta$_2$O$_5$ at various current densities are illustrated (Figure S4), TiN@C/S/Ta$_2$O$_5$ is capable of retaining the two-plateau discharge profile at a raised rate of up to 1 C without the severe electrochemical polarization-induced serious deformations of the voltage profile. The long-term cyclability performances are compared at a high current density of 0.5 C; the TiN@C/S/Ta$_2$O$_5$ cathode exhibits the optimal performances with an initial discharge capacity of 1175 mAh g^{-1} at 0.5 C, which is much higher than all the other cathodes. The TiN@C/S/Ta$_2$O$_5$ cathode still delivers a decent discharge capacity of 660 mAh g^{-1} and high Coulombic efficiency after 300 cycles (Figure 7d). The gentle capacity attenuation and prominent capacity retention of TiN@C/S/Ta$_2$O$_5$ cathode benefit from its abundant polysulfide-trapping and catalytic active sites: the out-coated Ta$_2$O$_5$ can serve as a covering layer to physically restrain part of LiPSs inside and chemically adsorb another part of out-diffused LiPSs on its polar active surface. Furthermore, TiN NPs presents satisfied catalysis ability to catalyze the conversion of LiPSs, originating from expediting electron transfer. With the synergistic and complementary roles of the cathode materials, the TiN@C/S/Ta$_2$O$_5$ cathode improves the efficient utilization of lithium polysulfides and promotes the chemical interaction with LiPSs and sulfur redox kinetics. It is worth noting that a distinct two-plateau discharge profile maintains well from the 1st to the 100 th cycle at a relatively high rate of 0.5 C (Figure S5), which is consistent with the results of cycling performance of TiN@C/S/Ta$_2$O$_5$ over 300 cycles at 0.5 C. This result is very competitive compared with the previously reported other electrodes (Table S1) [26–28] but it is under the condition of high S content up to 90 wt% and relatively low content of catalysts and carbon materials.

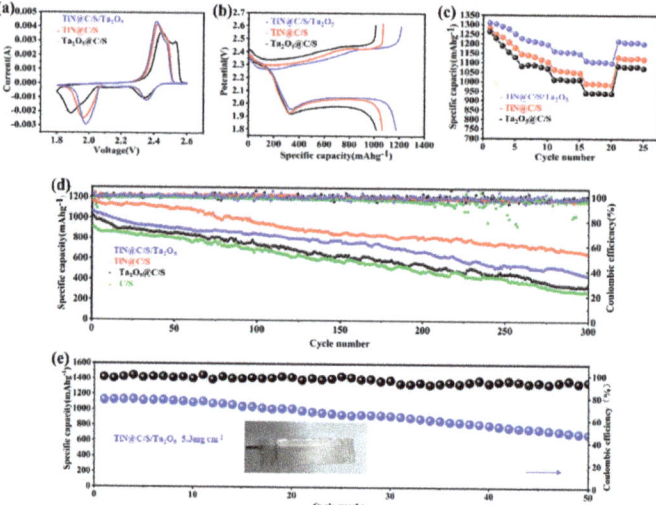

Figure 7. (a) CV curves of TiN@C/S/Ta$_2$O$_5$, Ta$_2$O$_5$@C/S, and TiN@C/S. (b) Galvanostatic charge–discharge curves of different cathodes at 0.5C. (c) rate performances of various sulfur electrodes. (d) Cycling performance of TiN@C/S/Ta$_2$O$_5$, Ta$_2$O$_5$@C/S, TiN@C/S and C/S cathodes at 0.5 C. (e) Cycling performances for TiN@C/Ta$_2$O$_5$/S pouch cells over 50 cycles at 0.2 C charge/discharge rate with a 5.3 mg cm^{-2}.

The wonderful performance of the TiN@C/S/Ta$_2$O$_5$ coin cell inspired us to fabricate a pouch cell with 200 mg sulfur loading in a single-piece cathode with dimensions of 75 mm × 50 mm. The pouch cell was cycled at a current density of 200 mA g^{-1}, the pouch cell showed a specific capacity of over 1100 mAh g^{-1} with a capacity retention rate of 61.63% for 50 cycles (Figure 7e). Even the uniform Li metal corrosion caused by the dissolution and diffusion of LiPSs in the electrolyte could result in the fluctuation of Coulombic efficiency with limited Li, but the pouch cell still shows high Coulombic efficiency of >96% and stable cycle life, further suggesting the function of the outer layer Ta_2O_5 in restraining the dissolution of LiPSs in organic electrolyte and eliminating Li metal corrosion. Our TiN@C/S/Ta_2O_5 pouch cell exhibits a superior electrocatalytic sulfur reduction reaction (SRR) and represents a significant advance in the light of specific capacity, good cycling life, and capacity retention when compared with some reported data (Table S2) [29–32]. Despite the substantial progress made in our work, there is much room to further enhance the energy densities by optimizing both the mass-production process and cell configuration for making the pouch cell.

4. Conclusions

In summary, the concept of combining the merits of rational materials to construct the high-energy-density lithium-sulfur battery has been introduced. The TiN@C/S/Ta_2O_5 composites with high sulfur fraction were synthesized via the co-precipitation method through a simple and low-cost preparation process. The novel design of TiN@C/S/Ta_2O_5 has demonstrated excellent cyclability and rate capability, owing to its potential for inhibiting the shuttle effect and facilitating LiPSs redox reaction in LSBs. Our work will guide the widespread commercialization of high-energy-density Li-S batteries.

Supplementary Materials: The following are available online at https://www.mdpi.com/article/10.3390/nano11112882/s1, Figure S1. XRD pattern of pure sulfur powder and TiN@C/S/Ta_2O_5 composite. Figure S2. EDS elemental mapping images of TiN@C/S/Ta_2O_5 material. Figure S3. CV profiles of TiN@C/S/Ta_2O_5 at a scan rate of 0.1 mV s^{-1}. Figure S4. Multi-rate discharge-charge profiles of TiN@C/S/Ta_2O_5. Figure S5. Discharge performance of TiN@C/S/Ta_2O_5 at different cycles at a rate of 0.5 C. Table S1. Performance Comparison with other sulfur electrodes. Table S2. Comparison of the pouch cell performance of our work with previously reported work focusing on sulfur cathodes.

Author Contributions: C.W.: design, analysis, investigation, writing—original draft. J.-H.L. and Z.-L.W.: software and investigation. A.-B.W. and H.Z.: investigation, methodology, formal analysis, and supervision. W.-K.W. and Z.-Q.J.: resources and supervision. L.-Z.F.: funding acquisition, resources, and supervision. All authors have read and agreed to the published version of the manuscript.

Funding: This research was supported by the National Natural Science Foundation of China (51872027) and the Beijing Natural Science Foundation (Z200011).

Conflicts of Interest: The authors declare no conflict of interest.

References

1. Peng, H.J.; Huang, J.Q.; Cheng, X.B.; Zhang, Q. Review on High-Loading and High-Energy Lithium–Sulfur Batteries. *Adv. Energy Mater.* **2017**, *7*, 1700260. [CrossRef]
2. Han, P.; Chung, S.H.; Manthiram, A. Recent Advances in Lithium Sulfide Cathode Materials and Their Use in Lithium Sulfur Batteries. *Energy Storage Mater.* **2019**, *17*, 317–324. [CrossRef]
3. Liu, D.; Zhang, C.; Zhou, G.; Lv, W.; Ling, G.; Zhi, L.; Yang, Q.H. Catalytic Effects in Lithium-Sulfur Batteries: Promoted Sulfur Transformation and Reduced Shuttle Effect. *Adv. Sci.* **2018**, *5*, 1700270. [CrossRef]
4. Xu, Z.L.; Lin, S.; Onofrio, N.; Zhou, L.; Shi, F.; Lu, W.; Kang, K.; Zhang, Q.; Lau, S.P. Exceptional catalytic effects of black phosphorus quantum dots in shuttling-free lithium sulfur batteries. *Nat. Commun.* **2018**, *9*, 4164. [CrossRef] [PubMed]
5. Tsao, Y.; Lee, M.; Miller, E.C.; Gao, G.; Park, J.; Chen, S.; Katsumata, T.; Tran, H.; Wang, L.-W.; Toney, M.F. Designing a Quinone-Based Redox Mediator to Facilitate Li2S Oxidation in Li-S Batteries. *Joule* **2019**, *3*, 872. [CrossRef]
6. Pang, Q.; Kundu, D.P.; Cuisinier, M.; Nazar, L.F. Surface-enhanced redox chemistry of polysulphides on a metallic and polar host for lithium-sulphur batteries. *Nat. Commun.* **2014**, *5*, 4759. [CrossRef]
7. Chao, D.; Zhu, C.; Yang, P.; Xia, X.; Liu, J.; Wang, J.; Fan, X.; Savilov, S.V.; Lin, J.; Fan, H.J.; et al. Array of nanosheets render ultrafast and high-capacity Na-ion storage by tunable pseudocapacitance. *Nat. Commun.* **2016**, *7*, 12122. [CrossRef]
8. Sun, Z.H.; Zhang, J.Q.; Yin, L.C.; Hu, G.J.; Fang, R.P.; Cheng, H.M.; Li, F. Conductive porous vanadium nitride/graphene composite as chemical anchor of polysulfides for lithium-sulfur batteries. *Nat. Commun.* **2017**, *8*, 14627. [CrossRef]
9. Chen, X.X.; Ding, X.Y.; Wang, C.S.; Feng, Z.Y.; Xu, L.Q.; Gao, X.; Zhai, Y.J.; Wang, D.B. A multi-shelled CoP nanosphere modified separator for highly efficient Li-S batteries. *Nanoscale* **2018**, *10*, 13694–13701. [CrossRef]

10. Xie, J.; Li, B.Q.; Peng, H.J.; Song, Y.W.; Zhao, M.; Chen, X.; Zhang, Q.; Huang, J.Q. Implanting Atomic Cobalt within Mesoporous Carbon toward Highly Stable Lithium-Sulfur Batteries. *Adv. Mater.* **2019**, *31*, 1903813. [CrossRef]
11. Zhou, T.H.; Lv, W.; Li, J.; Zhou, G.M.; Zhao, Y.; Fan, S.X.; Liu, B.L.; Li, B.H.; Kang, F.Y.; Yang, Q.H. Twinborn TiO_2–TiN heterostructures enabling smooth trapping–diffusion–conversion of polysulfides towards ultralong life lithium–sulfur batteries. *Energy Environ. Sci.* **2017**, *10*, 16941703. [CrossRef]
12. Hao, Z.X.; Yuan, L.X.; Chen, C.J.; Xiang, J.W.; Li, Y.Y.; Huang, Z.M.; Hu, P.; Huang, Y.H. TiN as a simple and efficient polysulfide immobilizer for lithium–sulfur batteries. *J. Mater. Chem. A* **2016**, *4*, 17711–17717. [CrossRef]
13. Cui, Z.M.; Zu, C.X.; Zhou, W.D.; Manthiram, A.; Goodenough, J.B. Mesoporous Titanium Nitride-Enabled Highly Stable Lithium-Sulfur Batteries. *Adv. Mater.* **2016**, *28*, 6926–6931. [CrossRef] [PubMed]
14. Xing, Z.; Li, G.; Sy, S.; Chen, Z. Recessed deposition of TiN into N-doped carbon as a cathode host for superior Li-S batteries performance. *Nano Energy* **2018**, *54*, 1–9. [CrossRef]
15. Choi, J.; Jeong, T.G.; Lee, D.; Oh, S.H.; Jung, Y.; Kim, Y.T. Enhanced rate capability due to highly active Ta_2O_5 catalysts for lithium sulfur batteries. *J. Power Sources* **2019**, *435*, 226707. [CrossRef]
16. Zhang, Z.; Luo, D.; Li, G.; Gao, R.; Li, M.; Li, S.; Zhao, L.; Dou, H.; Wen, G.; Sy, S.; et al. Tantalum-Based Electrocatalyst for Polysulfide Catalysis and Retention for High-Performance Lithium-Sulfur Batteries. *Matter* **2020**, *3*, 920. [CrossRef]
17. Peng, H.J.; Huang, J.Q.; Zhao, M.Q.; Zhao, M.Q.; Zhang, Q.; Cheng, X.B.; Liu, X.Y.; Qian, W.Z.; Wei, F. Nanoarchitectured Graphene/CNT@Porous Carbon with Extraordinary Electrical Conductivity and Interconnected Micro/Mesopores for Lithium-Sulfur Batteries. *Adv. Funct. Mater.* **2014**, *24*, 2772–2781. [CrossRef]
18. Xue, W.J.; Shi, Z.; Suo, L.M.; Wang, C.; Wang, Z.Q.; Wang, H.Z.; So, K.P.; Maurano, A.; Yu, D.W.; Chen, Y.M.; et al. Intercalation-conversion hybrid cathodes enabling Li-S full-cell architectures with jointly superior gravimetric and volumetric energy densities. *Nat. Energy* **2019**, *4*, 374–382. [CrossRef]
19. Li, Z.Q.; Xu, J.; Wang, J.; Niu, D.F.; Hu, S.Z.; Zhang, X.S. Well-dispersed Amorphous Ta_2O_5 Chemically Grafted onto Multi-Walled Carbon Nanotubes for High-performance Lithium Sulfur Battery. *Int. J. Electrochem. Sci.* **2019**, *14*, 6628–6642. [CrossRef]
20. Lim, W.G.; Jo, C.; Cho, A.; Hwang, J.; Kim, S.; Han, J.W.; Lee, J. Approaching Ultrastable High-Rate Li-S Batteries through Hierarchically Porous Titanium Nitride Synthesized by Multiscale Phase Separation. *Adv. Mater.* **2019**, *31*, 1806547. [CrossRef]
21. Shi, H.; Ren, X.; Lu, J.; Dong, C.; Liu, J.; Yang, Q.; Chen, J.; Wu, Z.-S. Dual-Functional Atomic Zinc Decorated Hollow Carbon Nanoreactors for Kinetically Accelerated Polysulfides Conversion and Dendrite Free Lithium Sulfur Batteries. *Adv. Energy Mater.* **2020**, *10*, 2002271. [CrossRef]
22. Shen, Z.; Cao, M.; Zhang, Z.; Pu, J.; Zhong, C.; Li, J.; Ma, H.; Li, F.; Zhu, J.; Pan, F.; et al. Efficient $Ni_2Co_4P_3$ Nanowires Catalysts Enhance Ultrahigh-Loading Lithium–Sulfur Conversion in a Microreactor-Like Battery. *Adv. Funct. Mater.* **2020**, *30*, 1906661. [CrossRef]
23. Chen, Y.; Choi, S.; Su, D.; Gao, X.; Wang, G. Self-standing sulfur cathodes enabled by 3D hierarchically porous titanium monoxide-graphene composite film for high-performance lithium-sulfur batteries. *Nano Energy* **2018**, *47*, 331. [CrossRef]
24. Li, Z.; Zhang, J.; Guan, B.; Wang, D.; Liu, L.M.; Lou, X.W. A sulfur host based on titanium monoxide@carbon hollow spheres for advanced lithium–sulfur batteries. *Nat. Commun.* **2016**, *7*, 13065. [CrossRef]
25. Wang, Y.; Huang, J.Y.; Lu, J.G.; Lu, B.; Ye, Z. Fabricating efficient polysulfide barrier via ultrathin tantalum pentoxide grown on separator for lithium–sulfur batteries. *J. Electroanal. Chem.* **2019**, *854*, 113539. [CrossRef]
26. Su, D.; Cortie, M.; Fan, H.; Wang, G. Prussian Blue Nanocubes with an Open Framework Structure Coated with PEDOT as High-Capacity Cathodes for Lithium–Sulfur Batteries. *Adv. Mater.* **2017**, *29*, 1700587. [CrossRef] [PubMed]
27. Bao, W.; Su, D.; Zhang, W.; Guo, X.; Wang, G. 3D Metal Carbide@Mesoporous Carbon Hybrid Architecture as a New Polysulfide Reservoir for Lithium-Sulfur Batteries. *Adv. Funct. Mater.* **2016**, *26*, 8746–8756. [CrossRef]
28. Sun, Q.; Xi, B.; Li, J.-Y.; Mao, H.; Ma, X.; Liang, J.; Feng, J.; Xiong, S. Nitrogen-Doped Graphene-Supported Mixed Transition-Metal Oxide Porous Particles to Confine Polysulfides for Lithium–Sulfur Batteries. *Adv. Energy Mater.* **2018**, *8*, 1800595. [CrossRef]
29. Luo, L.; Li, J.; Asl, H.Y.; Manthiram, A. In-Situ Assembled VS4 as a Polysulfide Mediator for High-Loading Lithium–Sulfur Batteries. *ACS Energy Lett.* **2020**, *5*, 1177–1185. [CrossRef]
30. Luo, L.; Chung, S.H.; Asl, H.Y.; Manthiram, A. Long-Life Lithium–Sulfur Batteries with a Bifunctional Cathode Substrate Configured with Boron Carbide Nanowires. *Adv. Mater.* **2018**, *30*, 1804149. [CrossRef] [PubMed]
31. Kim, M.S.; Kim, M.S.; Do, V.D.; Xia, Y.Y.; Kim, W.; Cho, W. Facile and scalable fabrication of high-energy-density sulfur cathodes for pragmatic lithium-sulfur batteries. *J. Power Sources* **2019**, *422*, 104–112. [CrossRef]
32. Zhang, J.; You, C.; Wang, J.; Xu, H.; Zhu, C.; Guo, S.; Zhang, W.; Yang, R.; Xu, Y. Confinement of sulfur species into heteroatom-doped, porous carbon container for high areal capacity cathode. *Chem. Eng. J.* **2019**, *368*, 340–349. [CrossRef]

Article

Enhancing the Capacity and Stability by CoFe$_2$O$_4$ Modified g-C$_3$N$_4$ Composite for Lithium-Oxygen Batteries

Xiaoya Li [1], Yajun Zhao [1], Lei Ding [1], Deqiang Wang [1], Qi Guo [1], Zhiwei Li [1], Hao Luo [1,*], Dawei Zhang [1] and Yan Yu [2,*]

1 School of Chemistry and Chemical Engineering, Hefei University of Technology, Hefei 230009, China; Sabrina_lxy@163.com (X.L.); zyj251264@163.com (Y.Z.); dlei1107@163.com (L.D.); wondreful26@163.com (D.W.); 18297919084@163.com (Q.G.); Zhiwei.Li@hfut.edu.cn (Z.L.); zhangdw@ustc.edu.cn (D.Z.)
2 Department of Materials of Science and Engineering, University of Science and Technology, Hefei 230026, China
* Correspondence: luohao@hfut.edu.cn (H.L.); yanyumse@ustc.edu.cn (Y.Y.)

Abstract: As society progresses, the task of developing new green energy brooks no delay. Li-O$_2$ batteries have high theoretical capacity, but are difficult to put into practical use due to problems such as high overvoltage, low charge-discharge efficiency, poor rate, and cycle performance. The development of high-efficiency catalysts to effectively solve the shortcomings of Li-O$_2$ batteries is of great significance to finding a solution for energy problems. Herein, we design CoFe$_2$O$_4$/g-C$_3$N$_4$ composites, and combine the advantages of the g-C$_3$N$_4$ material with the spinel-type metal oxide material. The flaky structure of g-C$_3$N$_4$ accelerates the transportation of oxygen and lithium ions and inhibits the accumulation of CoFe$_2$O$_4$ particles. The CoFe$_2$O$_4$ materials accelerate the decomposition of Li$_2$O$_2$ and reduce electrode polarization in the charge–discharge reaction. When CoFe$_2$O$_4$/g-C$_3$N$_4$ composites are used as catalysts in Li-O$_2$ batteries, the battery has a better discharge specific capacity of 9550 mA h g^{-1} (catalyst mass), and the cycle stability of the battery has been improved, which is stable for 85 cycles.

Keywords: Li-O$_2$ batteries; composite; ORR; OER

Citation: Li, X.; Zhao, Y.; Ding, L.; Wang, D.; Guo, Q.; Li, Z.; Luo, H.; Zhang, D.; Yu, Y. Enhancing the Capacity and Stability by CoFe$_2$O$_4$ Modified g-C$_3$N$_4$ Composite for Lithium-Oxygen Batteries. Nanomaterials **2021**, 11, 1088. https://doi.org/10.3390/nano11051088

Academic Editor: Christian M. Julien

Received: 18 March 2021
Accepted: 19 April 2021
Published: 22 April 2021

Publisher's Note: MDPI stays neutral with regard to jurisdictional claims in published maps and institutional affiliations.

Copyright: © 2021 by the authors. Licensee MDPI, Basel, Switzerland. This article is an open access article distributed under the terms and conditions of the Creative Commons Attribution (CC BY) license (https://creativecommons.org/licenses/by/4.0/).

1. Introduction

In the face of huge pressure from energy conservation and emission reduction advocates, it is necessary to replace traditional fuel vehicles with electric vehicles possessing high energy capacity and long range [1,2]. The theoretical energy density of Li-O$_2$ batteries is 3500 Wh kg^{-1}, much higher than other batteries. Li-O$_2$ batteries are expected to be used in electric vehicles and in large-scale productions [3–5]. However, so far, the actual specific energy of Li-O$_2$ batteries reported in literature is less than half the theoretical value. This is because during the charging and discharging processes, Li-O$_2$ batteries generate large polarization phenomena and high activation energy, leading to high energy loss [6]. One of the solutions to reduce energy loss for the Li-O$_2$ batteries is to develop new electrocatalysts with high activity. Metallic materials, such as platinum and iridium, or their alloys, have been shown to be the best electrode materials for oxygen reduction reaction. However, the high price of metallic materials hinders their commercial application [7,8]. Hence, more and more attention has been paid to the development of electrocatalysts prepared by some non-precious metals and non-metals [9].

Recently, spinel materials (such as Co$_3$O$_4$) with adjustable structure and stable chemical properties have attracted the attention of researchers [10,11]. However, compared with noble-metal-based catalysts, the catalytic performance of spinel material has a big gap. Studies have found that the structural stability and performance of the spinel material can be improved by replacing Co in Co$_3$O$_4$ with secondary metals such as Ni, Cu and Mn [12].

Unfortunately, when choosing this material as a catalyst, as the reaction begins, particles continue to aggregate, the number of active sites decreases, and cycle performance decreases. Choosing a suitable supporting substrate to form a stable structure can effectively reduce particle agglomeration and improve catalyst activity.

Graphite carbon nitride (g-C_3N_4) has a graphite-like planar phase, of which the nitrogen atoms have both three-fold coordination atoms and two-fold coordination atoms. They each also contain six nitrogen lone pairs of electrons [13–15]. This enables the g-C_3N_4-based catalyst to change the electronic structure and provides an ideal position for recombination [16,17]. Therefore, when employed as an ameliorative support for the nanoparticle surface, g-C_3N_4 plays a significant role in the catalysis process. It can not only restrain the migration of nanoparticles, but also acts as the second active-site supplier that imparts activity to the integral catalyst [18,19]. In addition, we know that rate capability, round-trip efficiency, and cycle life in Li-O_2 batteries are equally governed by parasitic reactions, which are now recognized to be caused by formation of the highly reactive singlet oxygen. Selected homogeneous catalyst approach to limit singlet oxygen release is a way to improve performance [20–22]. The uniformly dispersed particles supported on substrate are also beneficial to performance improvement due to more sufficient contact between the active constitutes and Li_2O_2 particles.

Taking these issues into account, we herein rationally design a scalable facile strategy for fabricating a $CoFe_2O_4$/g-C_3N_4 composite with the $CoFe_2O_4$ particles supported on the flaky g-C_3N_4. Thereinto, g-C_3N_4 not only provides rich catalytic sites, but also acts as a support for loading $CoFe_2O_4$ nanoparticles to restrain their aggregation. g-C_3N_4 and $CoFe_2O_4$ synergistically enhance the catalytic performance. The resultant $CoFe_2O_4$/g-C_3N_4 composite exhibits enhanced electrochemical performance of Li-O_2 batteries with respect to discharge capacity, voltage polarization, and cycling performance when compared with single $CoFe_2O_4$. This strategy may provide an efficient and versatile approach for designing spinel-based materials as efficient Li-O_2 batteries cathodes.

2. Materials and Methods

2.1. Synthesis

Preparation of g-C_3N_4: First, 4 g melamine was placed in a crucible with a cover. The crucible was annealed at 550 °C for 4 h in a muffle furnace, and the obtained product was grinded to obtain block C_3N_4. Next, 1 g C_3N_4 was dissolved in 35 mL hydrochloric acid and stirred for 30 min. The solution was then placed in a Teflon hydrothermal kettle at 110 °C for 300 min. Finally, the mixture was centrifuged, washed, and dried at 70 °C for 10 h.

Preparation of $CoFe_2O_4$/g-C_3N_4: First, 0.4 g g-C_3N_4 was dissolved in 30 mL ethylene glycol, then 0.0496 g cobalt nitrate and 0.138 g iron nitrate (n_{Co}:n_{Fe} = 1:2) were added and stirred vigorously for 30 min to achieve thorough mixing. The concentrated ammonia solution was added dropwise to the mixed solution to sustain the pH at 8, which was stirred for another 30 min. Next, the mixture was put in a 50 mL Teflon-lined autoclave and kept at the temperature of 160 °C for 20 h. The products were centrifuged, washed, and dried at 70 °C for 600 min. After grinding and annealing at 350 °C for 3 h, $CoFe_2O_4$/g-C_3N_4 composite was obtained.

Preparation of $CoFe_2O_4$: The procedure for the preparation of pure $CoFe_2O_4$ was the same as the preparation of $CoFe_2O_4$/g-C_3N_4, but without g-C_3N_4.

2.2. Material Characterization

The phase prepared in this experiment was tested with a D/MAX2500V X-ray diffractometer (XRD). The scan range was set to 10–90°, and the scan speed was 10° min^{-1}. For the observation and research of the microscopic morphology of the sample, the SU8020 field emission scanning electron microscope (SEM) produced by Japan's JOEL company was used. The surface microstructure and preliminary quantitative analysis of the sample was tested and analyzed by the JEM-2100F field emission transmission electron microscope

(TEM) and energy-dispersive spectrometer (EDS). A TriStar II 3020 V1.03 specific surface area tester was employed and the specific surface area was calculated by the BET method from the adsorption isotherm of the sample with nitrogen gas. The component elements of the sample and the analysis of element valence were performed by ESCALAB250 X-ray photoelectron spectrometer (XPS).

2.3. Electrochemical Performance Test

2.3.1. ORR/OER Performance Test

The ORR/OER electrocatalytic performance test was carried out by an ATA-1B rotating disc electrode (RDE). Preparation of working electrode: A mixture of 10 mg $CoFe_2O_4$/g-C_3N_4 and 2 mg Vulcan XC-72 were added to 40 µL Nafion solution, and 2 mL isopropanol water with the given fraction ($V_{isopropanol}:V_{Deionized\ water}$ = 1:5) was added into the solution. The ultrasonic treatment was then used to disperse the solution, then 5 µL was pipetted and mixed suspension added to the polished glassy carbon electrode surface. Finally, the electrode was dried by natural volatilization or low-temperature drying. Once the electrode was completely dried, it proceeded to the testing process. In the measurements, glassy carbon electrode (GCE) was used as the working electrode. Saturated calomel electrode (SCE) was used as the reference electrode. Graphite electrode was used as the auxiliary electrode.

For ORR/OER polarization curve testing, the scanning speed was set to 10 mV s^{-1} and the rotation speed was 400~2000 rpm. The electrolyte was 0.1 M oxygen-saturated KOH solution. The ORR potential scanning interval was set between 0 and -0.6 V. The electrode rotation speed tested in OER was 1600 rpm, and the potential scanning interval was 0~1 V.

2.3.2. Battery Performance Test

Preparation of oxygen electrode: First, 15 mg catalyst and 30 mg KB were grinded and mixed for 1 h. Next, 83.3 mg PVDF (6 wt%) and 10 drops of NMP were added to the grinded powder to form a uniform slurry without obvious particles, which was then coated on carbon paper and dried at 90 °C for 10 h.

The electrochemical performance of batteries was tested by using 2032-type coin cell. Each cell was composed of a lithium metal anode, a glass fiber separator (Whatman grade GF/D), an electrolyte containing 1 M $LiCF_3SO_3$ in TEGDME, an oxygen cathode, and two pieces of nickel foam (1 mm thick) as the filler and the current collector. This was assembled in an argon-filled glove box (M. Braun). The oxygen cathodes were prepared by coating catalyst ink onto carbon paper homogenously. The catalyst mass loading of the oxygen cathode is about 0.5 ± 0.1 mg cm^{-2}.

For the evaluation of the battery discharge–charge performance and overvoltage performance test, voltage range was 2.2–4.5 V (vs. Li^+/Li). The current density was 100 mA g^{-1}.

The voltage range for the battery cycle performance test was 2.0–4.5 V (vs. Li^+/Li). The current density was 500 mA g^{-1} and limited the capacity to 1000 mA h g^{-1}.

3. Results and Discussions

As shown in Scheme 1, the synthesis of $CoFe_2O_4$/g-C_3N_4 starts with a facile hydrothermal treatment of the solution containing cobalt nitrate, iron nitrate, and the prepared g-C_3N_4. The phase is studied by X-ray diffraction (XRD). In Figure 1a, g-C_3N_4 exhibits two characteristic diffraction peaks: A peak at 27.4° is formed by the stack of g-C_3N_4 rings, which is the (002) crystal plane [23]; another with relatively weak intensity is at 13.0°, which is a characteristic diffraction peak of Melamine substances and mainly refers to the in-plane nitrogen pores formed by the 3-s-triazine structure [24]. By comparing the two XRD patterns, before and after alkali treatment, the peak of the material does not change, which implies that the crystal structure of g-C_3N_4 remains unchanged. This confirms that carbon nitride possesses good chemical stability. For the XRD patterns of $CoFe_2O_4$/g-C_3N_4

composite, the peaks at 18.3°, 30.7°, 35.7°, 43.3°, 57.4°, and 63.2° are $CoFe_2O_4$ diffraction peaks. The peak at 2θ position of 27.7° correspond to the typical peak of g-C_3N_4, implying that the $CoFe_2O_4$/g-C_3N_4 material has been successfully synthesized.

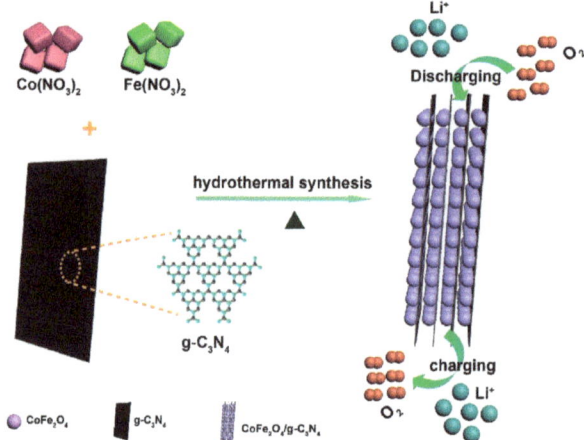

Scheme 1. Schematic illustration of the preparation of $CoFe_2O_4$/g-C_3N_4 composite.

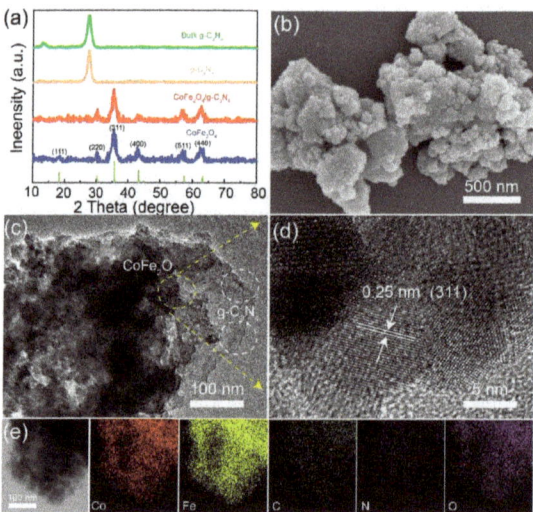

Figure 1. (a) XRD patterns of bulk g-C_3N_4, g-C_3N_4, $CoFe_2O_4$, and $CoFe_2O_4$/g-C_3N_4; images of $CoFe_2O_4$/g-C_3N_4 material: (b) SEM, (c) TEM, (d) HRTEM, and (e) EDS elemental mapping.

The surface and internal morphology of materials are observed with scanning electron microscopy (SEM) and transmission electron microscopy (TEM). Bulk g-C_3N_4 is composed of many layered nanosheets, which are stacked on each other (Figure S1). After alkali treatment, the g-C_3N_4 is stripped into flakes (Figure S2), which is conducive to the loading of spherical particles. Without g-C_3N_4, pure $CoFe_2O_4$ shows the particle morphology in Figure S3, while $CoFe_2O_4$/g-C_3N_4 exhibits stacked granular morphology connected with particles as shown in Figure 1b. TEM images indicate that $CoFe_2O_4$ nanoparticles are supported on the flaky g-C_3N_4 where the clear lattice fringes of 0.25 nm correspond well to the (311) plane of $CoFe_2O_4$, suggesting that these particles are $CoFe_2O_4$ (Figure 1c,d).

Moreover, the element distribution C, N, O, Co, and Fe of the material is observed by energy-dispersive spectroscopic (EDS) (Figure 1e), indicating successfully prepared material of $CoFe_2O_4/g\text{-}C_3N_4$ composite and uniformly distributed $CoFe_2O_4$ nanoparticles on the flat $g\text{-}C_3N_4$. The specific surface area and pore volume of $CoFe_2O_4/g\text{-}C_3N_4$ are determined to be 244.1 m^2 g^{-1} and 0.423 cm^3 g^{-1} by Braunauer–Emmett–Teller (BET) analysis (Figure S4), which is much larger than those of $g\text{-}C_3N_4$. Such a big increase for specific surface area and pore volume indicates that $CoFe_2O_4$ particles supported on the flaky $g\text{-}C_3N_4$ are effective for preventing $g\text{-}C_3N_4$ from stacking on each other. The increase in specific surface area and pore size can expose more active sites, increasing the ion migration rate during charging and discharging.

The elemental composition and chemical state of synthetic materials are analyzed by X-ray photoelectron spectroscopy (XPS). The full-spectrum scan result shows that the material contains C, N, O, Fe, and Co, which proves that the $CoFe_2O_4/g\text{-}C_3N_4$ material was successfully synthesized (Figure 2a). In the spectrum of C 1s (Figure 2b), two strong peaks are shown at 288.1 and 284.7 eV, which respectively belong to the sp^2 hybridized bounded carbon (C=N) and graphitic carbon (C-C) [25,26]. Figure 2c is the N 1s spectrum, which confirms that the $CoFe_2O_4/g\text{-}C_3N_4$ composite has three kinds of nitrogen. The 398.3 eV is the sp^2 hybrid nitrogen (N1), the 399.2 eV is the tertiary nitrogen (N2), and the 400.9 eV is the amino functional group (N3), respectively [27,28]. The high-resolution Co 2p$_{3/2}$ spectrum presents the signals of Co^{2+}, Co^{3+}, and satellite (Figure 2d), where the presence of Co^{3+}, Co^{2+}, and vibration satellite are respectively exhibited by the brown main peak (780.5 eV), the blue green peak at 782.2 eV, and the peak at 786.8 eV [29–31]. The Fe 2p$_{3/2}$ spectrum of the catalyst also exhibits Fe^{2+}, Fe^{3+}, and satellite peaks, which respectively appear at 710.7 eV, 712.2 eV, and 714.1 eV [31–33]. The above results indicate that two-electron pairs of Fe^{3+}/Fe^{2+} and Co^{3+}/Co^{2+} exist in the structure of $CoFe_2O_4/g\text{-}C_3N_4$ material. In addition, the O1s XPS peak at 530.2 eV is the intrinsic lattice, while another peak at 531.8 eV may be the adsorbed water [34,35].

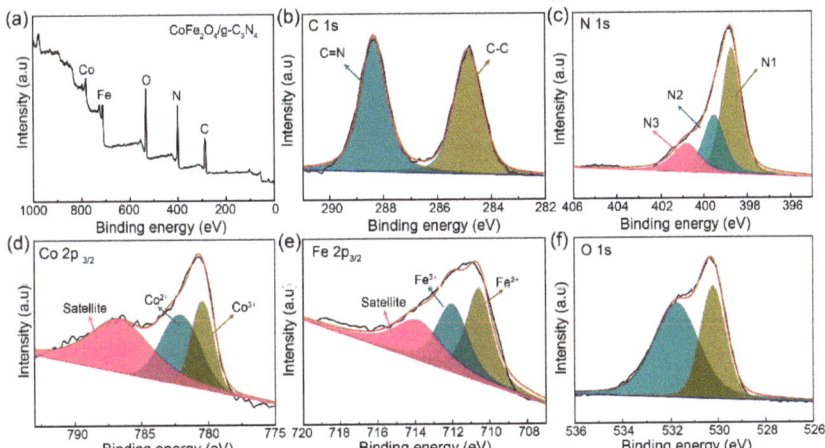

Figure 2. XPS survey spectra of $CoFe_2O_4/g\text{-}C_3N_4$: (**a**) full-spectrum scan, (**b**) C 1s, (**c**) N 1s, (**d**) Co 2p$_{3/2}$, (**e**) Fe 2p$_{3/2}$, and (**f**) O 1s.

The electrochemical catalytic activity of the sample was evaluated in 0.1M KOH solution by linear sweep voltammetry (LSV). In Figure 3a, the $CoFe_2O_4/g\text{-}C_3N_4$ displays an onset potential (E$_{onset}$) of 0.90 V and a half-wave potential (E$_{1/2}$) of 0.76 V, which are more positive than those of $CoFe_2O_4$ (0.67 V) and $g\text{-}C_3N_4$ (0.65 V) catalysts (Figure 3b), signifying the higher activity for ORR. Moreover, the $CoFe_2O_4/g\text{-}C_3N_4$ exhibits a larger diffusion-limited current, suggesting the material has strong mass transfer ability and fast electron transfer speed. The reaction kinetics of the material are calculated and analyzed by the LSV curve at different speeds, as shown in Figure 3c. The activity of all materials at different speeds is linearly related, and the calculated number of transferred electrons show Pt/C, $CoFe_2O_4$/g-C_3N_4, $CoFe_2O_4$, and g-C_3N_4 is 4.0, 3.8, 3.4, and 1.67, respectively. The ORR reaction of $CoFe_2O_4$/g-C_3N_4 composite is nearly a 4e$^-$ process, which indicates a better ORR catalytic performance and is close to commercial Pt/C. Furthermore, the OER curves of various catalysts show the $CoFe_2O_4$/g-C_3N_4 catalyst has the limiting current density of 49.3 mA cm^{-2}, higher than $CoFe_2O_4$ (~18.8 mA cm^{-2}), g-C_3N_4 (~1.93 mA cm^{-2}), and Pt/C (~9.87 mA cm^{-2}) (Figure 3d), indicating the $CoFe_2O_4$/g-C_3N_4 material has the best OER performance. This may be due to the synergies of g-C_3N_4 and $CoFe_2O_4$ which boost the performance.

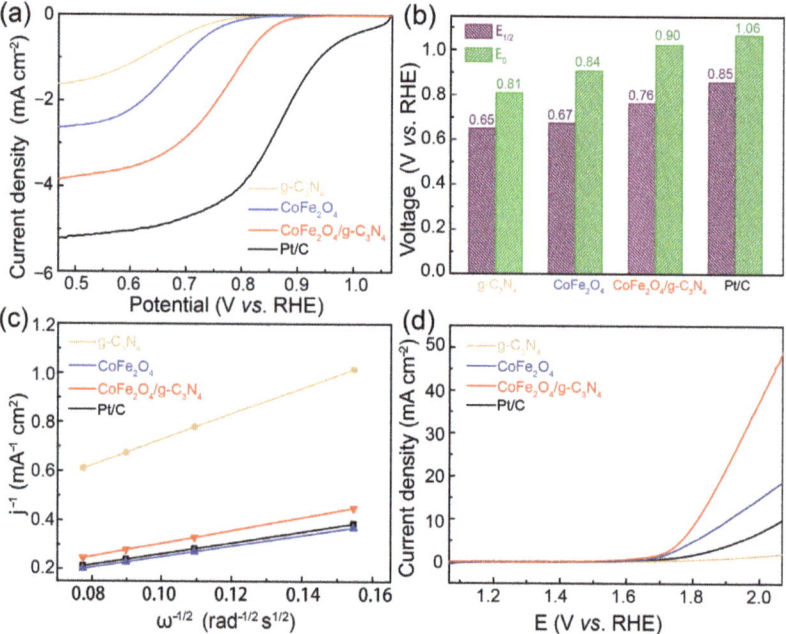

Figure 3. (a) ORR polarization curves; (b) the results summary of $E_{1/2}$ and E_0; (c) Koutecky-Levich plots; (d) OER curves of g-C_3N_4, $CoFe_2O_4$, $CoFe_2O_4$/g-C_3N_4, and Pt/C catalysts.

The electrochemical catalytic ability of the material was evaluated by assembling 2032 button batteries. In Figure 4a, the $CoFe_2O_4$/g-C_3N_4 composite exhibits a discharge specific capacity of 9550 mA h g^{-1}, significantly higher than $CoFe_2O_4$ and XC-72. The overpotential of the $CoFe_2O_4$/g-C_3N_4 composite as exhibited in Figure 4b is 1.21 V, which is lower than the catalysts of $CoFe_2O_4$ (1.33 V) and XC-72 (1.87 V). These results indicate that when $CoFe_2O_4$/g-C_3N_4 composite is used as catalyst for Li-O_2 batteries, the degree of polarization during the discharge–charge process is relatively lower than pure $CoFe_2O_4$. Figure 4c,d show the cycle performance of $CoFe_2O_4$ and $CoFe_2O_4$/g-C_3N_4. The $CoFe_2O_4$/g-C_3N_4 composite can stably cycle 85 times, which is significantly higher than pure $CoFe_2O_4$ cathodes (16 times). In the battery cycle reaction, the flake structure of g-C_3N_4 can supply sufficient space to store discharge product Li_2O_2 and accelerate the transportation of O_2 and Li$^+$; also, its larger specific surface area and rich N content provide more reactive sites for the discharge–charge reaction. In addition, g-C_3N_4 provides a stable support for restraining aggregation of $CoFe_2O_4$ nanoparticles due to its high chemical stability, which leads to an increase in the stability of the composite catalyst. Therefore, from the discussion, the synergistic effect between g-C_3N_4 and $CoFe_2O_4$ can effectively improve the electrocatalytic activity and stability of the catalyst in Li-O$_2$ batteries, which is much better than pure CoFe$_2$O$_4$ cathode. The capacity retention rate tests of Li-O$_2$ batteries of CoFe$_2$O$_4$/g-C$_3$N$_4$ material were carried out at four current densities (Figure 4e). Compared with the first discharge capacity, the corresponding capacity retention rates of the four current densities are 67.7%, 61.3%, 48.4%, and 42.3%, respectively (Figure 4f). All results prove that CoFe$_2$O$_4$/g-C$_3$N$_4$ composite can promote ORR/OER dynamics, thus improving the rate performance of the battery. Even at a high current, the battery keeps a good capacity retention rate.

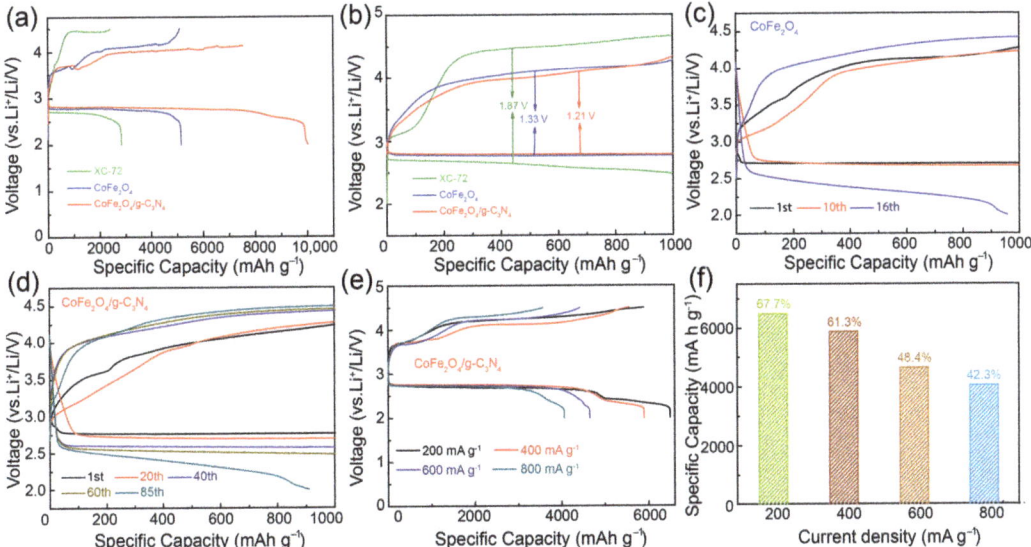

Figure 4. (**a**) First discharge–charge curves and (**b**) the overpotential curves of XC-72, CoFe$_2$O$_4$, and CoFe$_2$O$_4$/g-C$_3$N$_4$; cycle ability of (**c**) CoFe$_2$O$_4$ and (**d**) CoFe$_2$O$_4$/g-C$_3$N$_4$; (**e**) first discharge–charge plots at different current densities (200, 400, 600, and 800 mA g^{-1}) and (**f**) capacity retention plots of CoFe$_2$O$_4$/g-C$_3$N$_4$.

To investigate the composition and morphology changes of CoFe$_2$O$_4$/g-C$_3$N$_4$ electrodes at different stages after the discharge–charge process, Figure 5a displays XRD spectrum of batteries at different states. Compared with the catalyst in the initial state, after discharging, the surface of carbon sheet clearly exhibits characteristic peaks of Li$_2$O$_2$ at 33°, 35°, and 58°, and neither Li$_2$O nor LiOH is detected. After charging, no diffraction peak of Li$_2$O$_2$ is observed, which indicates that the battery has a good reversibility. Figure 5b–d are the SEM images of the battery after different processes. From Figure 5c, the morphology of the Li$_2$O$_2$ after the battery being deeply discharged is clearly observed. The lithium peroxide is mainly distributed uniformly on the carbon sheet in the form of filaments. After the charging process (Figure 5d), Li$_2$O$_2$ disappeared completely, which is similar with the initial state (Figure 5b) and consistent with XRD results, indicating that the battery has good reversibility.

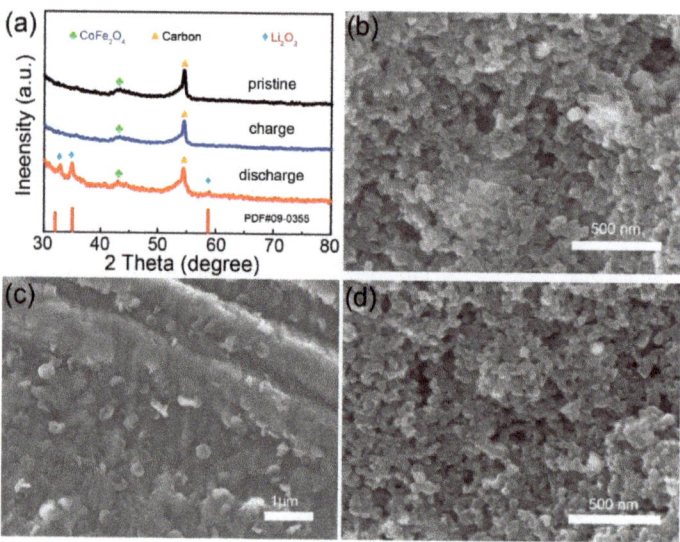

Figure 5. (a) XRD patterns of $CoFe_2O_4$/g-C_3N_4 electrode in different states; SEM images of $CoFe_2O_4$/g-C_3N_4 electrode in (b) pristine state, (c) 1st discharged state, and (d) 1st charged states.

4. Conclusions

In short, the g-C_3N_4/$CoFe_2O_4$ material is synthesized by simple methods. The flake structure of the $CoFe_2O_4$/g-C_3N_4 catalyst accelerates the transportation of O_2 and Li^+ and provides sufficient space to store the discharge product of Li_2O_2. Both g-C_3N_4 and $CoFe_2O_4$ can offer catalytic sites. In addition, g-C_3N_4 supplies a stable support for restraining the aggregation of $CoFe_2O_4$ nanoparticles. Therefore, the Li-O_2 batteries with such $CoFe_2O_4$ modified g-C_3N_4 composites as air cathodes deliver a discharge specific capacity of 9550 mA h g^{-1}, and the cycle stability has been enhanced more than pure $CoFe_2O_4$ cathodes. This strategy opens opportunities for rationally exploring different modified strategies on nanostructured electrocatalysts for diverse devices.

Supplementary Materials: The following are available online at https://www.mdpi.com/article/10.3390/nano11051088/s1, Figure S1: SEM image of bulk g-C_3N_4, Figure S2: SEM image of flake C_3N_4, Figure S3: SEM image of $CoFe_2O_4$, Figure S4: (a) N_2 adsorption-desorption isotherms and (b) specific surface area and pore volume of g-C_3N_4 and $CoFe_2O_4$/g-C_3N_4.

Author Contributions: Conceptualization, D.Z., Y.Y.; methodology, L.D.; software, Z.L.; formal analysis, Q.G.; data curation, X.L., D.W.; writing original draft preparation, Y.Z.; writing—review and editing, H.L. All authors have read and agreed to the published version of the manuscript.

Funding: This research was funded by the China Postdoctoral Science Foundation, grant number 172731 and National College Students' Innovative Entrepreneurship Training Scheme, grant number 201910359038.

Data Availability Statement: Data can be available upon request from the authors.

Conflicts of Interest: The authors declare no conflict of interest.

References

1. Wang, C.; Zhang, Z.; Liu, W.; Zhang, Q.; Wang, X.G.; Xie, Z.; Zhou, Z. Enzyme-Inspired Room-Temperature Lithium–Oxygen Chemistry via Reversible Cleavage and Formation of Dioxygen Bonds. *Angew. Chem. Int. Ed.* **2020**, *59*, 17856–17863. [CrossRef]
2. Zhang, X.; Yang, Y.; Zhou, Z. Towards practical lithium-metal anodes. *Chem. Soc. Rev.* **2020**, *49*, 3040–3071. [CrossRef]
3. Qiao, Y.; He, Y.; Wu, S.; Jiang, K.; Li, X.; Guo, S.; He, P.; Zhou, H. MOF-Based Separator in an Li–O2 Battery: An Effective Strategy to Restrain the Shuttling of Dual Redox Mediators. *ACS Energy Lett.* **2018**, *3*, 463–468. [CrossRef]

4. Mu, X.; Wen, Q.; Ou, G.; Du, Y.; He, P.; Zhong, M.; Zhu, H.; Wu, H.; Yang, S.; Liu, Y.; et al. A current collector covering nanostructured villous oxygen-deficient NiO fabricated by rapid laser-scan for Li-O$_2$ batteries. *Nano Energy* **2018**, *51*, 83–90. [CrossRef]
5. Akhtar, N.; Akhtar, W. Prospects, challenges, and latest developments in lithium-air batteries. *Int. J. Energy Res.* **2014**, *39*, 303–316. [CrossRef]
6. Ganesan, P.; Prabu, M.; Sanetuntikul, J.; Shanmugam, S. Cobalt sulfide nanoparticles grown on nitrogen and sulfur codo-ped graphene oxide: An efficient electrocatalyst for oxygen reduction and evolution reactions. *ACS Catal.* **2015**, *5*, 3625–3637. [CrossRef]
7. Zhao, W.; Wang, J.; Yin, R.; Li, B.; Huang, X.; Zhao, L.; Qian, L. Single-atom Pt supported on holey ultrathin g-C$_3$N$_4$ nanosheets as efficient catalyst for Li-O$_2$ batteries. *J. Colloid Interface Sci.* **2020**, *564*, 28–36. [CrossRef]
8. Hu, Y.; Zhang, T.; Cheng, F.; Zhao, Q.; Han, X.; Chen, J. Recycling application of Li-MnO$_2$ batteries as rechargeable lithi-um–air batteries. *Angew. Chem. Int. Ed.* **2015**, *54*, 4338–4343. [CrossRef]
9. Ionescu, M.I.; Laforgue, A. Synthesis of nitrogen-doped carbon nanotubes directly on metallic foams as cathode material with high mass load for lithium-air batteries. *Thin Solid Films* **2020**, *709*, 138211. [CrossRef]
10. Luo, H.; Jiang, W.-J.; Zhang, Y.; Niu, S.; Tang, T.; Huang, L.-B.; Chen, Y.-Y.; Wei, Z.; Hu, J.-S. Self-terminated activation for high-yield production of N,P-codoped nanoporous carbon as an efficient metal-free electrocatalyst for Zn-air battery. *Carbon* **2018**, *128*, 97–105. [CrossRef]
11. Li, S.; Hao, X.; Abudula, A.; Guan, G. Nanostructured Co-based bifunctional electrocatalysts for energy conversion and storage: Current status and perspectives. *J. Mater. Chem. A* **2019**, *7*, 18674–18707. [CrossRef]
12. Budnikova, Y.H. Recent advances in metal–organic frameworks for electrocatalytic hydrogen evolution and overall water splitting reactions. *Dalton Trans.* **2020**, *49*, 12483–12502. [CrossRef] [PubMed]
13. Luo, H.; Jiang, W.J.; Niu, S.; Zhang, X.; Zhang, Y.; Yuan, L.P.; He, C.X.; Hu, J.S. Self-Catalyzed Growth of Co-N-C Nano-brushes for Efficient Rechargeable Zn-Air Batteries. *Small* **2020**, *16*, 2001171. [CrossRef]
14. Duan, J.; Chen, S.; Jaroniec, M.; Qiao, S.Z. Porous C$_3$N$_4$Nanolayers@N-Graphene Films as Catalyst Electrodes for Highly Efficient Hydrogen Evolution. *ACS Nano* **2015**, *9*, 931–940. [CrossRef]
15. Dai, L.; Xue, Y.; Qu, L.; Choi, H.-J.; Baek, J.-B. Metal-Free Catalysts for Oxygen Reduction Reaction. *Chem. Rev.* **2015**, *115*, 4823–4892. [CrossRef]
16. Hang, Y.; Zhang, C.; Luo, X.; Xie, Y.; Xin, S.; Li, Y.; Zhang, D.W.; Goodenough, J.B. α-MnO$_2$ nanorods supported on porous graphitic carbon nitride as efficient electrocatalysts for lithium-air batteries. *J. Power Sources* **2018**, *392*, 15–22. [CrossRef]
17. Fu, J.; Yu, J.; Jiang, C.; Cheng, B. g-C$_3$N$_4$-Based Heterostructured Photocatalysts. *Adv. Energy Mater.* **2018**, *8*, 1701503. [CrossRef]
18. Wu, Y.; Wang, T.; Zhang, Y.; Xin, S.; He, X.; Zhang, D.; Shui, J. Electrocatalytic performances of g-C$_3$N$_4$-LaNiO$_3$ composite as bi-functional catalysts for lithium-oxygen batteries. *Sci. Rep.* **2016**, *6*, 24314. [CrossRef]
19. Wang, X.; Chen, X.; Thomas, A.; Fu, X.; Antonietti, M. Metal-containing carbon nitride compounds: A new functional or-ganic-metal hybrid material. *Adv. Mater.* **2009**, *21*, 1609–1612. [CrossRef]
20. Samojlov, A.; Schuster, D.; Kahr, J.; Freunberger, S.A. Surface and catalyst driven singlet oxygen formation in Li-O$_2$ cells. *Electrochim. Acta* **2020**, *362*, 137175. [CrossRef]
21. Zhang, J.; Sun, B.; Zhao, Y.; Tkacheva, A.; Liu, Z.; Yan, K.; Guo, X.; McDonagh, A.M.; Shanmukaraj, D.; Wang, C.; et al. A versatile functionalized ionic liquid to boost the solution-mediated performances of lithium-oxygen batteries. *Nat. Commun.* **2019**, *10*, 602. [CrossRef]
22. Qian, Z.; Li, X.; Sun, B.; Du, L.; Wang, Y.; Zuo, P.; Yin, G.; Zhang, J.; Sun, B.; Wang, G. Unraveling the Promotion Effects of a Soluble Cobaltocene Catalyst with Respect to Li–O$_2$ Battery Discharge. *J. Phys. Chem. Lett.* **2020**, *11*, 7028–7034. [CrossRef]
23. Sano, T.; Tsutsui, S.; Koike, K.; Hirakawa, T.; Teramoto, Y.; Negishi, N.; Takeuchi, K. Activation of graphitic carbon nitride (g-C$_3$N$_4$) by alkaline hydrothermal treatment for photocatalytic NO oxidation in gas phase. *J. Mater. Chem. A* **2013**, *1*, 6489–6496. [CrossRef]
24. Nie, H.; Ou, M.; Zhong, Q.; Zhang, S.; Yu, L. Efficient visible-light photocatalytic oxidation of gaseous NO with graphitic carbon nitride (g–C$_3$N$_4$) activated by the alkaline hydrothermal treatment and mechanism analysis. *J. Hazard. Mater.* **2015**, *300*, 598–606. [CrossRef] [PubMed]
25. Cai, B.; Zhao, M.; Ma, Y.; Ye, Z.; Huang, J. Bioinspired Formation of 3D Hierarchical CoFe2O4 Porous Microspheres for Magnetic-Controlled Drug Release. *ACS Appl. Mater. Interfaces* **2015**, *7*, 1327–1333. [CrossRef] [PubMed]
26. Xu, J.; Zhang, W.; Chen, Y.; Fan, H.; Su, D.; Wang, G. MOF-derived porous N-Co$_3$O$_4$@N-C nanododecahedra wrapped with reduced graphene oxide as a high capacity cathode for lithium–sulfur batteries. *J. Mater. Chem. A* **2018**, *6*, 2797–2807. [CrossRef]
27. Kong, H.J.; Won, D.H.; Kim, J.; Woo, S.I. Sulfur-Doped g-C$_3$N$_4$/BiVO$_4$ Composite Photocatalyst for Water Oxidation under Visible Light. *Chem. Mater.* **2016**, *28*, 1318–1324. [CrossRef]
28. Liu, Y.; Wang, Q.-L.; Chen, Z.; Li, H.; Xiong, B.-Q.; Zhang, P.-L.; Tang, K.-W. Visible-light photoredox-catalyzed dual C–C bond cleavage: Synthesis of 2-cyanoalkylsulfonylated 3,4-dihydronaphthalenes through the insertion of sulfur dioxide. *Chem. Commun.* **2020**, *56*, 3011–3014. [CrossRef]
29. Chen, D.; Dong, C.-L.; Zou, Y.; Su, D.; Huang, Y.-C.; Tao, L.; Dou, S.; Shen, S.; Wang, S. In situ evolution of highly dispersed amorphous CoO$_x$ clusters for oxygen evolution reaction. *Nanoscale* **2017**, *9*, 11969–11975. [CrossRef]

30. Feng, L.; Li, Y.; Sun, L.; Mi, H.; Ren, X.; Zhang, P. Heterostructured CoO-Co$_3$O$_4$ nanoparticles anchored on nitrogen-doped hollow carbon spheres as cathode catalysts for Li-O2 batteries. *Nanoscale* **2019**, *11*, 14769–14776. [CrossRef]
31. Biesinger, M.C.; Payne, B.P.; Grosvenor, A.P.; Lau, L.W.; Gerson, A.R.; Smart, R.S. Resolving surface chemical states in XPS analysis of first row transition metals, oxides and hydroxides: Cr, Mn, Fe, Co and Ni. *Appl. Surf. Sci.* **2011**, *257*, 2717–2730. [CrossRef]
32. Han, Q.; Wang, B.; Gao, J.; Cheng, Z.; Zhao, Y.; Zhang, Z.; Qu, L. Atomically Thin Mesoporous Nanomesh of Graphitic C$_3$N$_4$ for High-Efficiency Photocatalytic Hydrogen Evolution. *ACS Nano* **2016**, *10*, 2745–2751. [CrossRef]
33. Zhang, X.; Liang, C.; Qu, X.; Ren, Y.; Yin, J.; Wang, W.; Yang, M.S.; Huang, W.; Dong, X. Sandwich-Structured Fe-Ni$_2$P/MoSx/NF Bifunctional Electrocatalyst for Overall Water Splitting. *Adv. Mater. Interfaces* **2020**, *7*, 1901926. [CrossRef]
34. Lee, Y.J.; Kim, D.H.; Kang, T.-G.; Ko, Y.; Kang, K.; Lee, Y.J. Bifunctional MnO$_2$-Coated Co$_3$O$_4$ Hetero-structured Catalysts for Reversible Li-O2 Batteries. *Chem. Mater.* **2017**, *29*, 10542–10550. [CrossRef]
35. Wang, J.; Gao, R.; Zhou, D.; Chen, Z.; Wu, Z.; Schumacher, G.; Hu, Z.; Liu, X. Boosting the Electrocatalytic Activity of Co3O4 Nanosheets for a Li-O$_2$ Battery through Modulating Inner Oxygen Vacancy and Exterior Co^{3+}/Co^{2+} Ratio. *ACS Catal.* **2017**, *7*, 6533–6541. [CrossRef]

Article

Enhanced Electrochemical Behavior of Peanut-Shell Activated Carbon/Molybdenum Oxide/Molybdenum Carbide Ternary Composites

Ndeye F. Sylla [1], Samba Sarr [1], Ndeye M. Ndiaye [2], Bridget K. Mutuma [1], Astou Seck [3], Balla D. Ngom [2], Mohamed Chaker [3] and Ncholu Manyala [1,*]

[1] Department of Physics, Institute of Applied Materials, SARChI Chair in Carbon Technology and Materials, University of Pretoria, Pretoria 0028, South Africa; ntoufasylla@gmail.com (N.F.S.); ssarr3112@gmail.com (S.S.); bridgetmutuma@gmail.com (B.K.M.)

[2] Laboratoire de Photonique Quantique, d'Energie et de Nano-Fabrication, Faculté des Sciences et Techniques, Université Cheikh Anta Diop de Dakar (UCAD), Dakar-Fann Dakar B.P. 5005, Senegal; nmaty.ndiaye@gmail.com (N.M.N.); balla.ngom@ucad.edu.sn (B.D.N.)

[3] Institut National de la Recherche Scientifique Centre—Énergie Matériaux Télécommunications 1650, Boulevard Lionel Boulet, Varennes, QC J3X 1S2, Canada; astou.seck@emt.inrs.ca (A.S.); chaker@emt.inrs.ca (M.C.)

* Correspondence: ncholu.manyala@up.ac.za; Tel.: +27-12-420-3549; Fax: +27-12-420-2516

Citation: Sylla, N.F.; Sarr, S.; Ndiaye, N.M.; Mutuma, B.K.; Seck, A.; Ngom, B.D.; Chaker, M.; Manyala, N. Enhanced Electrochemical Behavior of Peanut-Shell Activated Carbon/Molybdenum Oxide/Molybdenum Carbide Ternary Composites. *Nanomaterials* 2021, 11, 1056. https://doi.org/10.3390/nano11041056

Academic Editors: Rongming Wang and Shuhui Sun

Received: 24 March 2021
Accepted: 16 April 2021
Published: 20 April 2021

Publisher's Note: MDPI stays neutral with regard to jurisdictional claims in published maps and institutional affiliations.

Copyright: © 2021 by the authors. Licensee MDPI, Basel, Switzerland. This article is an open access article distributed under the terms and conditions of the Creative Commons Attribution (CC BY) license (https://creativecommons.org/licenses/by/4.0/).

Abstract: Biomass-waste activated carbon/molybdenum oxide/molybdenum carbide ternary composites are prepared using a facile in-situ pyrolysis process in argon ambient with varying mass ratios of ammonium molybdate tetrahydrate to porous peanut shell activated carbon (PAC). The formation of MoO_2 and Mo_2C nanostructures embedded in the porous carbon framework is confirmed by extensive structural characterization and elemental mapping analysis. The best composite when used as electrodes in a symmetric supercapacitor (PAC/MoO_2/Mo_2C-1//PAC/MoO_2/Mo_2C-1) exhibited a good cell capacitance of 115 F g^{-1} with an associated high specific energy of 51.8 W h kg^{-1}, as well as a specific power of 0.9 kW kg^{-1} at a cell voltage of 1.8 V at 1 A g^{-1}. Increasing the specific current to 20 A g^{-1} still showcased a device capable of delivering up to 30 W h kg^{-1} specific energy and 18 kW kg^{-1} of specific power. Additionally, with a great cycling stability, a 99.8% coulombic efficiency and capacitance retention of ~83% were recorded for over 25,000 galvanostatic charge-discharge cycles at 10 A g^{-1}. The voltage holding test after a 160 h floating time resulted in increase of the specific capacitance from 74.7 to 90 F g^{-1} at 10 A g^{-1} for this storage device. The remarkable electrochemical performance is based on the synergistic effect of metal oxide/metal carbide (MoO_2/Mo_2C) with the interconnected porous carbon. The PAC/MoO_2/Mo_2C ternary composites highlight promising Mo-based electrode materials suitable for high-performance energy storage. Explicitly, this work also demonstrates a simple and sustainable approach to enhance the electrochemical performance of porous carbon materials.

Keywords: porous carbon; ternary composite; molybdenum oxide; molybdenum carbide; energy storage

1. Introduction

The high demand for energy in conjunction with the rapid depletion of fossil fuels has made it essential to develop alternative energy sources. Various researchers have shown an increased interest in the development of clean, sustainable and renewable energy sources such as solar, wind and geothermal [1,2]. However, there are still challenges linked to the production and continuous supply of energy in large quantities from these renewable energy systems [3]. Therefore, in order to supply energy needs on a long-term basis, in addition to it being affordable, sustainable and environmentally friendly, it is important to find diverse, efficient, safe and flexible methods for its simultaneous generation and storage [4–6]. Batteries and supercapacitors are the most renowned energy storage devices with supercapacitors being given gross attention by energy researchers globally due to their high performance features such as high specific power, long life cycle and quick charge-discharge dynamics [7,8].

Supercapacitors are typically categorized into electrical double layer-capacitor (EDLC) and pseudo capacitors. The former operates on a charge storage process that relies on an electrostatic charge accumulation formed at the interfacial region of electrode/electrolyte while the latter depends on reversible faradic-type redox reactions at the electrode surface material [9,10].

Porous carbon such as carbon nanotube, graphene, activated carbon (AC) and carbon onions owing to their large surface area, good electrical conductivity and great stability are mainly investigated as electrode materials for EDLC [11,12]. Porous AC materials obtained from biomass waste (peanut shell, walnut shell, pinecone and so on) have gained interest due to their distinctive features of well-developed surface area with hierarchical pore structure, abundant availability of precursors sources, environmental friendliness and low costs [13,14]. In addition, surface functional groups obtained from the AC can favor an easy ion adsorption/desorption at the electrode/electrolyte interface leading to an optimum electrochemical performance [15,16]. However, there is still a need to improve their performance to meet the high-energy demand.

To date, various methods have been reported to enhance the electrochemical performance of the porous ACs. For instance, the introduction of oxygen functional groups (ketone, ether, carboxylic acid, quinone, and so on) by oxidizing the porous carbon surface could promote the hydrophilicity and also surface reactivity of the carbon material [17–19]. Moreover, the presence of the surface oxygen-containing species not only provides some pseudo capacitance effect but also enriches the surface wetting capability which contributes a significant improvement in the capacitance and the overall specific energy/power of the carbon electrode material [19,20].

Song et al. [19] prepared O/S dual-modified nanoporous carbon (OSC) by a hydrothermal oxidation method using H_2O_2 (O) and H_2SO_4 (S) as oxidants. Their study revealed that the OSC electrode delivered a specific capacitance of 168 F g^{-1}, 3.5 times higher than the pristine nanoporous carbon in 6 M KOH aqueous electrolyte due to the introduction of the oxygen functional groups on the surface of nanoporous carbon [19]. Another promising method is the heteroatom-doping (nitrogen, sulphur, phosphorous, boron and so on) of a porous AC matrix, which enhances the capability to store charge in the material. Therefore, heteroatom doping into the porous carbon framework results in the modification of the electronic structure, which can remarkably improve the electrical conductivity, the surface wettability, properties of the electrons donor and hence the electrochemical features of the porous carbon materials [21–24]. For instance, we reported previously in our study on the synthesis of nitrogen-doped peanut shell activated carbon (NPAC) by chemical activation and nitrogen-post-doping processes with KOH and melamine respectively. The NPAC showed a considerable increase in the specific capacitance value (167 F g^{-1} to 216 F g^{-1}) for the non-doped PAC and NPAC electrodes in a 2.5 M KNO_3 aqueous electrolyte [24].

Incorporating transition metal oxides/hydroxides or conducting polymer into the porous carbon network is also an effective strategy that can improve the electrochemical properties of these carbon materials. The integration of porous carbon with transition metal oxides (MnO_2, NiO, Fe_3O_4, MoO_2, etc.) forming composites synergistically combines the advantages and mitigates the limitations of both materials [25–27]. MoO_2 is one of the most promising pseudocapacitive transition metal oxides which possesses several oxidation states (+2 to +6), high theoretical specific capacitance, excellent redox reaction capability, low electrical resistivity, good electrochemical activity and low cost [28,29].

Thus, the MoO_2 incorporated into the porous carbon may provide more electrochemical active sites that could influence additional surface pseudocapacitive effect endowing enhanced electrochemical performance of the composite [30–32]. Lina et al. [30] synthesized MoO_2 nanoparticles decorated into 3D porous graphene (MoO_2-rGO) using a hydrothermal route. Their work indicated that the MoO_2-rGO composite enhanced specific capacitance to 356 F g^{-1} as compared to the 3D porous graphene (244 F g^{-1}) in 6 M KOH aqueous electrolyte [30].

Other studies have also shown the incorporation of transition metal carbide (Mo$_2$C, W$_2$C, TiC, etc.) in the carbon framework could improve the performance metrics of the electrochemical capacitors [33–35]. Among the transition metal carbides, Mo$_2$C has recently attracted great interest due to the high specific conductance (1.02 × 10^2 S cm^{-1}), great conductivity (electrical and thermal) and good chemical stability [36–38]. Furthermore, these remarkable proprieties can offer additional actives sites, facilitate electron and ion transportation, good cyclic stability properties and a reduction of the charge resistance which could produce a relatively high electrochemical performance [38,39]. Hussain and co-workers [40] have prepared carbon nanotubes (CNTs) decorated with molybdenum carbide nanosheets (Mo$_2$C@CNT) by a chemical reduction approach followed by carbonization. In their report, the hybrid composite Mo$_2$C@CNT exhibited a specific capacitance value of 365 F g^{-1} which is 3.5 times higher than the CNTs (103 F g^{-1}) in KOH electrolyte owing to the synergy between the Mo$_2$C and CNTs [40].

Forming a ternary composite of metal oxide (MoO$_2$), metal carbide (Mo$_2$C) and porous carbon material result in a combination of the advantages of each component that could considerably enhance the electrochemical features of the energy storage device [38,41,42]. For instance, Ihsan et al. [36] have prepared a MoO$_2$/Mo$_2$C/C spheres by a two-step, hydrothermal process followed by a calcination procedure.

Yang et al. [42] have also synthesized MoO$_2$/Mo$_2$C/C hybrid microspheres by a template-free method. Both studies showed a great rate capability, cycling stability, good capacity as anode materials for Li-ion batteries. However, to the best of our knowledge no reports exist on the application of these Mo-based materials/porous carbon (MoO$_2$/Mo$_2$C/C) ternary composites as supercapacitor electrodes.

In this study, we have established a facile and low-cost approach of synthesizing a ternary composite (PAC/MoO$_2$/Mo$_2$C) for the first time as electrode material for supercapacitor by one-step pyrolysis route through varying mass ratios of ammonium molybdate tetrahydrate to porous peanut shell activated carbon (PAC) (1:0.5; 1:1; 1:2). The in-situ formation of MoO$_2$ and Mo$_2$C nanostructures incorporated into the PAC network were confirmed by the XRD, Raman, HRTEM, SAED, SEM, EDX mapping, as well as the XPS analysis. The obtained ternary composites portrayed interesting merits based on the existing incorporated nanostructures of the MoO$_2$-Mo$_2$C within the nanoporous PAC-based material including: (i) high specific surface area with hierarchically porous structure of the PAC, (ii) pseudocapacitive effect by the redox reaction of MoO$_2$ and (iii) superior electrical conductivity and stability of the Mo$_2$C. The best ternary composite (PAC/MoO$_2$/Mo$_2$C-1) exhibited superior capacitive performance in both half and full-cell test owing to the synergistic effect of the MoO$_2$ and Mo$_2$C nanostructures embedded into the PAC matrix. Moreover, our study demonstrates a simple and sustainable approach to enhance the electrochemical performance of porous carbon materials.

2. Experimental

2.1. Materials

In this study, all chemical reagents were used as obtained without any further purification. Ammonium molybdate tetrahydrate (NH$_4$)$_6$Mo$_7$O$_{24}$·4H$_2$O, 99.98%), potassium nitrate, (KNO$_3$, 99.99%), potassium hydroxide (KOH, 99%), polyvinylidene fluoride (PVDF, 99%), carbon acetylene black (CAB, 99.95%), hydrochloric acid (HCl, 37%) and N-methyl-2-pyrrolidone (NMP, 99%) were supplied from Merck (Johannesburg, South Africa). Argon gas (Ar, 99.99%) was purchased from Afrox (Johannesburg, South Africa) and polycrystalline nickel foam mesh (with 1.6 mm thickness, 420 g m^{-2} areal density) was obtained from Alantum (Munich, Germany).

2.2. Synthesis of the Peanut Shell Waste Derived Activated Carbon (PAC)

Peanut shell waste derived activated carbon (PAC) was prepared by a one-step chemical activation following a reported procedure from our previous study [24]. Briefly, the peanut shell waste raw material was mixed with KOH pellets in an optimized ratio by mass and then subjected to chemical activation at an elevated 850 °C temperature for 1 h to obtain the final product.

2.3. Synthesis of the Peanut Shell Derived Activated Carbon/Molybdenum Oxide/Molybdenum Carbide (PAC/MoO$_2$/Mo$_2$C) Ternary Composites

The synthesized PAC sample was mixed with ammonium molybdate tetrahydrate in mass ratios of 1:0.5, 1:1 and 1:2 for the PAC to the inorganic salt in an agate mortar. Few drops of deionized water were added to the as-prepared mixture to ensure thorough mixing of both materials which was then loaded onto a porcelain boat and air dried at 80 °C for 12 h in an electric oven.

The porcelain boat was transferred into a horizontal tube furnace and heated to 850 °C at a ramping rate of 5 °C min^{-1}. The furnace was kept constant at this temperature for 1 h under 250 sccm of argon gas flow. After cooling down to room temperature, the ternary composites were obtained and labelled PAC/MoO$_2$/Mo$_2$C-0.5, PAC/MoO$_2$/Mo$_2$C-1 and PAC/MoO$_2$/Mo$_2$C-2 corresponding to the mass ratio of 1:0.5, 1:1 and 1:2, of PAC to the inorganic salt, respectively. The schematic procedure for preparing the PAC/MoO$_2$/Mo$_2$C ternary composites is illustrated in Figure 1.

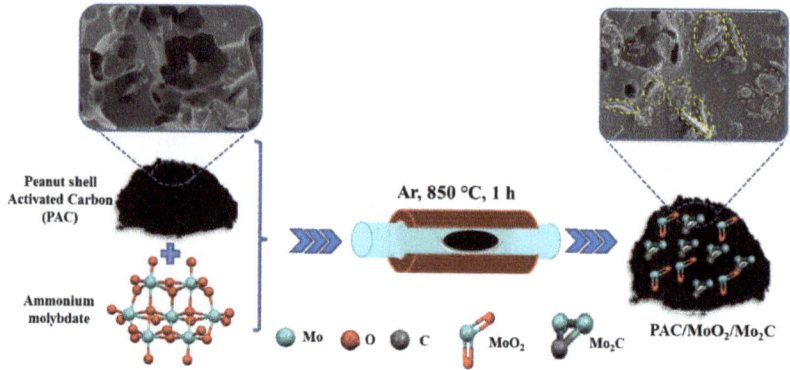

Figure 1. Schematic illustration for the synthesis method of the PAC/MoO$_2$/Mo$_2$C ternary composites.

2.4. Physical Characterization

Powder X-ray diffraction (XRD) analysis of the PAC/MoO$_2$/Mo$_2$C samples was determined using a Brucker D8 Advance diffractometer using Cu Kα (λ = 1.5406 Å) radiation operating in the 2θ range of 10–80°. Raman spectra of the ternary composite materials were characterized on WITec alpha300 RAS+ confocal Raman microscope (WITec, Ulm, Germany) using a 532 nm excitation laser at a power of 5 mW.

High-resolution transmission electron microscope (HRTEM) micrographs and selected area electron diffraction (SAED) patterns were obtained using a JEOL JEM-2100F field emission gun transmission electron microscope (FEG-TEM) operating at 200 kV. Scanning electron microscope (SEM) micrographs and energy dispersive X-ray (EDX) mappings were carried out using a Zeiss Ultra-plus 55 field emission scanning electron microscope (FE-SEM). The SEM images and EDX mapping images were operated at 1.0 and 10 kV accelerating voltage respectively. The surface area distribution was performed by Brunauer-Emmett-Teller (BET) and porosity pore size by Barrett-Joyner-Halenda (BJH) methods on the Micrometrics TriStar II 3020 (version 2.00) system in a relative pressure (P/P$_0$) range of 0.01–1.0 at 77 K. X-ray photoelectron spectroscopy (XPS) of the samples was obtained by a

VG Escalab 220i-XL instrument equipped with a monochromatic Al-Kα (1486.6 eV) source of radiation.

2.5. Electrochemical Characterization

The PAC/MoO$_2$/Mo$_2$C electrodes were prepared by mixing the active material, carbon acetylene black (CAB) as conductive additive and polyvinylidene difluoride (PVDF) as binder in a weight ratio of 8:1:1, respectively in an agate mortar. Few drops of N-methylpyrrolidone (NMP) as solvent was added to the mixture to obtain slurry which was uniformly coated on nickel foam (NF) as current collector followed by drying at 80 °C for a period of 12 h in an electric oven.

The cyclic voltammetry (CV), galvanostatic charge-discharge (GCD) and electrochemical impedance spectroscopy (EIS) were investigated using VMP-300 16-channel potentiostat (Bio-Logic, Knoxville, USA) associated with EC-Lab® (V11.33) software. The three electrode (half-cell) test was performed using the as-prepared electrode as working electrode, the glassy carbon as counter electrode (CE) and Ag/AgCl (in saturated 3M KCl) as reference electrode (RE).

For the two-electrode measurement (full-cell), a Swagelok cell and a microfiber filter paper (separator) were used to assemble the symmetric device. All electrochemical tests were performed in 2.5 M KNO$_3$ aqueous electrolyte at room temperature.

The specific capacitance C_s (F g^{-1}) of the half-cell was obtained from the GCD profiles using the following Equation [43]:

$$C_s = \frac{I \Delta t}{m \Delta V} \quad (1)$$

where I represents the current (mA), Δt is the time (s) of the discharge slop from GCD, m is the mass (mg) of the active electrode and ΔV is the operating potential (V).

The specific capacitance C_s (F g^{-1}), specific energy E (W h kg^{-1}) and the specific power P (W kg^{-1}) for the symmetric device were calculated using the mass total m_T (mg) of the positive and negative electrode from the Equations (2)–(4) [44]:

$$C_s = \frac{I \Delta t}{m_T \Delta V} \quad (2)$$

$$E = \frac{C_s \Delta V^2}{7.2} \quad (3)$$

$$P = 3600 \frac{E_d}{\Delta t} \quad (4)$$

3. Results and Discussion

3.1. Structural, Morphological and Textural Characterization

XRD patterns of the as-prepared PAC/MoO$_2$/Mo$_2$C ternary composites are displayed in Figure 2a. These XRD spectra reveal diffraction peaks matching with the monoclinic MoO$_2$ (ICSD card No. 86-0135, space group: P21/c, cell parameters: a = 5.6096 Å, b = 4.8570 Å, c = 5.6259 Å) and the hexagonal Mo$_2$C (ICSD card No. 35-0787, space group: P63/mmc, with cell parameters: a = 3.0124 Å, b = 3.0124 Å, c = 4.7352 Å) in PAC/MoO$_2$/Mo$_2$C-0.5, PAC/MoO$_2$/Mo$_2$C-1 and PAC/MoO$_2$/Mo$_2$C-2 samples. All diffraction peaks located with approximate 2θ values of 26.1°, 37.1°, 53.7°, 60.7° and 66.9° corresponding to (011), (211), (220), (310) and (131) crystallographic planes, respectively can be assigned to the monoclinic MoO$_2$ [45,46].

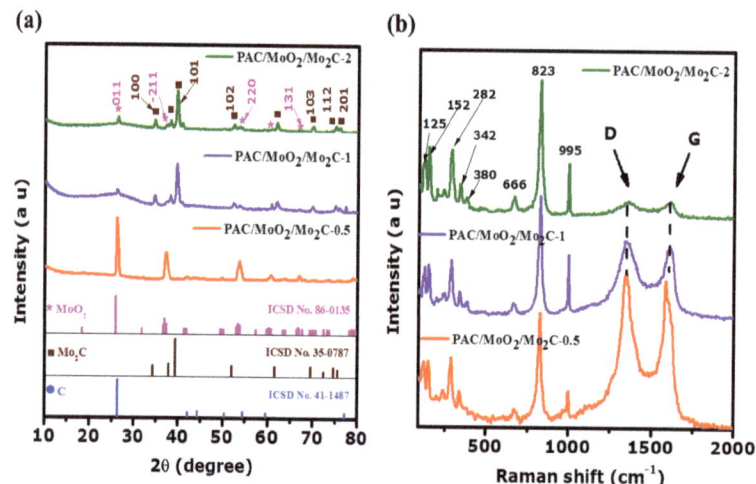

Figure 2. (a) XRD patterns and (b) Raman spectra of the PAC/MoO$_2$/Mo$_2$C-0.5, PAC/MoO$_2$/Mo$_2$C-1 and PAC/MoO$_2$/Mo$_2$C-2.

Therefore, the other featured peaks located at 2θ of 34.3°, 37.9°, 39.4°, 52.1°, 61.5°, 69.5°, 74.6° and 75.5° can be ascribed to the (100), (002), (101), (102), (110), (103), (112) and (201) crystallographic planes from the hexagonal Mo$_2$C, respectively [47,48]. Additionally, the broad peak around 2θ of 26.5° corresponding to (002) diffraction of graphite (ICSD card No. 41-1487) could be attributed to the presence of the amorphous carbon domains of the PAC which overlaps with MoO$_2$ peak at 26.1° [45,48].

The Raman spectra of the as-prepared PAC/MoO$_2$/Mo$_2$C porous ternary composites are shown in Figure 2b. The characteristic peaks at around 125, 152, 200, 342, 381 and 666 cm^{-1} bands could be attributed to the vibration modes of the monoclinic MoO$_2$. The Raman active modes at 285, 825 and 998 cm^{-1} bands could be assigned to vibrational features of the Mo$_2$C [49–52]. Two other characteristics peaks are also observed at (1341–1360 cm^{-1}) and (1583–1609 cm^{-1}) which correspond to the D and G bands, respectively (as shown in Table 1). The D band is associated to the disordered graphitic structure in carbon matrix while the G is due to the sp^2-hybridized graphitic carbon [43]. The intensity ratio of D and G bands (I_D/I_G ratio) recorded in Table 1 reveals the graphitization degree of the as-synthesized ternary composites [42]. The I_D/I_G ratio values decreased from 1.03 to 0.97 with increasing mass ratio of the molybdenum precursor to the porous carbon. This indicates a balanced amorphous carbon to the graphitic carbon in the PAC/MoO$_2$/Mo$_2$C composites resulting from the MoO$_2$ and Mo$_2$C nanoparticles embedded into the porous carbon network [53,54].

Table 1. Raman data of the PAC/MoO$_2$/Mo$_2$C ternary composites.

Samples	D-Band (cm^{-1})	G-Band (cm^{-1})	I_D/I_G Ratio
PAC/MoO$_2$/Mo$_2$C-0.5	1349	1583	1.03
PAC/MoO$_2$/Mo$_2$C-1	1341	1607	1.01
PAC/MoO$_2$/Mo$_2$C-2	1360	1610	0.97

High-resolution transmission electron microscopic (HRTEM) micrographs and selected area electron diffraction (SAED) patterns were further performed to provide more crystal structural information of the PAC/MoO$_2$/Mo$_2$C-0.5, PAC/MoO$_2$/Mo$_2$C-1 and PAC/MoO$_2$/Mo$_2$C-2 ternary composites as shown in Figure 3. Figure 3a–c revealed the HRTEM micrographs of the ternary composites in which the lattice fringes are highlighted in yellow arrow and the layer of amorphous PAC in orange arrow. The lattice fringes with an inter-planar spacing (d) approximate values of 0.340 nm, 0.219 nm and 0.283 nm are corresponding to the crystallographic planes (011), (−212) and (−102) of the monoclinic

MoO$_2$ (ICSD card No. 86-0135), respectively. The other d-spacing values of 0.237 nm and 0.227 nm are assigned to the crystallographic planes (002) and (101) of the hexagonal Mo$_2$C (ICSD card No. 35-0787), respectively. The as-obtained SAED patterns of the PAC/MoO$_2$/Mo$_2$C ternary composites are exhibited in Figure 3d–f. The SAED patterns show the bright diffraction rings which are attributed to (302), (310) and (220) planes of MoO$_2$ and those (002), (100), (101) and (201) planes to Mo$_2$C.

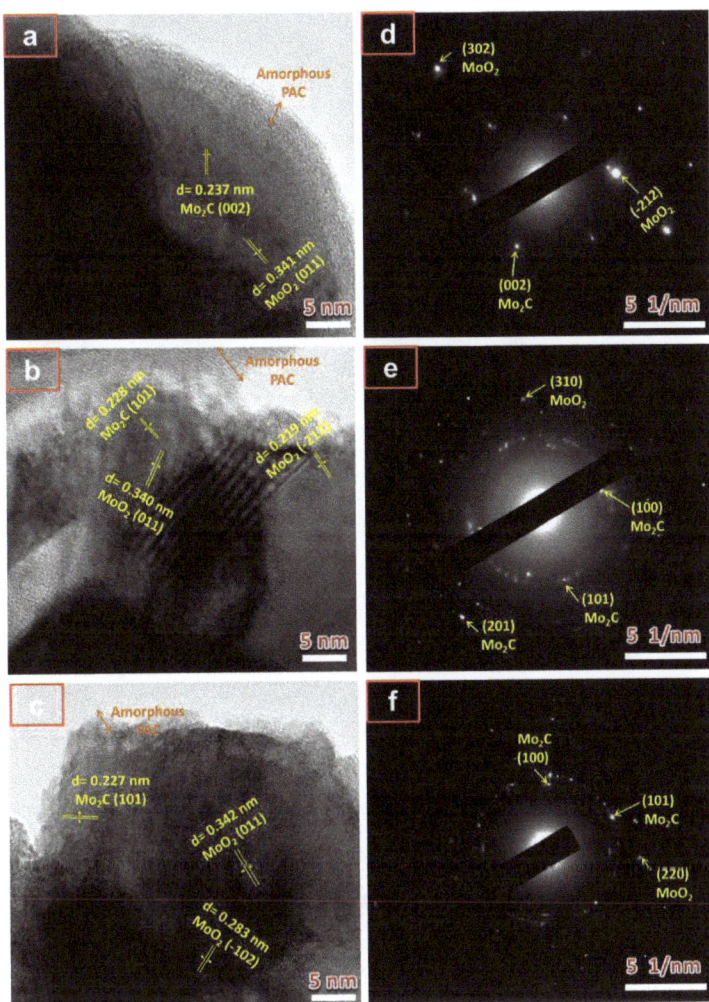

Figure 3. HRTEM micrographs and SAED patterns of (**a,d**) PAC/MoO$_2$/Mo$_2$C-0.5, (**b,e**) PAC/MoO$_2$/Mo$_2$C-1 and (**c,f**) PAC/MoO$_2$/Mo$_2$C-2 ternary composites, respectively.

These results imply a successful incorporation of the MoO$_2$ and Mo$_2$C heterostructures into the amorphous PAC matrix through a pyrolysis process which are consistent with the XRD analysis in Figure 2a.

The SEM micrographs of PAC/MoO$_2$/Mo$_2$C ternary composites prepared at different mass ratios of molybdenum precursor to PAC are displayed in Figure 4. The morphology of the ternary composites reveals the formation of a mixture of agglomerated nanoparticles and nanoplates embedded into the interconnected porous structure of the PAC at low magnification (Figure 4a,c,e).

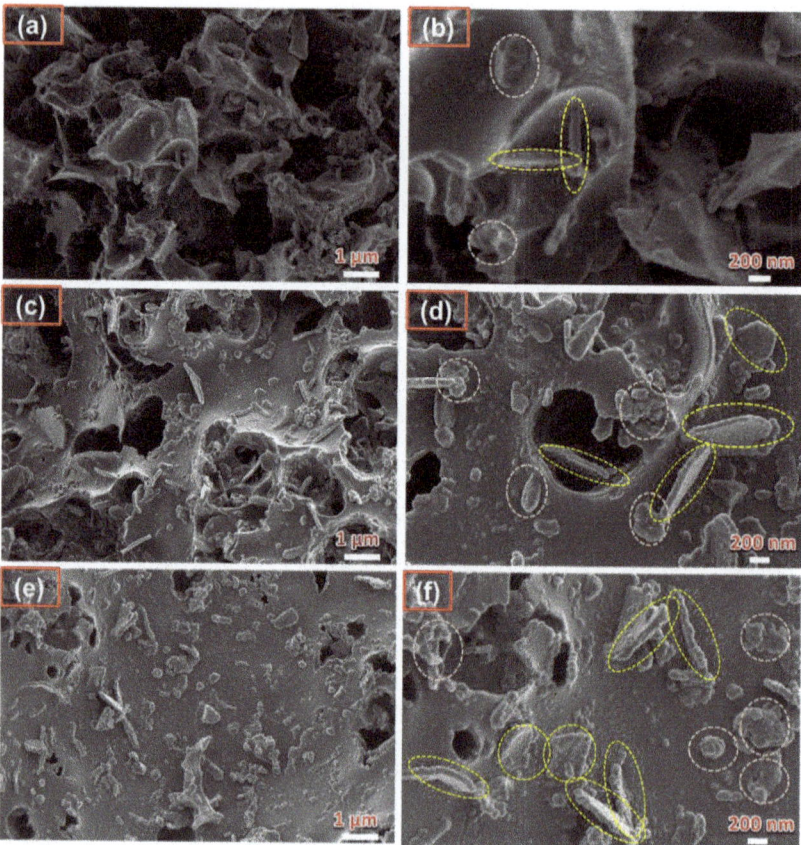

Figure 4. SEM micrographs at low and high magnification of (**a,b**) PAC/MoO$_2$/Mo$_2$C-0.5, (**c,d**) PAC/MoO$_2$/Mo$_2$C-1 and (**e,f**) PAC/MoO$_2$/Mo$_2$C-2 ternary composites.

An increase of molybdenum content reveals an increased tendency of both agglomerated nanoparticles and nanoplates morphologies as highlighted in circles in Figure 4b,d,f (high magnification). These two different morphologies are perhaps due to the presence MoO$_2$ and Mo$_2$C in the composites, but it is not easy to identify which of these belongs to a specific morphology.

EDX mapping was also applied to determine the elemental distribution of the PAC/MoO$_2$/Mo$_2$C-0.5, PAC/MoO$_2$/Mo$_2$C-1 and PAC/MoO$_2$/Mo$_2$C-2 ternary composites as seen in Figure 5a–i. It is observed that the Mo, O and C elements are uniformly distributed throughout the interconnected porous carbon structure. This suggests that the agglomerated nanoparticles and nanoplates were composed of MoO$_2$ and Mo$_2$C embedded into the carbon matrix.

N$_2$ adsorption/desorption experiment was conducted to investigate the textural properties of the PAC/MoO$_2$/Mo$_2$C ternary composites as shown in Figure 6 and Table 2. Figure 6a,b presents the sorption isotherms and the pore size distribution curves, respectively, of the ternary composites. All isotherms depicted a type IV features associated with a H4 hysteresis loop which indicates the coexistence of the micropores and mesopores structures in the ternary composites [55,56]. The BET specific surface area (SSA) values of PAC/MoO$_2$/Mo$_2$C-

0.5, PAC/MoO$_2$/Mo$_2$C-1 and PAC/MoO$_2$/Mo$_2$C-2 samples are 804, 711 and 301 m^2 g^{-1}, respectively. A decrease in SSA was observed upon increasing the ammonium molybdate precursor loading from 0.5 to 2. For the total pore volume and micropore area (Table 2), the same trend is also identified for all samples from with the decrease from 0.44 to 0.20 cm^3 g^{-1} and 670 to 165 m^2 g^{-1}, respectively.

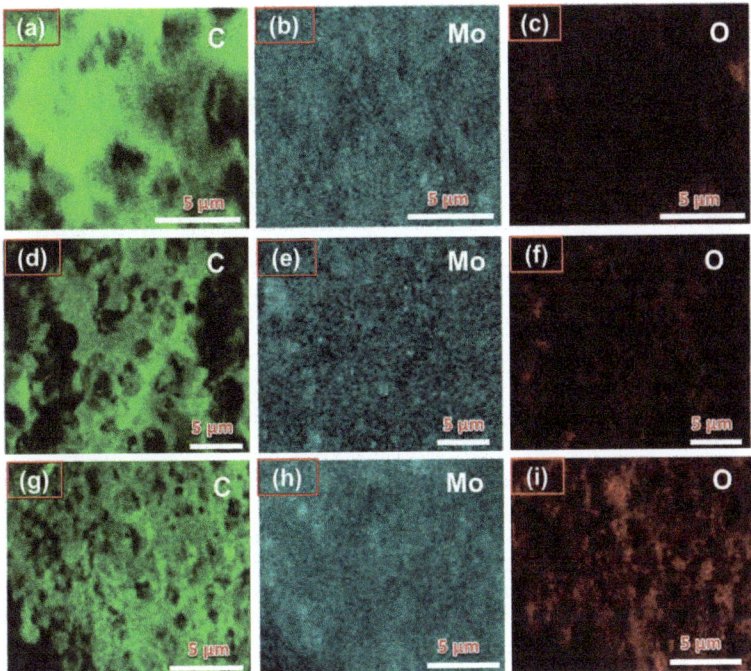

Figure 5. EDX mapping images showing the distribution of C, Mo and O individual elements in: (**a–c**) PAC/MoO$_2$/Mo$_2$C-0.5, (**d–f**) PAC/MoO$_2$/Mo$_2$C-1 and (**g–i**) PAC/MoO$_2$/Mo$_2$C-2 ternary composites.

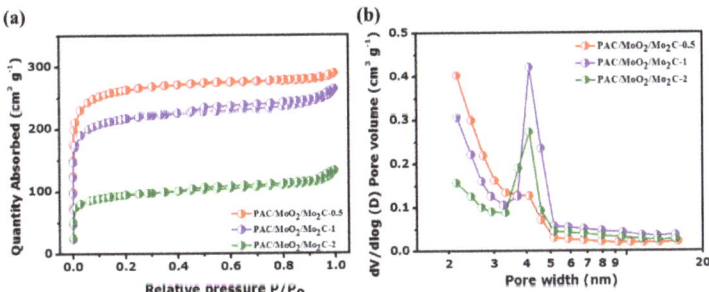

Figure 6. (**a**) N$_2$ absorption-desorption isotherms, (**b**) pore size distribution of the PAC/MoO$_2$/Mo$_2$C-0.5, PAC/MoO$_2$/Mo$_2$C-1 and PAC/MoO$_2$/Mo$_2$C-2.

Table 2. Textural properties of the PAC/MoO$_2$/Mo$_2$C ternary composites.

Samples	BET SSA (m^2 g^{-1})	Total Pore Volume (cm^3 g^{-1})	Micropore Volume (cm^3 g^{-1})	Micropore SSA (m^2 g^{-1})	Mesopore Volume (cm^3 g^{-1})
PAC/MoO$_2$/Mo$_2$C-0.5	804	0.44	0.23	670	0.21
PAC/MoO$_2$/Mo$_2$C-1	711	0.40	0.19	575	0.21
PAC/MoO$_2$/Mo$_2$C-2	301	0.20	0.07	165	0.13

The decrease in the SSA and total pore volume could be ascribed to the embedding of the MoO_2 and Mo_2C nanoparticles into the porous PAC during the pyrolysis process which can block some pores [57].

However, the SSA of the as-synthesized samples are much higher than that reported for similar materials such as $MoO_2/Mo_2C/C$ composite prepared via ion-exchange method (73.4 $m^2\ g^{-1}$) [58], $MoO_2/Mo_2C/C$ spheres by hydrothermal and calcination processes (159.6 $m^2\ g^{-1}$) [36] and $MoO_2/Mo_2C/C$ microspheres obtained by a mild polymer regulation procedure followed by calcination treatment (57.6 $m^2\ g^{-1}$) [59].

The formation of the nanoparticles MoO_2 and Mo_2C in the ternary composite emanates from the interaction between the ammonium molybdate tetrahydrate (($NH_4)_6Mo_7O_{24} \cdot 4H_2O$) and the porous PAC (denoted as C) at elevated temperature (≥ 800 °C) under argon atmosphere. It is good to mention that typically activated carbon (PAC) comprised of OH and COOH groups on the surface, which makes it acidic, favors a thermal reduction of the ammonium molybdate precursor to MoO_2 instead of MoO_3. The MoO_2 could react with carbon at high temperature and under inert atmosphere (Ar) to give Mo_2C. This process can be described with the following Equations [42,60]:

$$(NH_4)_6Mo_7O_{24} \cdot 4H_2O \rightarrow MoO_2 + H_2O + NH_3 \uparrow \quad (5)$$

$$2C + 4MoO_2 \rightarrow 2Mo_2C + 4O_2 \quad (6)$$

During the pyrolysis, the ammonium molybdenum precursor decomposes to form MoO_2, H_2O and ammonia gas (NH_3) being released at high temperature. In addition, the generated MoO_2 nanoparticles could react with the porous carbon (PAC) leading to the formation of Mo_2C. The formation of a **ternary composite** comprising of PAC, MoO_2 and Mo_2C could possibly enhance ion intercalation as well as create an interconnected porous network. This might promote an easy diffusion of the electrolyte's ions through the electrode materials and further enhance the fast transport of the ions which are beneficial for the electrochemical analysis.

3.2. XPS Analysis

The surface chemistry property and the elemental composition of the as-synthesized ternary composites were determined using X-ray photoelectron spectroscopy (XPS). The wide survey scan spectrum depicted the distinctive peaks of the carbon (C 1s), molybdenum ($Mo3p_{1/2}$, $Mo3p_{3/2}$ and Mo 3d) and oxygen (O 1s) elements in $PAC/MoO_2/Mo_2C$-0.5, $PAC/MoO_2/Mo_2C$-1 and $PAC/MoO_2/Mo_2C$-2 composites as illustrated in Figure 7a.

Table 3 presents the atomic percentage (at.%) of C, Mo and O elements in the as-synthesized ternary composites. From these samples, it can be seen that the carbon content decreases from 73.9 to 54.9 at.% as the yield of the molybdenum increases. However, the molybdenum and oxygen contents were found to increase from 8.6 to 20.2 at.% and 17.5 to 24.9 at.%, respectively. Notably, the $PAC/MoO_2/Mo_2C$-2 material exhibited the smallest carbon content and highest molybdenum and oxygen contents. This could be due to the formation of MoO_2 and Mo_2C nanoparticles during the pyrolysis process. On the other hand, the elemental composition in the ternary composites is significantly influenced by the mass loading of molybdenum precursor into the PAC.

The high-resolution XPS spectra of the Mo 3d split into $3d_{5/2}$ and $3d_{3/2}$ spin-orbit components in the binding energy range of 226–241 eV as presented in Figure 7b–d. In Figure 7b, the deconvolution of the core level Mo 3d spectrum exhibits six sets of peaks which indicate the presence of four oxidation states Mo^{2+}, Mo^{4+}, Mo^{5+} and Mo^{6+} in $PAC/MoO_2/Mo_2C$-0.5 ternary composite. The peak located at 228.8 eV (Mo^{2+} $3d_{5/2}$) is associated to Mo-C bond in Mo_2C while the pair of peaks located at 229.6 and 232.9 eV (Mo^{4+} $3d_{5/2}/3d_{3/2}$) are attributed to the formation of MoO_2 [42,61–63]. The pair of peaks at binding energies of 231.0 and 234.4 eV (Mo^{5+} $3d_{5/2}/3d_{3/2}$) and that located at 232.5 and 235.9 eV (Mo^{6+} $3d_{5/2}/3d_{3/2}$) are the characteristics of the MoO_3 which could be assigned to the surface oxidation and sample oxidation in air of the metastable phase of the MoO_2 [64–67]. Figure 7c,d presents the fitting of the core level Mo3d in PAC/MoO$_2$/Mo$_2$C-1 and PAC/MoO$_2$/Mo$_2$C-2 ternary composites. In comparison with PAC/MoO$_2$/Mo$_2$C-0.5 ternary composite, there are no changes in the number of peaks deconvoluted which means that all the composites have similar oxidation states (Mo^{2+}, Mo^{4+}, Mo^{5+} and Mo^{6+}). Table S1 presents the atomic percentage (at.%) of all deconvoluted peaks of the ternary composites. What is noticeable is that PAC/MoO$_2$/Mo$_2$C-1 has high at.% of Mo$_2$C and MoO$_2$ as compared to the other composites.

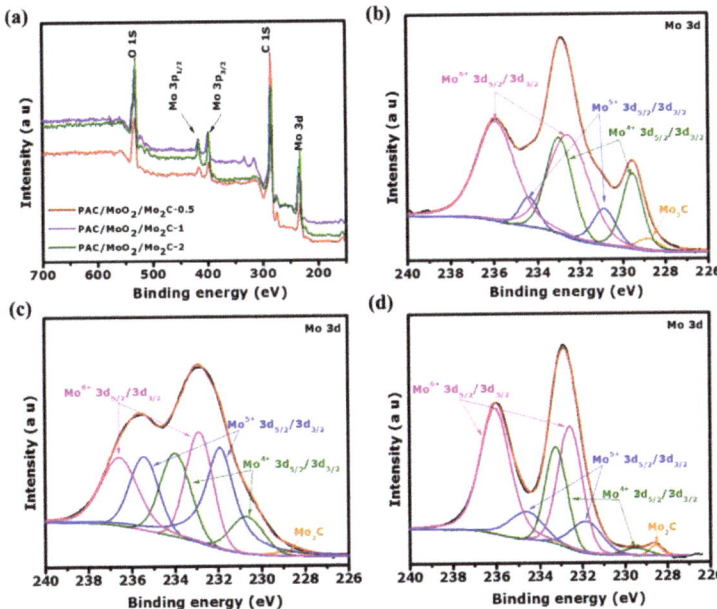

Figure 7. (a) XPS survey spectra of the as-synthesized ternary composites and high resolution of Mo 3d, (b) PAC/MoO$_2$/Mo$_2$C-0.5, (c) PAC/MoO$_2$/Mo$_2$C-1 and (d) PAC/MoO$_2$/Mo$_2$C-2 ternary composites.

Table 3. Elemental composite of the PAC/MoO$_2$/Mo$_2$C ternary composites.

Samples	Elemental Composition (at.%)		
	C 1s	O 1s	Mo 3d
PAC/MoO$_2$/Mo$_2$C-0.5	73.9	17.5	8.6
PAC/MoO$_2$/Mo$_2$C-1	62.4	22.6	15.0
PAC/MoO$_2$/Mo$_2$C-2	54.9	24.9	20.2

This could be beneficial in the electrochemical measurements of these composites because these two materials are expected to improve the electrochemical properties of PAC where MoO$_2$ is expected to contribute pseudocapacitive behavior, while Mo$_2$C will contribute stability and conductivity.

The high-resolution C 1s and O 1s core levels of the ternary composites are shown in Figure S1. The fitting of the C 1s spectrum (Figure S1a–c) shows the characteristic peak of Mo-C bond in Mo$_2$C at 283.3 ± 0.2 eV [68,69].

The other four peaks are attributed to the C=C (sp^2 hybridized), C-C (sp^3 hybridized), C-OH and O-C=O bonds corresponding to the binding energies at 284.4 ± 0.3eV, 285.2 ± 0.2 eV, 287.5 ± 0.6 and 290.7 ± 0.3 eV, respectively [70–72]. The deconvolution of the O 1s feature (Figure S1d–f) exhibits a peak located at 530.6 ± 0.2 eV linked to Mo-O bond, the two other peaks at 533.3 ± 0.2 eV and 536.4 ± 0.2 eV are associated to C-O and O-C=O bonds, respectively [73–75]. The results of the XPS analysis confirm the formation of the MoO$_2$ and

Mo$_2$C nanoparticles in all ternary composites which are consistent with the XRD, Raman, HRTEM and SAED results.

3.3. Electrochemical Characterization

All measurements of the ternary composite electrodes with different molybdenum precursor content were done first in a three-electrode configuration using 2.5 M KNO$_3$ aqueous electrolyte. Figure 8a,b displays the comparative cyclic voltammetry (CV) profiles of the PAC/MoO$_2$/Mo$_2$C-0.5, PAC/MoO$_2$/Mo$_2$C-1 and PAC/MoO$_2$/Mo$_2$C-2 electrodes at a constant scan rate of 50 mV s^{-1} in both negative and positive potential windows of −0.9–0 V and 0–0.9 V vs. Ag/AgCl, respectively. A quasi-rectangular characteristic was observed for all CV profiles indicating an electrical double layer capacitor (EDLC) behavior for these samples [24]. It can be seen that the CV profile of PAC/MoO$_2$/Mo$_2$C-1 electrode displays a higher current response than other electrode materials. This superior current response could be assigned to the moderate loading of molybdenum precursor into the porous carbon network, which provided more active sites enhancing the fast ion diffusion and good interaction between the interface of electrode/KNO$_3$ electrolyte.

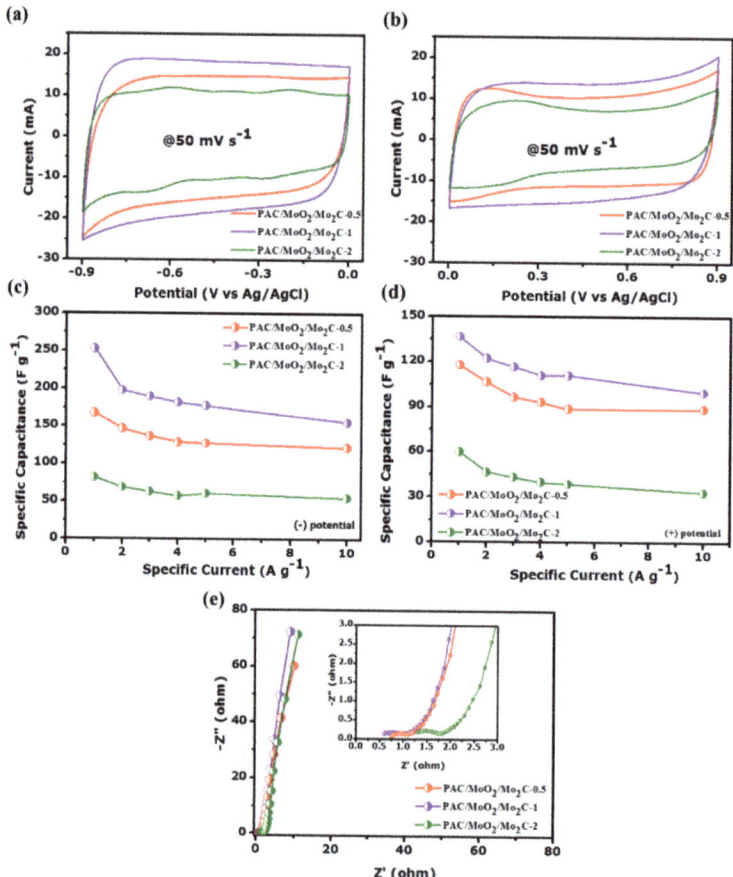

Figure 8. (**a**,**b**) Cyclic voltammetry at scan rate of 50 mV s^{-1}, (**c**,**d**) specific capacitance at different specific current values in (−0.9–0.0 V) and (0.0–0.9 V) operating potential windows and (**e**) Nyquist plots of the PAC/MoO$_2$/Mo$_2$C ternary composites in a three-electrode configuration.

Figure S2a,b display the galvanostatic charge-discharge (GCD) comparison curves of the PAC/MoO$_2$/Mo$_2$C ternary composites.

The GCD curves of all electrodes are performed within both negative (−0.9–0.0 V vs. Ag/AgCl) and positive (0.0–0.9 V vs. Ag/AgCl) operating potential windows, respectively at a constant specific current of 1 A g^{-1}.

The GCD curves depicted a symmetrical triangular profile that confirms the electrical double layer capacitor nature of the ternary composites electrodes supported by the CV curves [76]. It is also observed that the GCD curve of the PAC/MoO$_2$/Mo$_2$C-1 electrode has a longer discharge time as compared to PAC/MoO$_2$/Mo$_2$C-0.5 and PAC/MoO$_2$/Mo$_2$C-2 electrodes agreeing with the CV results. Furthermore, the details of the CV curves at different scan rates from 10 to 100 mV s^{-1} and GCD profiles at specific currents ranging from 1 to 10 A g^{-1} of the PAC/MoO$_2$/Mo$_2$C-1 ternary composite because of superior electrochemical properties are shown in Figure S3a–d, respectively. The corresponding specific capacitance (C_s) as a function of the specific current in the range of 1–10 A g^{-1} is presented in Figure 8c,d in the negative (−0.9–0.0 V vs. Ag/AgCl) and positive (0.0–0.9 V vs. Ag/AgCl) operating potential windows for all three composites, respectively. The C_s of the ternary composites was determined from the discharge period of GCD patterns using Equation (1). From both Figure 8c,d, it is observed that the PAC/MoO$_2$/Mo$_2$C-1 electrode depicted the highest C_s value in both potential windows reflecting longer charge-discharge pattern which is consistent with the highest current response from the CV curve. However, the smallest C_s value of PAC/MoO$_2$/Mo$_2$C-2 in both potential windows can be explained by the fact that further increasing the molybdenum precursor content could lead to the blockage of some pores in the porous carbon matrix as evidenced by a decrease in SSA of 301 m^2 g^{-1} and thus limit the ion diffusion at the electrode/electrolyte interface [77].

Electrochemical impedance spectroscopy (EIS) analysis of the as-synthesized PAC/MoO$_2$/Mo$_2$C ternary composites was evaluated at open circuit voltage (V_{OC}) in the frequency range of 100 kHz to 10 mHz. The EIS data are provided using Nyquist plot which illustrates the variation of the impedance as a function of the frequency as shown Figure 8e.

The Nyquist plots of all ternary composites exhibit a semi-circle at high to medium frequency region corresponding to the charge transfer resistance (R_{ct}) and a quasi-straight line slightly tilted to Z" imaginary axis at low frequency region indicating the ion diffusion throughout the electrolyte [27,78]. The Nyquist curves of the ternary composites further confirms the capacitive characteristic. The intercept of the Z' real axis (beginning of the arc) at high frequency depicts the equivalent series resistance (ESR) which represents the combination of resistance at electrode/electrolyte and electrode/current collector interfaces [79]. As seen in the inset to Figure 8e, PAC/MoO$_2$/Mo$_2$C-1 electrode depicts smaller ESR and R_{ct} values of 0.59 and 1.01 Ω as compared to PAC/MoO$_2$/Mo$_2$C-0.5 (0.76 and 1.10 Ω) and PAC/MoO$_2$/Mo$_2$C-2 electrodes (1.31 and 1.84 Ω), respectively. In addition, PAC/MoO$_2$/Mo$_2$C-1 electrode also has the shortest diffusion path length and closest to the Z" axis suggesting a quicker ion diffusion of the interfacial electrode and KNO$_3$ electrolyte and shows better capacitive features among the ternary composites.

Considering all the above results, the as-prepared PAC/MoO$_2$/Mo$_2$C-1 electrode recorded superior electrochemical performance among the other ternary composites. This might be ascribed to the synergistic effect of the MoO$_2$, MO$_2$C and porous carbon obtained after the reaction of ammonium molybdate and porous carbon at equal mass ratios during the pyrolysis route. This is also supported by the at.% of all the deconvoluted peaks in Table S1 where for this particular sample MoO$_2$ and Mo$_2$C show higher at.% as compared to the rest of the composites. The formation of the MoO$_2$ and Mo$_2$C nanoparticles embedded into the porous PAC provided abundant electro-actives sites for the charge transfer ability and quick ion diffusion of the electrolyte, which improve the wettability and the electrical conductivity, thus the charge storage of the electrode.

The electrochemical measurement of the PAC/MoO$_2$/Mo$_2$C-1 ternary composite electrode was further performed in a two-cell configuration by assembling a symmetric device using identical electrolyte. Figure 9a,b displays the CV features of the as-fabricated symmetric supercapacitor (PAC/MoO$_2$/Mo$_2$C-1//PAC/MoO$_2$/Mo$_2$C-1) under an operating cell potential of 0–1.8 V. The CV features of the ternary composite device reveals a quasi-rectangular behavior whereas the current response increased upon increasing the scan rates from 10 to 400 mV s^{-1} (Figure 9a) suggesting quasi-reversible electron transfer kinetics and dominated electrical double layer.

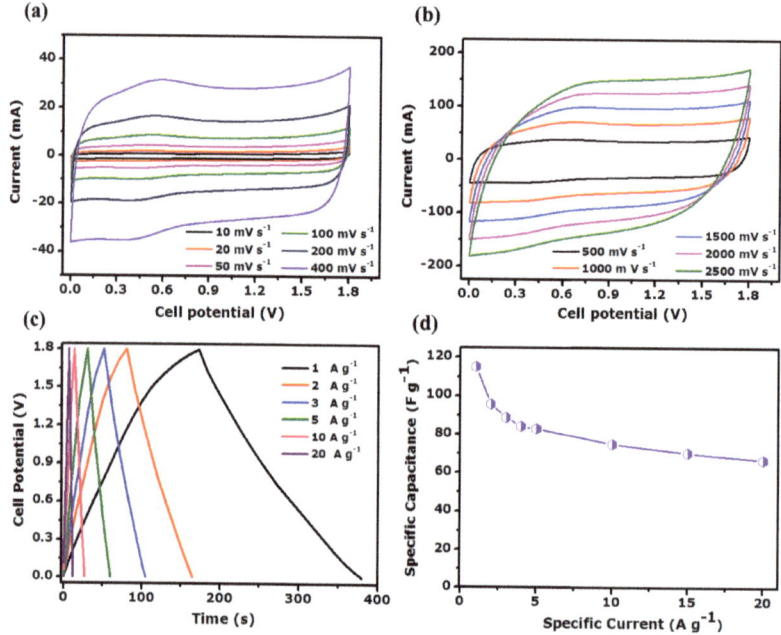

Figure 9. (a,b) Cyclic voltammetry at scan rates variant from 10–400 mV s^{-1} and from 500–2500 mV s^{-1}, respectively, (c) galvanostatic charge-discharge profiles at various specific currents from 1 to 20 A g^{-1} and (d) specific capacitance as a function of specific current for PAC/MoO$_2$/Mo$_2$C-1//PAC/MoO$_2$/Mo$_2$C-1 symmetric cell.

A slight redox peak was observed in the CV curves due to the pseudocapacitive contribution from the molybdenum oxide. The minor redox reactions could emanate from the insertion of K$^+$ ions into the MoO$_2$ containing electrodes. A similar observation was reported by Wang et al. [80], on the lithium-ion insertion onto MoO$_2$ that is associated with the monoclinic–orthorhombic–monoclinic phase transition of MoO$_2$ [80]. As such a transition from MoO$_2$ to K$_X$MoO$_2$ by the insertion of K$^+$ ions to the molybdenum oxide can be postulated [81]. The CV curves of the PAC/MoO$_2$/Mo$_2$C-1//PAC/MoO$_2$/Mo$_2$C-1 device still maintain the rectangular-like feature upon increasing the scan rate to high rate from 0.5 to 2.5 V s^{-1} (as seen in Figure 9b) which demonstrated a high rate capability [82]. GCD plots of the assembled device in the operating potential of 0–1.8 V are shown in Figure 9c. A typical triangular behavior was recorded for all GCD plots at various specific currents from 1 to 20 A g^{-1} confirming the capacitive charge storage mechanism of the symmetric ternary composite device [24]. The plot of the obtained specific capacitance (C_s) calculated using Equation (2) against the specific current from 1 to 20 A g^{-1} is depicted in Figure 9d. The recorded value C_s of the ternary composite device was found to be 115 F g^{-1} at 1 A g^{-1} specific current. The ternary composite PAC/MoO$_2$/Mo$_2$C-1//PAC/MoO$_2$/Mo$_2$C-1 device still delivered a high C_s of 67 F g^{-1} even after a twentyfold increase of specific current which confirms the good rate capability of 58.3% obtained from the symmetric supercapacitor.

Figure 10a depicts the specific energy against specific power (Ragone plot) measured at various specific currents (1–20 A g^{-1}). The symmetric ternary composite device recorded high specific energy of 51.8 W h kg^{-1} with an associated power of 0.9 kW kg^{-1} at 1 A g^{-1}. Interestingly, the symmetric ternary composite device can maintain up to 30 W h kg^{-1} of specific energy with a corresponding specific power of 18 kW kg^{-1} even at 20 A g^{-1} increase of specific current. These specific energy/power values recorded for the PAC/MoO$_2$/Mo$_2$C-1//PAC/MoO$_2$/Mo$_2$C-1 symmetric cell are better than reports on Mo-based/C composites for supercapacitors applications as shown in Table S2.

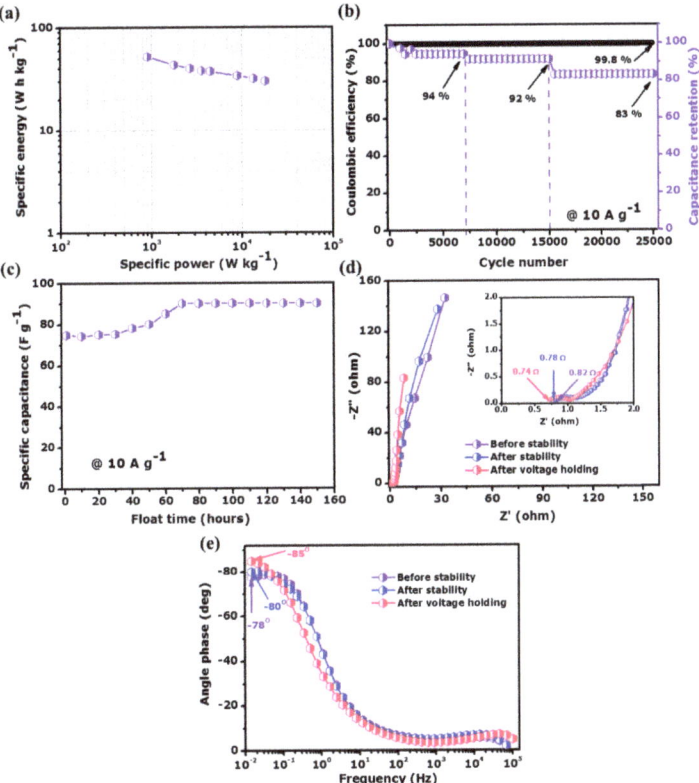

Figure 10. (a) Ragone plot, (b) cycling stability up to 25,000 cycles, (c) specific capacitance as function of voltage holding time up to 180 h, (d) Nyquist plots and (e) Bode plot of PAC/MoO$_2$/Mo$_2$C-1//PAC/MoO$_2$/Mo$_2$C-1 symmetric cell.

To investigate the stability of the PAC/MoO$_2$/Mo$_2$C-1//PAC/MoO$_2$/Mo$_2$C-1 device, the cycling test based on the long-term galvanostatic charge-discharge was performed at 10 A g^{-1} specific current as shown in Figure 10b.

The ternary composite device recorded a columbic efficiency of 99.8% up to 25,000 charge-discharge cycles and a capacitance retention found to be 94%, 92% and 83% after 7000, 15,000 and 25,000 constant charge-discharge cycles, respectively. These results indicate that even up to 25,000 continuous cycling the ternary composite device still maintains good stability with a specific capacitance loss of 17% as compared to the initial value. The good long-term stability of the symmetric device is owed to the rapid electron transfer kinetics offered by the ternary composite electrode.

An additional stability performance, floating test (or voltage holding) has been investigated on the PAC/MoO$_2$/Mo$_2$C-1//PAC/MoO$_2$/Mo$_2$C-1 symmetric supercapacitor. Figure 10c displays the variation of the specific capacitance versus the floating time of each 10 h during 150 h at a maximum potential cell of 1.8 V at 10 A g^{-1}. An increase of 21% from the initial value of the specific capacitance is observed during the first 60 h of floating subsequently stabilizing up to 150 h floating time. The specific capacitance of the ternary composite device was enhanced from 74.7 to 90 F g^{-1} after the floating time which also highlights an improvement of the specific energy from 33.7 to 40.2 W h kg^{-1}. The improvement of the PAC/MoO$_2$/Mo$_2$C-1//PAC/MoO$_2$/Mo$_2$C-1 device in the specific capacitance and specific energy could be ascribed to more penetration of ions electrolyte into the network of MoO$_2$ and Mo$_2$C nanoparticles embedded in the porous carbon. This could consequently increase the electrode wettability and enable faster diffusion of electrolyte ions at the electrode/KNO$_3$ electrolyte interface, hence enhancing the charge storage [83,84]. In brief, the as-fabricated PAC/MoO$_2$/Mo$_2$C-1//PAC/MoO$_2$/Mo$_2$C-1 device revealed a good stability performance in terms of long-term cycling up 25,000 cycles and floating time over 150 h thereby implying a superior electrochemical performance of the entire device.

Figure 10d illustrates the Nyquist plots of PAC/MoO$_2$/Mo$_2$C-1//PAC/MoO$_2$/Mo$_2$C-1 symmetric supercapacitor before and both after long-term cycling and holding test. All Nyquist plots exhibit a nearly vertical feature at low frequency referring to the ideal capacitive characteristic and the great electrical conductivity of the PAC/MoO$_2$/Mo$_2$C-1 ternary composite. A slight decrease of the ESR values was observed from the original value of 0.82 Ω to 0.78 Ω and 0.74 Ω for both after 25,000 cycles and 150 h floating test respectively as shown in the inset to Figure 10d. Similarly, the R_{ct} value depicted also a small decrease from the initial value of 1.14 Ω to 1.11 Ω and 1.04 Ω after cycling and holding tests, respectively. These small ESR and R_{ct} values demonstrate a fast charge transport capability and rapid ion diffusion through the full symmetric device which got improved by the stability tests as indication that the electrode had better wettability after stability [85].

The Bode plot of the as-fabricated-symmetric device which defines the plot of the angle phase versus frequency is shown in Figure 10e before and after cycling stability and voltage holding. The phase angle values increased from −78° to −80° and −85° after cycling and holding test, respectively. These values are close to −90° which confirm the ideal capacitive behavior of the PAC/MoO$_2$/Mo$_2$C-1 ternary composite [24].

According to these results, the as-fabricated symmetric supercapacitor demonstrated a superior electrochemical performance after cycling and holding test in aqueous electrolyte which could be due to the fact that the electrodes wettability had been improved and hence ions have more access to the pores. The high performance of the PAC/MoO$_2$/Mo$_2$C-1//PAC/MoO$_2$/Mo$_2$C-1 symmetric supercapacitor is based on the synergistic effect of the ternary composite with the following benefits: (i) high electrical conductivity of Mo$_2$C, (ii) pseudo capacitor effect of MoO$_2$ and (iii) large surface area of porous carbon (PAC). Owing to these advantages, the PAC/MoO$_2$/Mo$_2$C-1//PAC/MoO$_2$/Mo$_2$C-1 can be used as an excellent charge storage device.

4. Conclusions

A ternary peanut shell activated carbon/molybdenum oxide/molybdenum carbide (PAC/MoO$_2$/Mo$_2$C) composite was successfully synthesized via a facile in-situ pyrolysis route of ratio of porous carbon to different mass loading of ammonium molybdate (1:0.5; 1:1; 1:2). All as-synthesized materials display the in-situ formation of MoO$_2$ and Mo$_2$C nanostructures into the porous carbon based on the XRD, Raman, HRTEM, SAED, EDX mapping and XPS analysis. The ternary composite with the mass ratio of 1 to 1 (PAC/MoO$_2$/Mo$_2$C-1) provided a superior electrochemical characteristic in a neutral 2.5 M KNO$_3$ electrolyte. The as-assembled PAC/MoO$_2$/Mo$_2$C-1//PAC/MoO$_2$/Mo$_2$C-1 symmetric device delivered an excellent specific capacitance of 115 F g^{-1} at 1 A g^{-1} with a good rate capability (58% at 20 A g^{-1}) and cycling stability (99.8% columbic efficiency after 25,000 cycles). Moreover, a specific energy of 51.8 W h kg^{-1} with a corresponding power of 0.9 kW kg^{-1} was recorded for the symmetric device within an operating potential window of 1.8 V at 1 A g^{-1} specific current. Interestingly, the electrochemical results of the device show a significant enhancement of 21% from its initial specific capacitance value after 160 h holding test. These remarkable performances are linked to the great synergistic effect of the different components into the ternary composite by supplying favorable properties: Pseudo capacitor behavior of the MoO_2, highly conductive Mo_2C and high surface area of the porous carbon. This study provides a simple and low-cost way to enhance the electrochemical performance of carbons by incorporating Mo-based components into porous activated carbon.

Supplementary Materials: The following are available online at https://www.mdpi.com/article/10.3390/nano11041056/s1, Table S1: Composition (at.%) of the PAC/MoO_2/Mo_2C ternary composites, Figure S1: High resolution XPS spectra C 1S and O 1 S of (a,d) PAC/MoO_2/Mo_2C-0.5 (b,e) PAC/MoO_2/Mo_2C-1 (c,f) PAC/MoO_2/Mo_2C-2 ternary composites, Figure S2: Galvanostatic charge-discharge curves at 1 g^{-1} (a) in -0.9–0.0 V negative and (b) 0.0–0.9 V positive potential windows of the PAC/MoO_2/Mo_2C ternary composites in three-electrode configuration, Figure S3: (a,b) CV plots at different scan rate from 10 to 100 mV s^{-1} in (-0.9–0.0 V) and (0.0–0.9 V) operating potential, (c,d) GCD curves at different specific currents ranging from 1 to 10 A g^{-1} in (-0.9–0.0 V) and (0.0–0.9 V) operating potential of the PAC/MoO_2/Mo_2C-1 ternary composite in three-electrode configuration, Table S2: Comparison of electrochemical performance of Mo-based composite with carbon material in aqueous electrolyte.

Author Contributions: N.F.S. and N.M. conceived and designed the concept of all the experiments. N.F.S., B.D.N., N.M.N. and A.S. carried out, performed the experiments and data analysis. N.F.S., S.S., N.M.N. and B.K.M. studied and drafted the manuscript. B.D.N., N.M. and M.C. edited the manuscript. All authors have read and agreed to the published version of the manuscript.

Funding: This research was supported by the South African Research Chairs Initiative (SARChI) of the Department of Science and Technology through the National Research Foundation (NRF) of South Africa (Grant No. 61056).

Data Availability Statement: Data can be available upon request from the authors.

Acknowledgments: This research was supported by the South African Research Chairs Initiative (SARChI) of the Department of Science and Technology through the National Research Foundation (NRF) of South Africa (Grant No. 61056). Any idea, finding, conclusion or recommendation expressed in this publication is that of the author(s). The NRF does not accept any liability in this regard. N.F.S. acknowledges NRF through SARChI in Carbon Technology and Materials and the University of Pretoria for financial support.

Conflicts of Interest: The authors declare no conflict of interest.

References

1. Owusu, P.A.; Asumadu-Sarkodie, S. A review of renewable energy sources, sustainability issues and climate change mitigation. *Cogent Eng.* **2016**, *3*, 1167990. [CrossRef]
2. Xing, H.; Long, G.; Zheng, J.; Zhao, H.; Zong, Y.; Li, X.; Wang, Y.; Zhu, X.; Zhang, M.; Zheng, X. Interface engineering boosts electrochemical performance by fabricating CeO_2@CoP Schottky conjunction for hybrid supercapacitors. *Electrochim. Acta* **2020**, *337*, 135817. [CrossRef]
3. Yekini Suberu, M.; Wazir Mustafa, M.; Bashir, N. Energy storage systems for renewable energy power sector integration and mitigation of intermittency. *Renew. Sustain. Energy Rev.* **2014**, *35*, 499–514. [CrossRef]
4. Chu, S.; Cui, Y.; Liu, N. The path towards sustainable energy. *Nat. Mater.* **2016**, *16*, 16–22. [CrossRef] [PubMed]
5. Momodu, D.; Madito, M.; Barzegar, F.; Bello, A.; Khaleed, A.; Olaniyan, O.; Dangbegnon, J.; Manyala, N. Activated carbon derived from tree bark biomass with promising material properties for supercapacitors. *J. Solid State Electrochem.* **2017**, *21*, 859–872. [CrossRef]
6. Seh, Z.W.; Kibsgaard, J.; Dickens, C.F.; Chorkendorff, I.; Nørskov, J.K.; Jaramillo, T.F. Combining theory and experiment in electrocatalysis: Insights into materials design. *Science* **2017**, *355*. [CrossRef]
7. Ndiaye, N.M.; Ngom, B.D.; Sylla, N.F.; Masikhwa, T.M.; Madito, M.J.; Momodu, D.; Ntsoane, T.; Manyala, N. Three dimensional vanadium pentoxide/graphene foam composite as positive electrode for high performance asymmetric electrochemical supercapacitor. *J. Colloid Interface Sci.* **2018**, *532*, 395–406. [CrossRef]
8. Kang, L.; Huang, C.; Zhang, J.; Zhang, M.; Zhang, N.; Liu, S.; Ye, Y.; Luo, C.; Gong, Z.; Wang, C.; et al. Effect of fluorine doping and sulfur vacancies of $CuCo_2S_4$ on its electrochemical performance in supercapacitors. *Chem. Eng. J.* **2020**, *390*, 124643. [CrossRef]

9. Simon, P.; Gogotsi, Y.; Dunn, B. Where do batteries end and supercapacitors begin? *Science* **2014**, *343*, 1210–1211. [CrossRef] [PubMed]
10. Ngom, B.D.; Ndiaye, N.M.; Sylla, N.F.; Mutuma, B.K.; Manyala, N. Sustainable development of vanadium pentoxide carbon composites derived from Hibiscus sabdariffa family for application in supercapacitors. *Sustain. Energy Fuels* **2020**, *4*, 4814–4830. [CrossRef]
11. Dubey, R.; Guruviah, V. Review of carbon-based electrode materials for supercapacitor energy storage. *Ionics* **2019**, *25*, 1419–1445. [CrossRef]
12. Wang, Y.; Zhang, L.; Hou, H.; Xu, W.; Duan, G.; He, S.; Liu, K.; Jiang, S. Recent progress in carbon-based materials for supercapacitor electrodes: A review. *J. Mater. Sci.* **2021**, *56*, 173–200. [CrossRef]
13. Abioye, A.M.; Ani, F.N. Recent development in the production of activated carbon electrodes from agricultural waste biomass for supercapacitors: A review. *Renew. Sustain. Energy Rev.* **2015**, *52*, 1282–1293. [CrossRef]
14. Dubey, P.; Shrivastav, V.; Maheshwari, P.H.; Sundriyal, S. Recent advances in biomass derived activated carbon electrodes for hybrid electrochemical capacitor applications: Challenges and opportunities. *Carbon N. Y.* **2020**, *170*, 1–29. [CrossRef]
15. Wang, K.; Zhao, N.; Lei, S.; Yan, R.; Tian, X.; Wang, J.; Song, Y.; Xu, D.; Guo, Q.; Liu, L. Promising biomass-based activated carbons derived from willow catkins for high performance supercapacitors. *Electrochim. Acta* **2015**, *166*, 1–11. [CrossRef]
16. Musyoka, N.M.; Mutuma, B.K.; Manyala, N. Onion-derived activated carbons with enhanced surface area for improved hydrogen storage and electrochemical energy application. *RSC Adv.* **2020**, *10*, 26928–26936. [CrossRef]
17. Zuliani, J.E.; Tong, S.; Jia, C.Q.; Kirk, D.W. Contribution of surface oxygen groups to the measured capacitance of porous carbon supercapacitors. *J. Power Sources* **2018**, *395*, 271–279. [CrossRef]
18. Zhai, D.D.; Liu, H.; Wang, M.; Wu, D.; Chen, X.Y.; Zhang, Z.J. Integrating surface functionalization and redox additives to improve surface reactivity for high performance supercapacitors. *Electrochim. Acta* **2019**, *323*, 134810. [CrossRef]
19. Song, W.; Zhang, Z.; Wan, P.; Wang, M.; Chen, X.; Mao, C. Low temperature and highly efficient oxygen/sulfur dual-modification of nanoporous carbon under hydrothermal conditions for supercapacitor application. *J. Solid State Electrochem.* **2020**, *24*, 761–770. [CrossRef]
20. Liu, C.; Koyyalamudi, B.B.; Li, L.; Emani, S.; Wang, C.; Shaw, L.L. Improved capacitive energy storage via surface functionalization of activated carbon as cathodes for lithium ion capacitors. *Carbon N. Y.* **2016**, *109*, 163–172. [CrossRef]
21. Yaglikci, S.; Gokce, Y.; Yagmur, E.; Aktas, Z. The performance of sulphur doped activated carbon supercapacitors prepared from waste tea. *Environ. Technol.* **2020**, *41*, 36–48. [CrossRef]
22. Lin, G.; Wang, Q.; Yang, X.; Cai, Z.; Xiong, Y.; Huang, B. Preparation of phosphorus-doped porous carbon for high performance supercapacitors by one-step carbonization. *RSC Adv.* **2020**, *10*, 17768–17776. [CrossRef]
23. Muduli, S.; Naresh, V.; Martha, S.K. Boron, Nitrogen-Doped Porous Carbon Derived from Biowaste Orange Peel as Negative Electrode Material for Lead-Carbon Hybrid Ultracapacitors. *J. Electrochem. Soc.* **2020**, *167*, 090512. [CrossRef]
24. Sylla, N.F.; Ndiaye, N.M.; Ngom, B.D.; Mutuma, B.K.; Momodu, D.; Chaker, M.; Manyala, N. Ex-situ nitrogen-doped porous carbons as electrode materials for high performance supercapacitor. *J. Colloid Interface Sci.* **2020**, *569*, 332–345. [CrossRef] [PubMed]
25. Lee, H.; Park, I.-S.; Park, Y.-K.; An, K.-H.; Kim, B.-J.; Jung, S.-C. Facile Preparation of Ni-Co Bimetallic Oxide/Activated Carbon Composites Using the Plasma in Liquid Process for Supercapacitor Electrode Applications. *Nanomaterials* **2019**, *10*, 61. [CrossRef]
26. Sinha, P.; Banerjee, S.; Kar, K.K. Transition Metal Oxide/Activated Carbon-Based Composites as Electrode Materials for Supercapacitors. In *Handbook of Nanocomposite Supercapacitor Materials II*; Springer: Cham, Switzerland, 2020; pp. 145–178.
27. Yang, Y.; Niu, H.; Qin, F.; Guo, Z.; Wang, J.; Ni, G.; Zuo, P.; Qu, S.; Shen, W. MnO_2 doped carbon nanosheets prepared from coal tar pitch for advanced asymmetric supercapacitor. *Electrochim. Acta* **2020**, *354*, 136667. [CrossRef]
28. Li, X.; Shao, J.; Li, J.; Zhang, L.; Qu, Q.; Zheng, H. Ordered mesoporous MoO_2 as a high-performance anode material for aqueous supercapacitors. *J. Power Sources* **2013**, *237*, 80–83. [CrossRef]
29. Wu, K.; Zhao, J.; Zhang, X.; Zhou, H.; Wu, M. Hierarchical mesoporous MoO_2 sphere as highly effective supercapacitor electrode. *J. Taiwan Inst. Chem. Eng.* **2019**, *102*, 212–217. [CrossRef]
30. Zhang, L.; Lin, H.; Zhai, L.; Nie, M.; Zhou, J.; Zhuo, S. Enhanced supercapacitor performance based on 3D porous graphene with MoO_2 nanoparticles. *J. Mater. Res.* **2017**, *32*, 292–300. [CrossRef]
31. Si, H.; Sun, L.; Zhang, Y.; Zhang, Y.; Bai, L.; Zhang, Y. Carbon-coated MoO_2 nanoclusters anchored on RGO sheets as high-performance electrodes for symmetric supercapacitors. *Dalt. Trans.* **2019**, *48*, 285–295. [CrossRef] [PubMed]
32. Yuan, X.; Yan, X.; Zhou, C.; Wang, D.; Zhu, Y.; Wang, J.; Tao, X.; Cheng, X. Promising carbon nanosheets decorated by self-assembled MoO_2 nanoparticles: Controllable synthesis, boosting performance and application in symmetric coin cell supercapacitors. *Ceram. Int.* **2020**, *46*, 19981–19989. [CrossRef]
33. Weng, Y.T.; Tsai, C.B.; Ho, W.H.; Wu, N.L. Polypyrrole/carbon supercapacitor electrode with remarkably enhanced high-temperature cycling stability by TiC nanoparticle inclusion. *Electrochem. Commun.* **2013**, *27*, 172–175. [CrossRef]
34. Xiao, Y.; Hwang, J.Y.; Sun, Y.K. Transition metal carbide-based materials: Synthesis and applications in electrochemical energy storage. *J. Mater. Chem. A* **2016**, *4*, 10379–10393. [CrossRef]
35. Tian, J.; Shi, Y.; Fan, W.; Liu, T. Ditungsten carbide nanoparticles embedded in electrospun carbon nanofiber membranes as flexible and high-performance supercapacitor electrodes. *Compos. Commun.* **2019**, *12*, 21–25. [CrossRef]

36. Ihsan, M.; Wang, H.; Majid, S.R.; Yang, J.; Kennedy, S.J.; Guo, Z.; Liu, H.K. MoO$_2$/Mo$_2$C/C spheres as anode materials for lithium ion batteries. *Carbon N. Y.* **2016**, *96*, 1200–1207. [CrossRef]
37. Zhang, W.-B.; Ma, X.-J.; Kong, L.-B.; Liu, M.-C.; Luo, Y.-C.; Kang, L. Intermetallic Molybdenum Carbide for Pseudocapacitive Electrode Material. *J. Electrochem. Soc.* **2016**, *163*, A2441–A2446. [CrossRef]
38. Hou, C.; Wang, J.; Du, W.; Wang, J.; Du, Y.; Liu, C.; Zhang, J.; Hou, H.; Dang, F.; Zhao, L.; et al. One-pot synthesized molybdenum dioxide-molybdenum carbide heterostructures coupled with 3D holey carbon nanosheets for highly efficient and ultrastable cycling lithium-ion storage. *J. Mater. Chem. A* **2019**, *7*, 13460–13472. [CrossRef]
39. Yan, Q.; Yang, X.; Wei, T.; Zhou, C.; Wu, W.; Zeng, L.; Zhu, R.; Cheng, K.; Ye, K.; Zhu, K.; et al. Porous β-Mo$_2$C nanoparticle clusters supported on walnut shell powders derived carbon matrix for hydrogen evolution reaction. *J. Colloid Interface Sci.* **2020**, *563*, 104–111. [CrossRef]
40. Hussain, S.; Rabani, I.; Vikraman, D.; Feroze, A.; Karuppasamy, K.; Haq, Z.U.; Seo, Y.-S.; Chun, S.-H.; Kim, H.-S.; Jung, J. Hybrid Design Using Carbon Nanotubes Decorated with Mo$_2$C and W$_2$C Nanoparticles for Supercapacitors and Hydrogen Evolution Reactions. *ACS Sustain. Chem. Eng.* **2020**, *8*, 12248–12259. [CrossRef]
41. Yang, L.; Li, X.; Ouyang, Y.; Gao, Q.; Ouyang, L.; Hu, R.; Liu, J.; Zhu, M. Hierarchical MoO$_2$/Mo$_2$C/C Hybrid Nanowires as High-Rate and Long-Life Anodes for Lithium-Ion Batteries. *ACS Appl. Mater. Interfaces* **2016**, *8*, 19987–19993. [CrossRef]
42. Yang, X.; Li, Q.; Wang, H.; Feng, J.; Zhang, M.; Yuan, R.; Chai, Y. In-situ carbonization for template-free synthesis of MoO$_2$-Mo$_2$C-C microspheres as high-performance lithium battery anode. *Chem. Eng. J.* **2018**, *337*, 74–81. [CrossRef]
43. Momodu, D.; Sylla, N.F.; Mutuma, B.; Bello, A.; Masikhwa, T.; Lindberg, S.; Matic, A.; Manyala, N. Stable ionic-liquid-based symmetric supercapacitors from Capsicum seed porous carbons. *J. Electroanal. Chem.* **2019**, *838*, 119–128. [CrossRef]
44. Men, B.; Guo, P.; Sun, Y.; Tang, Y.; Chen, Y.; Pan, J.; Wan, P. High-performance nitrogen-doped hierarchical porous carbon derived from cauliflower for advanced supercapacitors. *J. Mater. Sci.* **2019**, *54*, 2446–2457. [CrossRef]
45. Liu, X.; Ji, W.; Liang, J.; Peng, L.; Hou, W. MoO$_2$@carbon hollow microspheres with tunable interiors and improved lithium-ion battery anode properties. *Phys. Chem. Chem. Phys.* **2014**, *16*, 20570–20577. [CrossRef] [PubMed]
46. Bao, S.; Luo, S.H.; Yan, S.X.; Wang, Z.Y.; Wang, Q.; Feng, J.; Wang, Y.L.; Yi, T.F. Nano-sized MoO$_2$ spheres interspersed three-dimensional porous carbon composite as advanced anode for reversible sodium/potassium ion storage. *Electrochim. Acta* **2019**, *307*, 293–301. [CrossRef]
47. Li, H.; Ye, H.; Xu, Z.; Wang, C.; Yin, J.; Zhu, H. Freestanding MoO$_2$/Mo$_2$C imbedded carbon fibers for Li-ion batteries. *Phys. Chem. Chem. Phys.* **2017**, *19*, 2908–2914. [CrossRef]
48. Kumar, R.; Ahmed, Z.; Kumar, R.; Jha, S.N.; Bhattacharyya, D.; Bera, C.; Bagchi, V. In-situ modulation of silica-supported MoO$_2$/Mo$_2$C heterojunction for enhanced hydrogen evolution reaction. *Catal. Sci. Technol.* **2020**, *10*, 4776. [CrossRef]
49. Frauwallner, M.L.; López-Linares, F.; Lara-Romero, J.; Scott, C.E.; Ali, V.; Hernández, E.; Pereira-Almao, P. Toluene hydrogenation at low temperature using a molybdenum carbide catalyst. *Appl. Catal. A Gen.* **2011**, *394*, 62–70. [CrossRef]
50. Zhou, E.; Wang, C.; Zhao, Q.; Li, Z.; Shao, M.; Deng, X.; Liu, X.; Xu, X. Facile synthesis of MoO$_2$ nanoparticles as high performance supercapacitor electrodes and photocatalysts. *Ceram. Int.* **2016**, *42*, 2198–2203. [CrossRef]
51. Huo, J.; Xue, Y.; Liu, Y.; Ren, Y.; Yue, G. Polyvinyl alcohol-assisted synthesis of porous MoO$_2$/C microrods as anodes for lithium-ion batteries. *J. Electroanal. Chem.* **2020**, *857*, 113751. [CrossRef]
52. Zheng, J.; Ren, P.; Peng, Y.; Zhou, W.; Yin, Y.; Wu, H.; Gong, W.; Wang, W.; Tang, D.; Zou, B. In-Plane Anisotropic Raman Response and Electrical Conductivity with Robust Electron−Photon and Electron−Phonon Interactions of Air Stable MoO$_2$ Nanosheets. *J. Phys. Chem. Lett.* **2019**, *10*, 2182–2190. [CrossRef] [PubMed]
53. Zhang, J.; Liang, W.; Zhi, L.; Kang, D.; Liu, J.; Su, Z. Study on the structure and electrochemical performances of amphiphilic carbonaceous material-based porous carbon electrode materials. *Mater. Lett.* **2020**, *278*, 128430. [CrossRef]
54. Cheng, Y.; Wu, L.; Fang, C.; Li, T.; Chen, J.; Yang, M.; Zhang, Q. Synthesis of porous carbon materials derived from laminaria japonica via simple carbonization and activation for supercapacitors. *J. Mater. Res. Technol.* **2020**, *9*, 3261–3271. [CrossRef]
55. Seaton, N.A. Determination of the connectivity of porous solids from nitrogen sorption measurements. *Chem. Eng. Sci.* **1991**, *46*, 1895–1909. [CrossRef]
56. Lv, Y.; Ding, L.; Wu, X.; Guo, N.; Guo, J.; Hou, S.; Tong, F.; Jia, D.; Zhang, H. Coal-based 3D hierarchical porous carbon aerogels for high performance and super-long life supercapacitors. *Sci. Rep.* **2020**, *10*, 1–11. [CrossRef] [PubMed]
57. An, K.; Xu, X.; Liu, X. Mo$_2$C-Based Electrocatalyst with Biomass-Derived Sulfur and Nitrogen Co-Doped Carbon as a Matrix for Hydrogen Evolution and Organic Pollutant Removal. *ACS Sustain. Chem. Eng* **2018**, *6*, 1446–1455. [CrossRef]
58. Zhu, Y.; Wang, S.; Zhong, Y.; Cai, R.; Li, L.; Shao, Z. Facile synthesis of a MoO$_2$-Mo$_2$C-C composite and its application as favorable anode material for lithium-ion batteries. *J. Power Sources* **2016**, *307*, 552–560. [CrossRef]
59. Li, X.; Xiao, Q.; Zhang, H.; Xu, H.; Zhang, Y. Fabrication and application of hierarchical mesoporous MoO$_2$/Mo$_2$C/C microspheres. *J. Energy Chem.* **2018**, *27*, 940–948. [CrossRef]
60. Hou, M.; Lan, R.; Hu, Z.; Chen, Z. The preparation of Ni/Mo-based ternary electrocatalysts by the self-propagating initiated nitridation reaction and their application for efficient hydrogen production. *Nanoscale* **2019**, *11*, 17093–17103. [CrossRef] [PubMed]
61. Li, J.S.; Wang, Y.; Liu, C.H.; Li, S.L.; Wang, Y.G.; Dong, L.Z.; Dai, Z.H.; Li, Y.F.; Lan, Y.Q. Coupled molybdenum carbide and reduced graphene oxide electrocatalysts for efficient hydrogen evolution. *Nat. Commun.* **2016**, *7*, 1–8. [CrossRef] [PubMed]
62. Lei, C.; Zhou, W.; Feng, Q.; Lei, Y.; Zhang, Y.; Chen, Y.; Qin, J. Charge Engineering of Mo$_2$C@Defect-Rich N-Doped Carbon Nanosheets for Efficient Electrocatalytic H2 Evolution. *Nano-Micro Lett.* **2019**, *11*, 1–10. [CrossRef]

63. Zhu, Y.; Ji, X.; Yang, L.; Jia, J.; Cheng, S.; Chen, H.; Wu, Z.-S.; Passarello, D.; Liu, M. Targeted synthesis and reaction mechanism discussion of Mo_2C based insertion-type electrodes for advanced pseudocapacitors †. *J. Mater. Chem. A* **2020**, *8*, 7819–7827. [CrossRef]
64. Baltrusaitis, J.; Mendoza-Sanchez, B.; Fernandez, V.; Veenstra, R.; Dukstiene, N.; Roberts, A.; Fairley, N. Generalized molybdenum oxide surface chemical state XPS determination via informed amorphous sample model. *Appl. Surf. Sci.* **2015**, *326*, 151–161. [CrossRef]
65. Xie, X.; Lin, L.; Liu, R.Y.; Jiang, Y.F.; Zhu, Q.; Xu, A.W. The synergistic effect of metallic molybdenum dioxide nanoparticle decorated graphene as an active electrocatalyst for an enhanced hydrogen evolution reaction. *J. Mater. Chem. A* **2015**, *3*, 8055–8061. [CrossRef]
66. Yang, X.; Feng, X.; Tan, H.; Zang, H.; Wang, X.; Wang, Y.; Wang, E.; Li, Y. N-Doped graphene-coated molybdenum carbide nanoparticles as highly efficient electrocatalysts for the hydrogen evolution reaction. *J. Mater. Chem. A* **2016**, *4*, 3947–3954. [CrossRef]
67. Devina, W.; Hwang, J.; Kim, J. Synthesis of $MoO_2/Mo_2C/RGO$ composite in supercritical fluid and its enhanced cycling stability in Li-ion batteries. *Chem. Eng. J.* **2018**, *345*, 1–12. [CrossRef]
68. Murugappan, K.; Anderson, E.M.; Teschner, D.; Jones, T.E.; Skorupska, K.; Román-Leshkov, Y. Operando NAP-XPS unveils differences in MoO_3 and Mo_2C during hydrodeoxygenation. *Nat. Catal.* **2018**, *1*, 960–967. [CrossRef]
69. Zhang, Q.; Pastor-Pérez, L.; Jin, W.; Gu, S.; Reina, T.R. Understanding the promoter effect of Cu and Cs over highly effective $β-Mo_2C$ catalysts for the reverse water-gas shift reaction. *Appl. Catal. B Environ.* **2019**, *244*, 889–898. [CrossRef]
70. Anandan, C.; Mohan, L.; Babu, P.D. Electrochemical studies and growth of apatite on molybdenum doped DLC coatings on titanium alloy β-21S. *Appl. Surf. Sci.* **2014**, *296*, 86–94. [CrossRef]
71. Liu, C.; Lin, M.; Fang, K.; Meng, Y.; Sun, Y. Preparation of nanostructured molybdenum carbides for CO hydrogenation. *RSC Adv.* **2014**, *4*, 20948–20954. [CrossRef]
72. Song, X.; Ma, X.; Li, Y.; Ding, L.; Jiang, R. Tea waste derived microporous active carbon with enhanced double-layer supercapacitor behaviors. *Appl. Surf. Sci.* **2019**, *487*, 189–197. [CrossRef]
73. Payne, B.P.; Biesinger, M.C.; McIntyre, N.S. The study of polycrystalline nickel metal oxidation by water vapour. *J. Electron Spectros. Relat. Phenomena* **2009**, *175*, 55–65. [CrossRef]
74. Gao, Q.; Zhao, X.; Xiao, Y.; Zhao, D.; Cao, M. A mild route to mesoporous Mo_2C-C hybrid nanospheres for high performance lithium-ion batteries. *Nanoscale* **2014**, *6*, 6151–6157. [CrossRef]
75. Xiao, Y.; Liu, Y.; Qin, G.; Han, P.; Guo, X.; Cao, S.; Liu, F. Building $MoSe_2$-Mo_2C incorporated hollow fluorinated carbon fibers for Li-S batteries. *Compos. Part B Eng.* **2020**, *193*, 108004. [CrossRef]
76. Sylla, N.F.; Ndiaye, N.M.; Ngom, B.D.; Momodu, D.; Madito, M.J.; Mutuma, B.K.; Manyala, N.; Manyala, N. Effect of porosity enhancing agents on the electrochemical performance of high-energy ultracapacitor electrodes derived from peanut shell waste. *Sci. Rep.* **2019**, *9*, 1–15. [CrossRef]
77. Li, L.; Chen, C.; Chen, X.; Zhang, X.; Huang, T.; Yu, A. Structure and Catalyst Effects on the Electrochemical Performance of Air Electrodes in Lithium-Oxygen Batteries. *ChemElectroChem* **2018**, *5*, 2666–2671. [CrossRef]
78. Wang, Y.; Zhang, D.; Lu, Y.; Wang, W.; Peng, T.; Zhang, Y.; Guo, Y.; Wang, Y.; Huo, K.; Kim, J.K.; et al. Cable-like double-carbon layers for fast ion and electron transport: An example of $CNT@NCT@MnO_2$ 3D nanostructure for high-performance supercapacitors. *Carbon N. Y.* **2019**, *143*, 335–342. [CrossRef]
79. Arvani, M.; Keskinen, J.; Lupo, D.; Honkanen, M. Current collectors for low resistance aqueous flexible printed supercapacitors. *J. Energy Storage* **2020**, *29*, 101384. [CrossRef]
80. Wang, Z.; Chen, J.S.; Zhu, T.; Madhavi, S.; Lou, X.W. One-pot synthesis of uniform carbon-coated MoO_2 nanospheres for high-rate reversible lithium storage. *Chem. Commun.* **2010**, *46*, 6906–6908. [CrossRef] [PubMed]
81. Giardi, R.; Porro, S.; Topuria, T.; Thompson, L.; Pirri, C.F.; Kim, H.C. One-pot synthesis of graphene-molybdenum oxide hybrids and their application to supercapacitor electrodes. *Appl. Mater. Today* **2015**, *1*, 27–32. [CrossRef]
82. Vikraman, D.; Hussain, S.; Karuppasamy, K.; Feroze, A.; Kathalingam, A.; Sanmugam, A.; Chun, S.H.; Jung, J.; Kim, H.S. Engineering the novel $MoSe_2$-Mo_2C hybrid nanoarray electrodes for energy storage and water splitting applications. *Appl. Catal. B Environ.* **2020**, *264*, 118531. [CrossRef]
83. Nugent, J.M.; Santhanam, K.S.V.; Rubio, A.; Ajayan, P.M. Fast Electron Transfer Kinetics on Multiwalled Carbon Nanotube Microbundle Electrodes. *NANO Lett.* **2001**, *1*, 87–91. [CrossRef]
84. Ndiaye, N.M.; Sylla, N.F.; Ngom, B.D.; Mutuma, B.K.; Dangbegnon, J.K.; Ray, S.C.; Manyala, N. Nitridation Temperature Effect on Carbon Vanadium Oxynitrides for a Symmetric Supercapacitor. *Nanomaterials* **2019**, *9*, 1762. [CrossRef] [PubMed]
85. Javed, M.S.; Chen, J.; Chen, L.; Xi, Y.; Zhang, C.; Wan, B.; Hu, C. Flexible full-solid state supercapacitors based on zinc sulfide spheres growing on carbon textile with superior charge storage. *J. Mater. Chem. A* **2015**, *4*, 667–674. [CrossRef]

Article

Self-Templating Synthesis of N/P/Fe Co-Doped 3D Porous Carbon for Oxygen Reduction Reaction Electrocatalysts in Alkaline Media

Yan Rong [1,*] and Siping Huang [2]

[1] College of Resources & Environment and Historical Culture, Xianyang Normal University, 43 Wenlin Road, Weicheng District, Xianyang 712000, China
[2] College of Chemistry and Chemical Engineering, Xianyang Normal University, 43 Wenlin Road, Weicheng District, Xianyang 712000, China; huangsiping1971@163.com
* Correspondence: allenry@126.com

Abstract: The development of low-cost, highly active, and stable oxygen reduction reaction (ORR) catalysts is of great importance for practical applications in numerous energy conversion devices. Herein, iron/nitrogen/phosphorus co-doped carbon electrocatalysts (NPFe-C) with multistage porous structure were synthesized by the self-template method using melamine, phytic acid and ferric trichloride as precursors. In an alkaline system, the ORR half-wave potential is 0.867 V (vs. RHE), comparable to that of platinum-based catalysts. It is noteworthy that NPFe-C performs better than the commercial Pt/C catalyst in terms of power density and specific capacity. Its unique structure and the feature of heteroatom doping endow the catalyst with higher mass transfer ability and abundant available active sites, and the improved performance can be attributed to the following aspects: (1) Fe-, N-, and P triple doping created abundant active sites, contributing to the higher intrinsic activity of catalysts. (2) Phytic acid was crosslinked with melamine to form hydrogel, and its carbonized products have high specific surface area, which is beneficial for a large number of active sites to be exposed at the reaction interface. (3) The porous three-dimensional carbon network facilitates the transfer of reactants/intermediates/products and electric charge. Therefore, Fe/N/P Co-doped 3D porous carbon materials prepared by a facile and scalable pyrolysis route exhibit potential in the field of energy conversion/storage.

Keywords: N/P/Fe co-doped carbon; self-templating synthesis; 3D porous structure; oxygen reduction reaction electrocatalysts

Citation: Rong, Y.; Huang, S. Self-Templating Synthesis of N/P/Fe Co-Doped 3D Porous Carbon for Oxygen Reduction Reaction Electrocatalysts in Alkaline Media. *Nanomaterials* **2022**, *12*, 2106. https://doi.org/10.3390/nano12122106

Academic Editor: Justo Lobato

Received: 26 May 2022
Accepted: 15 June 2022
Published: 19 June 2022

Publisher's Note: MDPI stays neutral with regard to jurisdictional claims in published maps and institutional affiliations.

Copyright: © 2022 by the authors. Licensee MDPI, Basel, Switzerland. This article is an open access article distributed under the terms and conditions of the Creative Commons Attribution (CC BY) license (https://creativecommons.org/licenses/by/4.0/).

1. Introduction

Oxygen reduction reaction (ORR) kinetics pose a major challenge to the development of energy conversion devices, especially metal-air batteries and fuel cells [1–4]. Platinum based noble metal catalysts are recognized as the best catalysts for ORR. However, its large-scale commercial application is greatly limited by scarce resources, high cost and poor durability [5–7]. Therefore, there is an urgent need to develop non-precious metal-based catalysts apply earth-rich materials [8,9]. Among existing non-noble metal based catalysts, the incorporation of heteroatoms into carbon based materials can change the charge distribution (charge distribution or spin density redistribution), which can introduce abundant active sites and thus enhance the adsorption of and reduction in oxygen [10–12]. Researchers have confirmed that the structure design of carbon-based materials is of great significance to the improvement of properties [6,13–15]. In addition, researchers have found that active species such as transition metals (Mn, Fe, Co, Ni, Cu, etc.) and heteroatoms (B, N, F, P, S, etc.) in carbon-based nanomaterials have synergistic effects to change the topical work function of materials, resulting in enhanced ORR performance [16,17].

Generally, the catalytic performance of carbon-based doped catalysts depends on the specific surface area and pore structure, the composition, content, and distribution of doped elements [18–21]. Conventional synthetic techniques are mainly obtained by porous templates at high carbonization temperatures [22,23]. However, template-assisted methods often require additional complex processes to synthesize and remove templates (for example, silica, polystyrene spheres or metal oxides), during which toxic reagents may be introduced, causing the catalyst to be poisoned. Therefore, it is of great importance to develop a simple process to prepare efficient carbon-based catalysts with good doping conditions and three-dimensional porous structure for ORR catalysis [6,10,24–27].

Herein, the 3D N/P/Fe co-doped material with reasonable pore size and heteroatom distribution was prepared by self-template assisted preparation. The porous structure and surface area were constructed to enrich the active sites exposed to the catalytic reaction interface [28]. Melamine coordinates with Fe and crosslinks with phytic acid (PA) to form a hydrogel. In the pyrolysis process, the decomposition of melamine with PA will generate a large number of micropores and small-sized mesopores, so that the doped atoms are exposed at the three-phase boundary, generating a large number of available active sites (Figure 1). It is expected to promote the formation of carbon porous structures, increase the surface area, and expose more active sites. Three-dimensional interconnected porous networks contribute to the diffusion and transport of substances involved in the reaction (such as O_2 and H_2O) in the process of oxygen reduction. Benefiting from the well controlled distribution and content of N, P, and Fe atoms, as well as the synergistic effect of three-dimensional hierarchical structures, the NPFe-C exhibit much better ORR catalytic activity and durability than the currently advanced Pt/C catalysts. In addition, compared with the traditional templating method, PA cross mixing with melamine hydrogel serves as a self sacrificial template, avoiding the unfavorable process of template removal.

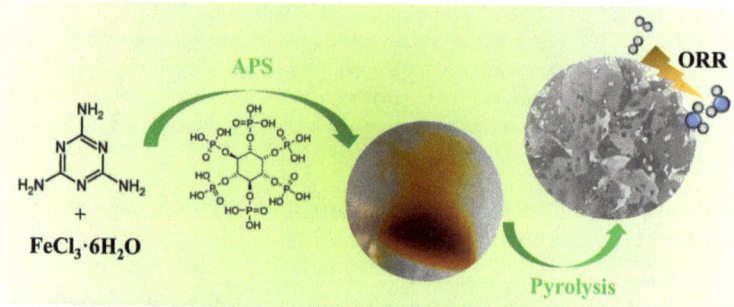

Figure 1. Schematic diagram of the preparation of the NPFe-C.

2. Materials and Methods

2.1. Reagents and Chemicals

Iron trichloride ($FeCl_3$), melamine ($C_3H_6N_6$), phytic acid (C_6HOP_6) were purchased from Sinopharm Chemical Reagent Co., Ltd. (Shanghai, China). Commercial 20 wt% Pt/C elelctrocatalyst was supplied by E-TEK, Inc (Shenzhen, China). Potassium hydroxide (KOH) and hydrochloric acid (HCl) were supplied by Aladdin Reagent Co., Ltd. (Shanghai, China).

2.2. Preparation of NPFe-C

A total of 0.5 g of melamine and 0.1 g of $FeCl_3$ were added in 200 mL water. Then, 0.1 mL of PA was added in the mixture. After stirring for 3 h, the gel product was recovered, dried, and carbonized at 800 °C for 2 h under a N_2 flow with a heating rate of 1 °C/min. Finally, the NPFe-C were obtained after immersed in 0.5 M HCl solution for 3 h. For comparison, the Fe-free catalyst (NP-C) was synthesized by similar processes.

2.3. Physical Characterization

The crystalline structure, morphology and surface composition of the sample were physically characterized by X-ray diffraction (XRD), Raman Spectrometer, scan electron microscopy (SEM) equipped with energy-dispersive spectrometer (EDS), transmission electron microscopy (TEM), and X-ray photoelectron spectroscopy (XPS). The surface area and pore volume of the samples were measured on a physical adsorption instrument.

2.4. Electrochemical Measurement

The electrochemical experiments were carried out on a computer-controlled CHI 760E electrochemical workstation. The standard 3-electrode system was used at 30 °C, a rotating disk electrode (RDE, 0.196 cm^2) or a rotating ring disk electrode (RRDE, 0.2475 cm^2) coated with the electrocatalyst was used as working electrode, a carbon rod served as counter electrode, and a saturated calomel electrode acted as reference electrode. The electrocatalyst ink was obtained by mixing 10 mg of electrocatalyst and 5 mL of water/ethanol/Nafion solution, and sonicating for 60 min. Then, 10 µL electrocatalyst ink was dropped onto the polished glassy carbon electrode surface and dried at room temperature. The catalyst loading is 0.10 mg cm^{-2}. The CV curve and LSV curve were recorded in the N_2/O_2-saturated 0.1 M KOH solution. The rotating speed and potential sweep rate were set to 1600 rpm and 10 mV s^{-1}, respectively [29].

2.5. Zn−Air Battery Measurements

The electrochemical properties of the catalysts were evaluated in the simulated Zn-air battery devices. Polarization curves and galvanostatic discharge tests were carried out on a CHI 760E electrochemical workstation and LAND testing system, respectively. The as prepared electrocatalyst ink was coated on PTFE treated carbon pentafiber paper as cathode, and zinc foil and 6 M KOH were used as anode and electrolyte [25].

3. Results

3.1. Characterization of NPFe-C

The samples at different synthesis stages were characterized by XRD (Figure 2A). The samples before carbonization had wide C (002) peaks at 25°. After carbonization at 800 °C, the sample shows 3 sets of X-ray diffraction peaks, which are Fe (PDF#52-0513), Fe (PDF#65-4899), Fe_3O_4 (JCPDS No.11-0614) and Fe_3P (PDF#19-0617). For NPFe-C, there are a few weak and widened diffraction peaks, which may be due to the existence of trace iron-based nanocrystals. In addition, 2 broad characteristic peaks can be observed after carbonization of polymer samples, which correspond to the (002) and (100)/(101) crystal planes of carbon, about 25° and 42°, respectively [30]. The corresponding (002) diffracted graphite plane indicates a highly crystalline structure. Raman spectroscopy (Figure 2B) confirms this assertion by using the intensity ratios of D to G bands (I_D/I_G) caused by Raman vibrations of disordered (sp3) and ordered (sp2) carbon, respectively. I_D/I_G of D-band (1337 cm^{-1}) and G-band (1587 cm^{-1}) with relative peak strength of NPFe-C is 0.93, it indicates that the material has both amorphous carbon and graphitized carbon with high degree of graphitization. In general, carbon-based electrocatalysts with a high graphitization degree increase the electrical conductivity, thus effectively improving the activity of the ORR [31].

NPFe-C was analyzed by SEM (Figure 3A) and TEM (Figure 3B), and the pyrolyzed NPFe-C catalysts exhibited typical interconnected 3D macroporous networks. High resolution TEM (HRTEM) images of the NPFe-C reveal that the material is not dense and contains numerous pores (Figure 3C). The mapping results show that N, Fe and P elements are evenly distributed in the three-dimensional carbon network (Figure 3D).

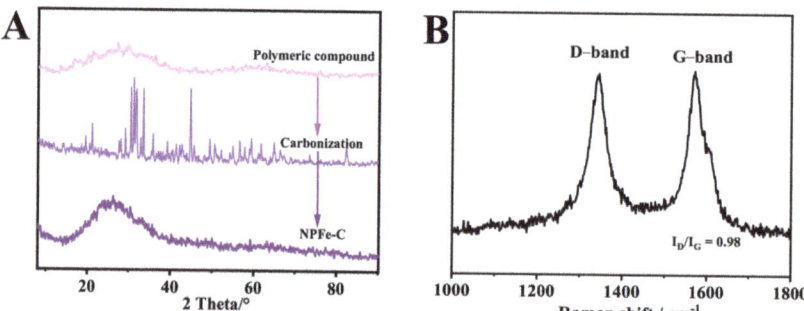

Figure 2. (**A**) XRD patterns and (**B**) Raman spectrum of the NPFe-C.

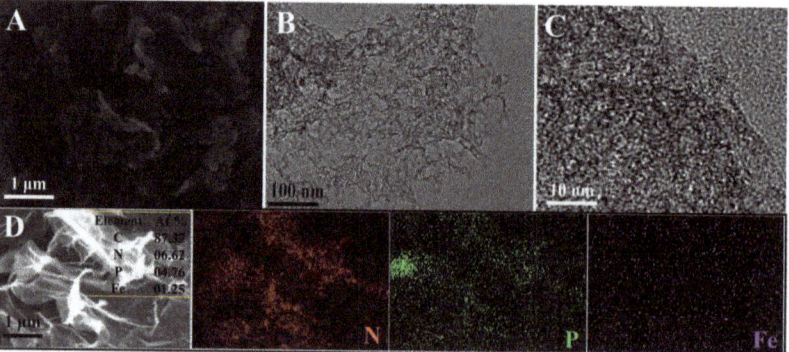

Figure 3. (**A**) SEM images, (**B**) TEM images, (**C**) HRTEM image and (**D**) elemental mapping of the NPFe-C. The insets are the corresponding EDS data.

To further investigate the pore structure of the materials, the pore structure of NPFe-C was characterized by N_2 adsorption desorption isotherms (Figure 4A). The isothermal curves showed a type I/IV mixing shape. The 2 steep adsorption processes occurred at low (<0.05) and high pressures (>0.9) and showed the coexistence of micropores, mesopores, and macropores. The specific surface area was calculated to be 774.5 m^2 g^{-1}. The simulation results based on QSDFT model show that the micropore size in NPFe-C is mainly 0.9 nm and the mesoporous size is widely distributed in the range of 2–35 nm. The micro/mesopores are formed during carbonization, in which the polymer is pyrolyzed to generate volatile gas, forming porous carbon. The hierarchical pore structure may facilitate the diffusion of substances during the ORR process. The larger BET surface area and abundant microporous structures is conducive to expose the abundant ORR active sites to the reaction interface and promote the promotion of the activity of NPFe-C catalysts. In addition, constructing a reasonable 3D microporous structure can provide efficient mass transport channels for reactants, intermediates, and products. All of these features are useful for improving ORR performance.

The surface composition of NPFe-C was further analyzed by X-ray photoelectron spectroscopy (XPS). The presence of C, N, O, P, Fe in the doped carbon materials was confirmed by XPS survey spectra (Figure 4B), indicating the successful introduction of heteroatoms into the carbon substrate. The N, P, and Fe contents in NPFe-C were 2.12, 1.24, and 0.31 at%, respectively (Figure 4C). The high-resolution N1s spectra of the heteroatom doped catalysts (Figure 4D) are classified into 5 types, including pyridinic-N (398.2 eV, 26.7%), Fe-Nx (399.5 eV, 19.8%), pyrrolic-N (401.1 eV, 35.6%), graphitic-N (402.4 eV, 9.2%), and oxidized N (404.7 eV, 8.7%), some studies have demonstrated that pyridinic-N, Fe-Nx and graphitic-N are active sites for the ORR [32–35]. Some researchers have proposed that the N atom can reduce the band gap and increase the charge mobility of graphite lattices. Quantum chemical calculations show that the N-containing group at the edge has the highest charge mobility. In addition, these changes in the carbon band structure ultimately reduce the electronic work function at the interface, which has an important impact on the ORR performance [6,36]. The P 2p high resolution spectrum has two characteristic peaks at 132.9 eV and 131.2 eV, which are assigned to P-C and P-O species respectively. Doped P mostly exists in P-C (Figure 4E). Because the electronegativity of P (2.19) is lower than C (2.55), P atom with positive charge will be induced and new ORR active sites may be generated in alkaline medium [2]. Moreover, the presence of phosphorus atoms in P-O is beneficial for enhancing the electronegativity of the O atom, thereby promoting the charge delocalization of nearby C atoms, increasing the adsorption on O_2 [37–39]. Moreover, P atom doping in the carbon lattice induces distortion of its structure, and generates edge sites, which further enhance its electrochemical activity [40]. As identified by XPS, the high-resolution Fe 2p spectrum (Figure 4F) has obvious peaks at 725.7 eV and 721.3 eV, which are the Fe 2p 1/2 of Fe (III) and Fe (II), respectively. The peak at 709.2 eV is Fe 2p 3/2, while the peak at 716.3 eV is probably a satellite peak. According to previous studies, Fe sites are favorable for the adsorption of O_2 and oxygenated intermediates, thereby significantly promoting the activity of ORR [41–43].

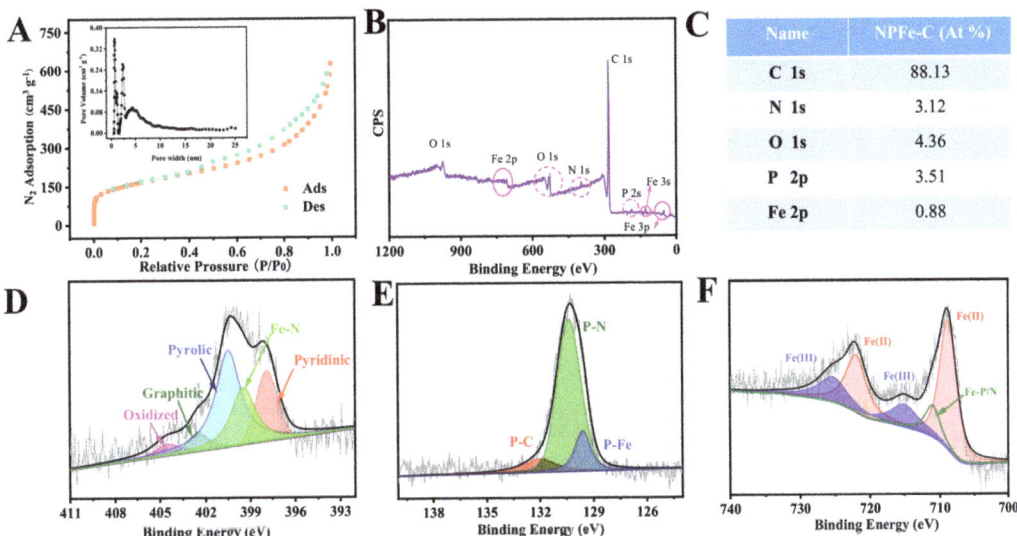

Figure 4. (**A**) N_2 adsorption–desorption isotherm curve (inset: pore size distribution) of NPFe-C, (**B,C**) XPS spectrum and the corresponding data of the Polymeric compound, the Carbonization and the NPFe-C, (**D–F**) high-resolution XPS spectra of N 1s, P 2p, Fe 2p of NPFe-C.

3.2. ORR Performance and the Zn-Air Battery Tests of NPFe-C

The ORR activity of NPFe-C was first measured by cyclic voltammetry (CV) in 0.1 M KOH electrolyte saturated with N_2/O_2 (Figure 5A). In oxygen saturated electrolyte, a cathode peak was observed at 0.81 V. In contrast, such cathode peaks were not found in N_2-saturated electrolytes. This indicates that NPFe-C has significant electrocatalytic activity for ORR. The ORR performance of NPFe-C electrocatalysts was investigated by rotating disk electrode tests (Figure 5B). The polarization curves show that the NPFe-C catalysts exhibit the most superior electrocatalytic activity for half wave potential of 0.867 V compared with NP-C ($E_{1/2}$ = 0.712 V), even comparable to that of the commercial catalyst. We observed that the annealing temperature significantly affected the ORR activity of N/P/Fe-C (Figure S1A). According to the $E_{1/2}$ value of ORR and the limiting current density, the catalyst with the best performance was obtained at 800 °C. In addition, the formation of N/P/Fe-C and the ORR performance were significantly affected by different ratio of reagents. The ORR activity of the samples with the ratio of melamine to phytic acid of 0.5 g:0.1 mL is significantly better than that of 0.5 g:0.04 mL, 0.5 g:0.06 mL, 0.5 g:0.15 mL and 0.5 g:0.2 mL (Figure S1B). When changing the amount of $FeCl_3$, the ORR activity tends to increase with the amount of $FeCl_3$ added, but the hydrogel can not be obtained when the amount is more than 0.12 g (Figure S1C).

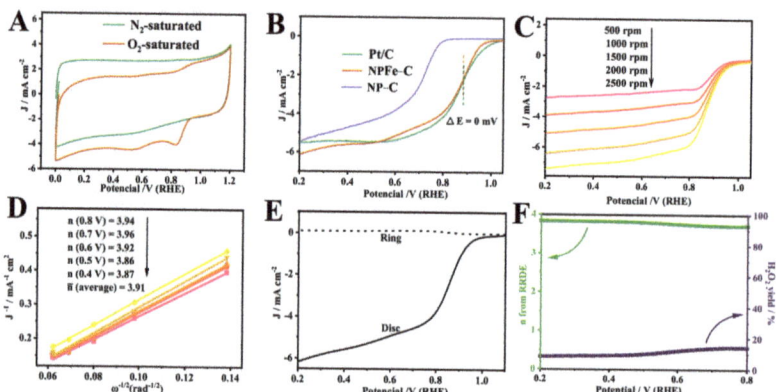

Figure 5. (**A**) CV curves of NPFe-C in an N_2- and O_2-saturated 0.1 M KOH solution. (**B**) LSV of NPFe-C, NP-C and commercial Pt/C in an O_2-saturated 0.1 M KOH solution (sweep rate: 10 mV s^{-1}; electrode rotation speed: 1600 rpm). (**C**) LSV of NPFe-C at different rotation rates. (**D**) Koutecky-Levich of NPFe-C. (**E,F**) Rotating ring-disk electrode voltammograms obtained at 1600 rpm in O_2 saturated 0.1 M KOH solutions.

To further investigate the electron transfer process of NPFe-C catalysts, the ORR polarization curves at different rotating speeds were recorded in saturated 0.1 M KOH solution. The kinetic parameters were analyzed using the koutecky Levich (K-L) equation (Figure 5C,D), which revealed that the electron transfer number was close to 4, indicating high selectivity for the complete four electron oxygen reduction reaction. The selectivity of ORR was then determined by using the rotating ring disc electrode (RRDE) (Figure 5E,F). The disc current density (I_D) sharply increased over 0.8 V. Similar to what RDE observed, the disc current (I_R) was negligible in the range of 0.2–0.8 V, corresponding to a lower hydrogen peroxide yield of 7–12%. Therefore, in this potential range, the value of n is about 3.88, which is consistent with the results of RDE.

In order to further explore the active sites of NPFe-C, the method of SCN^- poisoning catalyst was adopted. It is well known that SCN^- poisons the active site of Fe-N/Fe-P and inactivates its active site because of the strong action of SCN^- ions on Fe. As shown in Figure 6A, after the introduction of 0.1 M KSCN, $E_{1/2}$ was negatively shifted by 72 mV, consistent with the results reported in the literature [43–45]. Therefore, the presence of Fe-N/Fe-P species is essential to provide high performance. Moreover, the $E_{1/2}$ of the catalyst still reaches 0.78 V after the Fe-N/Fe-P species is toxified, indicating that Fe-N is not the only active site and part of the activity of the catalyst originates from the doping of heteroatoms. Therefore, it can be concluded that both Fe-N/Fe-P species and doped nitrogen atoms contribute to the improvement of ORR performance of NPFe-C. In addition, the electrical conductivity of the materials significantly affects the catalytic performance. The conductivity of catalysts is evaluated by EIS test. The charge transfer resistance (Rct = 193.62 Ω) of NPFe-C is similar to that of Pt/C (Rct = 137.75 Ω) (Figure S2). Since the electrical conductivity of carbon materials is related to their graphitization degree, combined with Raman measurements, it is concluded that the graphitization degree of N/P/Fe-C contributes to the kinetic improvement of ORR.

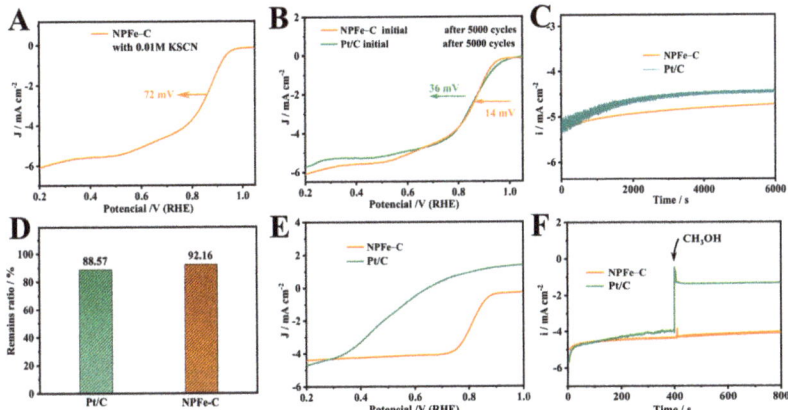

Figure 6. (**A**) ORR polarization curves of Fe-NCCs electrocatalysts in 0.1 M KOH + 0.01 M KSCN solution, (**B**) ORR curves before and after 5000 CV cycles (CV potential range: 0 V–1.2 V, RHE), (**C,D**) Chronoamperometric curves and the corresponding data of NPFe-C and Pt/C electrocatalyst, (**E**) ORR polarization curves of Fe-NCCs in 0.1 M KOH + 0.2 M methanol solution, (**F**) Chronoamperometric curves of the Pt/C electrocatalyst and NPFe-C before and after adding 1 M methanol.

In addition to its high catalytic activity, the long-term stability of the electrocatalysts must also be considered. The stability of NPFe-C was evaluated using cycling experiments (Figure 6B) and chronoamperometry (Figure 6C). After 5000 cycles, the $E_{1/2}$ of NPFe-C only decreased by 14 mV. In contrast, the $E_{1/2}$ of the Pt/C electrocatalyst is negatively shifted by 36 mV. The i-t curves show that the NPFe-C decays more slowly than the Pt/C electrocatalyst at 0.8 V (Figure 6C). Among them, NPFe-C maintained 92.16% of the initial current value after 6000 s; this was much better than the commercial Pt/C electrocatalyst (88.57%). These results demonstrate that the NPFe-C catalysts exhibit remarkable electrocatalytic activity and long-lasting stability in alkaline medium, serving as a cost-effective electrocatalyst for ORR.

Over the past 30 years, direct methanol fuel cells (DMFCs) have emerged as a highly promising energy conversion device. In the DMFC system, the methanol fuel of the anode causes the cathodic Pt/C electrocatalyst to generate mixed potentials and CO poisoning. Therefore, the cathode electrocatalyst with high ORR selectivity has an important promoting effect on the commercialization of DMFCs. To examine the ORR selectivity of the cathode electrocatalyst, the ORR performance of the NPFe-C and commercial Pt/C in the presence of methanol was investigated (Figure 5F,F) [29]. As shown in Figure 6E, the $E_{1/2}$ value of the ORR for NPFe-C is more positive than that of commercial Pt/C in 0.1 M KOH + 0.2 M methanol solution. The i-t curves show that the ORR current of NPFe-C during the first 400 s is comparable to that of the Pt/C electrocatalyst in the absence of methanol, and after 1 M methanol was injected into the electrolyte at 400 s, there was no obvious decrease in the current of NPFe-C but not that of the Pt/C electrocatalyst. Therefore, with the addition of 1 M methanol, the electrocatalytic activity of NPFe-C for ORR is much higher than that of Pt/C electrocatalysts, indicating the high selectivity of NPFe-C for ORR.

In order to further evaluate the ORR activity of NPFe-C catalyst under actual battery operating conditions, NPFe-C and Pt/C were assembled to fabricate Zn-air batteries, the 6 M KOH solution was used as the electrolyte. As shown in Figure 7A, the Zn-air battery with NPFe-C as catalyst exhibits higher discharge current and power density than that with Pt/C. The maximum power density of NPFe-C catalyst is 79.6 mWcm^{-2}, which is significantly higher than that of commercialized Pt/C (68.5 mWcm^{-2}). In addition, typical galvanostatic discharge tests were performed at different current densities (Figure 7B), which revealed that the NPFe-C based cells exhibited a slightly higher voltage (1.23 V) and a longer discharge time (74 h) compared to the Pt/C electrocatalyst at 5 mA cm^{-2}

current density (1.22 V, 68 h). When normalized to the mass of consumed Zn (Figure 7C), the calculated cell specific capacity is 752 mAh g^{-1} at 5 mA cm^{-2} for the NPFe-C catalyzed air cathode, superior to that of commercialized Pt/C (707 mAh g^{-1}). In particular, the NPFe-C still exhibits excellent electrocatalytic performance for Zn-air batteries under high current conditions.

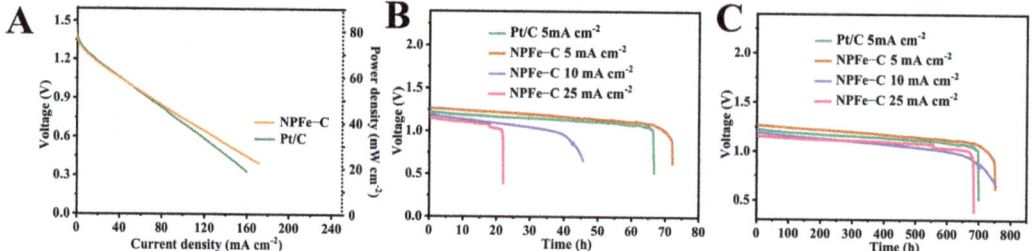

Figure 7. (**A**) Polarization and power-density curves, (**B**) galvanostatic discharge curves; (**C**) specific capacities at various current densities (5, 10, and 25 mA cm^{-2}) of the Zn-air batteries using NPFe-C and Pt/C as electrocatalysts.

4. Conclusions

In summary, N-, P-, and Fe triple doped three-dimensional nanoporous carbon materials synthesized by a simple self templating strategy, with large specific surface area, optimal porous structure, and well-designed distribution of doped atoms, effectively enhance the number of active sites and mass transfer efficiency. Compared with commercial Pt/C catalysts, NPFe-C catalysts show better electrocatalytic activity and long-term durability for ORR. Assembled in the Zn-air cell, the catalyst exhibits superior power density and specific capacity over the Pt/C catalyst. These results indicate that NPFe-C as ORR electrocatalysts can be used as potential substitutes for precious metal catalysts in energy conversion and storage devices.

Supplementary Materials: The following supporting information can be downloaded at: https://www.mdpi.com/article/10.3390/nano12122106/s1, Figure S1: ORR LSV curves of a series of NPFe-Cs samples with (A) different annealing temperature; (B) different ratio of melamine to phytic acid and (C) different ratio of melamine to FeCl$_3$.; Figure S2: EIS curves of the N/P/Fe-C and Pt/C in 0.1 M KOH electrolyte. Potential: at open circuit potential from 0.01 to 100,000 Hz with a modulation amplitude: 10 mV.

Author Contributions: Data curation, Y.R.; Formal analysis, Y.R.; Funding acquisition, Y.R.; Methodology, S.H.; Writing—original draft, Y.R. All authors have read and agreed to the published version of the manuscript.

Funding: This research was sponsored by Shaanxi Social Science Foundation Project (2021G008); Key projects of Shaanxi Provincial Department of Education (17JZ080); Key cultivation and scientific research projects of Xianyang Normal University (XSYK21035); "Blue talents" project of Xianyang Normal University (XSYQL201903); Supported by "academic leader" of Xianyang Normal University (XSYXSDT202114).

Data Availability Statement: Not applicable.

Acknowledgments: Gratefully acknowledges methodology provided by Siping Huang, and thanks for the fund support.

Conflicts of Interest: The authors declare no conflict of interest.

References

1. Kostuch, A.; Jarczewski, S.; Surówka, M.K.; Kuśtrowski, P.; Sojka, Z.; Kruczała, K. The joint effect of electrical conductivity and surface oxygen functionalities of carbon supports on the oxygen reduction reaction studied over bare supports and Mn–Co spinel/carbon catalysts in alkaline media. *Catal. Sci. Technol.* **2021**, *11*, 7578–7591. [CrossRef]
2. Zhan, X.; Tong, X.; Gu, M.; Tian, J.; Gao, Z.; Ma, L.; Xie, Y.; Chen, Z.; Ranganathan, H.; Zhang, G.; et al. Phosphorus-doped graphene electrocatalysts for oxygen reduction reaction. *Nanomaterials* **2022**, *12*, 1141. [CrossRef] [PubMed]
3. Jia, Y.; Shi, C.; Zhang, W.; Xia, W.; Hu, M.; Huang, R.; Qi, R. Iron single atoms anchored on nitrogen-doped carbon matrix/nanotube hybrid supports for excellent oxygen reduction properties. *Nanomaterials* **2022**, *12*, 1593. [CrossRef] [PubMed]
4. Huang, Z.-F.; Wang, J.; Peng, Y.; Jung, C.-Y.; Fisher, A.; Wang, X. Design of efficient bifunctional oxygen reduction/evolution electrocatalyst: Recent advances and perspectives. *Adv. Energy Mater.* **2017**, *7*, 1700544. [CrossRef]
5. Jafari, M.; Gharibi, H.; Parnian, M.J.; Nasrollahpour, M.; Vafaee, M. Iron-nanoparticle-loaded nitrogen-doped carbon nanotube/carbon sheet composites derived from mof as electrocatalysts for an oxygen reduction reaction. *ACS Appl. Nano Mater.* **2021**, *4*, 459–477. [CrossRef]
6. Wang, D.-W.; Su, D. Heterogeneous nanocarbon materials for oxygen reduction reaction. *Energ. Environ. Sci.* **2014**, *7*, 576. [CrossRef]
7. Labata, M.F.; Li, G.; Ocon, J.; Chuang, P.-Y.A. Insights on platinum-carbon catalyst degradation mechanism for oxygen reduction reaction in acidic and alkaline media. *J. Power Sources* **2021**, *487*, 229356. [CrossRef]
8. Alemany-Molina, G.; Quílez-Bermejo, J.; Navlani-García, M.; Morallón, E.; Cazorla-Amorós, D. Efficient and cost-effective orr electrocatalysts based on low content transition metals highly dispersed on c3n4/super-activated carbon composites. *Carbon* **2022**, *196*, 378–390. [CrossRef]
9. Liu, D.; Tong, Y.; Yan, X.; Liang, J.; Dou, S.X. Recent advances in carbon-based bifunctional oxygen catalysts for zinc-air batteries. *Batter. Supercaps* **2019**, *2*, 743–765. [CrossRef]
10. Yang, L.; Shui, J.; Du, L.; Shao, Y.; Liu, J.; Dai, L.; Hu, Z. Carbon-based metal-free orr electrocatalysts for fuel cells: Past, present, and future. *Adv. Mater.* **2019**, *31*, e1804799. [CrossRef]
11. Wang, X.X.; Cullen, D.A.; Pan, Y.-T.; Hwang, S.; Wang, M.; Feng, Z.; Wang, J.; Engelhard, M.H.; Zhang, H.; He, Y.; et al. Nitrogen-coordinated single cobalt atom catalysts for oxygen reduction in proton exchange membrane fuel cells. *Adv. Mater.* **2018**, *30*, 1706758. [CrossRef]
12. Tao, L.; Wang, Y.; Zou, Y.; Zhang, N.; Zhang, Y.; Wu, Y.; Wang, Y.; Chen, R.; Wang, S. Charge transfer modulated activity of carbon-based electrocatalysts. *Adv. Energy Mater.* **2019**, *10*, 1901227. [CrossRef]
13. Bakoglidis, K.D.; Palisaitis, J.; dos Santos, R.B.; Rivelino, R.; Persson, P.O.Å.; Gueorguiev, G.K.; Hultman, L. Self-healing in carbon nitride evidenced as material inflation and superlubric behavior. *ACS Appl. Mater. Inter.* **2018**, *10*, 16238–16243. [CrossRef]
14. Gueorguiev, G.K.; Broitman, E.; Furlan, A.; Stafström, S.; Hultman, L. Dangling bond energetics in carbon nitride and phosphorus carbide thin films with fullerene-like and amorphous structure. *Chem. Phys. Lett.* **2009**, *482*, 110–113. [CrossRef]
15. Seh, Z.W.; Kibsgaard, J.; Dickens, C.F.; Chorkendorff, I.; Norskov, J.K.; Jaramillo, T.F. Combining theory and experiment in electrocatalysis: Insights into materials design. *Science* **2017**, *355*, eaad4998. [CrossRef]
16. Ge, H.; Zhang, H. Fungus-based mno/porous carbon nanohybrid as efficient laccase mimic for oxygen reduction catalysis and hydroquinone detection. *Nanomaterials* **2022**, *12*, 1596. [CrossRef]
17. Hu, C.; Dai, L. Carbon-based metal-free catalysts for electrocatalysis beyond the orr. *Angew. Chem. Int. Edit.* **2016**, *55*, 11736–11758. [CrossRef]
18. Liu, Z.; Zhao, Z.; Wang, Y.; Dou, S.; Yan, D.; Liu, D.; Xia, Z.; Wang, S. In situ exfoliated, edge-rich, oxygen-functionalized graphene from carbon fibers for oxygen electrocatalysis. *Adv. Mater.* **2017**, *29*, 1606207. [CrossRef]
19. Lu, Z.; Chen, G.; Siahrostami, S.; Chen, Z.; Liu, K.; Xie, J.; Liao, L.; Wu, T.; Lin, D.; Liu, Y.; et al. High-efficiency oxygen reduction to hydrogen peroxide catalysed by oxidized carbon materials. *Nat. Catal.* **2018**, *1*, 156–162. [CrossRef]
20. Cieluch, M.; Podleschny, P.Y.; Kazamer, N.; Wirkert, F.J.; Rost, U.W.; Brodmann, M. Development of a bifunctional Ti-based gas diffusion electrode for ORR and OER by one- and two-step Pt-Ir electrodeposition. *Nanomaterials* **2022**, *12*, 1233. [CrossRef]
21. Hu, Y.; Jensen, J.O.; Zhang, W.; Cleemann, L.N.; Xing, W.; Bjerrum, N.J.; Li, Q. Hollow spheres of iron carbide nanoparticles encased in graphitic layers as oxygen reduction catalysts. *Angew. Chem. Int. Edit.* **2014**, *53*, 3675–3679. [CrossRef]
22. Hasche, F.; Oezaslan, M.; Strasser, P.; Fellinger, T.-P. Electrocatalytic hydrogen peroxide formation on mesoporous non-metal nitrogen-doped carbon catalyst. *J. Energy Chem.* **2016**, *25*, 251–257. [CrossRef]
23. Rybarczyk, M.K.; Cysewska, K.; Yuksel, R.; Lieder, M. Microporous n-doped carbon obtained from salt melt pyrolysis of chitosan toward supercapacitor and oxygen reduction catalysts. *Nanomaterials* **2022**, *12*, 1162. [CrossRef]
24. Wang, X.X.; Hwang, S.; Pan, Y.-T.; Chen, K.; He, Y.; Karakalos, S.; Zhang, H.; Spendelow, J.S.; Su, D.; Wu, G. Ordered Pt3Co intermetallic nanoparticles derived from metal-organic frameworks for oxygen reduction. *Nano Lett.* **2018**, *18*, 4163–4171. [CrossRef] [PubMed]
25. Zheng, X.; Wu, J.; Cao, X.; Abbott, J.; Jin, C.; Wang, H.; Strasser, P.; Yang, R.; Chen, X.; Wu, G. N-, P-, and S-doped graphene-like carbon catalysts derived from onium salts with enhanced oxygen chemisorption for Zn-air battery cathodes. *Appl. Catal. B Environ.* **2019**, *241*, 442–451. [CrossRef]
26. Dai, L. Carbon-based catalysts for metal-free electrocatalysis. *Curr. Opin. Electrochem.* **2017**, *4*, 18–25. [CrossRef]

27. Zhang, T.; Asefa, T. Heteroatom-doped carbon materials for hydrazine oxidation. *Adv. Mater.* **2018**, *31*, 1804394. [CrossRef] [PubMed]
28. Contreras, E.; Dominguez, D.; Tiznado, H.; Guerrero-Sanchez, J.; Takeuchi, N.; Alonso-Nunez, G.; Contreras, O.E.; Oropeza-Guzman, M.T.; Romo-Herrera, J.M. N-doped carbon nanotubes enriched with graphitic nitrogen in a buckypaper configuration as efficient 3D electrodes for oxygen reduction to H_2O_2. *Nanoscale* **2019**, *11*, 2829–2839. [CrossRef]
29. Jia, N.; Weng, Q.; Shi, Y.; Shi, X.; Chen, X.; Chen, P.; An, Z.; Chen, Y. N-doped carbon nanocages: Bifunctional electrocatalysts for the oxygen reduction and evolution reactions. *Nano Res.* **2018**, *11*, 1905–1916. [CrossRef]
30. He, X.; Xia, Y.; Liang, C.; Zhang, J.; Huang, H.; Gan, Y.; Zhao, C.; Zhang, W. A flexible non-precious metal Fe-N/C catalyst for highly efficient oxygen reduction reaction. *Nanotechnology* **2019**, *30*, 144001. [CrossRef]
31. Wang, T.; Ola, O.; Dapaah, M.F.; Lu, Y.; Niu, Q.; Cheng, L.; Wang, N.; Zhu, Y. Preparation and characterization of multi-doped porous carbon nanofibers from carbonization in different atmospheres and their oxygen electrocatalytic properties research. *Nanomaterials* **2022**, *12*, 832. [CrossRef]
32. Jiang, W.-J.; Hu, W.-L.; Zhang, Q.-H.; Zhao, T.-T.; Luo, H.; Zhang, X.; Gu, L.; Hu, J.-S.; Wan, L.-J. From biological enzyme to single atomic fe-n-c electrocatalyst for efficient oxygen reduction. *Chem. Commun.* **2018**, *54*, 1307–1310. [CrossRef]
33. Li, Z.; Wei, L.; Jiang, W.-J.; Hu, Z.; Luo, H.; Zhao, W.; Xu, T.; Wu, W.; Wu, M.; Hu, J.-S. Chemical state of surrounding iron species affects the activity of Fe-Nx for electrocatalytic oxygen reduction. *Appl. Catal. B Environ.* **2019**, *251*, 240–246. [CrossRef]
34. Gong, K.; Du, F.; Xia, Z.; Durstock, M.; Dai, L. Nitrogen-doped carbon nanotube arrays with high electrocatalytic activity for oxygen reduction. *Science* **2009**, *323*, 760–764. [CrossRef]
35. Sa, Y.J.; Seo, D.-J.; Woo, J.; Lim, J.T.; Cheon, J.Y.; Yang, S.Y.; Lee, J.M.; Kang, D.; Shin, T.J.; Shin, H.S.; et al. A general approach to preferential formation of active fe-n-x sites in Fe-N/C electrocatalysts for efficient oxygen reduction reaction. *J. Am. Chem. Soc.* **2016**, *138*, 15046–15056. [CrossRef]
36. Strelko, V.V.; Kuts, V.S.; Thrower, P.A. On the mechanism of possible influence of heteroatoms of nitrogen, boron and phosphorus in a carbon matrix on the catalytic activity of carbons in electron transfer reactions. *Carbon* **2000**, *38*, 1499–1503. [CrossRef]
37. Esrafili, M.D.; Mohammad-Valipour, R.; Mousavi-Khoshdel, S.M.; Nematollahi, P. A comparative study of co oxidation on nitrogen- and phosphorus-doped graphene. *Chemphyschem* **2015**, *16*, 3719–3727. [CrossRef]
38. Ma, T.Y.; Ran, J.; Dai, S.; Jaroniec, M.; Qiao, S.Z. Phosphorus-doped graphitic carbon nitrides grown in situ on carbon-fiber paper: Flexible and reversible oxygen electrodes. *Angew. Chem. Int. Edit.* **2015**, *54*, 4646–4650. [CrossRef]
39. Shinde, S.S.; Sami, A.; Lee, J.-H. Nitrogen- and phosphorus-doped nanoporous graphene/graphitic carbon nitride hybrids as efficient electrocatalysts for hydrogen evolution. *ChemCatChem* **2015**, *7*, 3873–3880. [CrossRef]
40. Banhart, F.; Kotakoski, J.; Krasheninnikov, A.V. Structural defects in graphene. *ACS Nano* **2011**, *5*, 26–41. [CrossRef]
41. Liang, J.; Zhou, R.F.; Chen, X.M.; Tang, Y.H.; Qiao, S.Z. Fe-N decorated hybrids of cnts grown on hierarchically porous carbon for high-performance oxygen reduction. *Adv. Mater.* **2014**, *26*, 6074–6079. [CrossRef]
42. Liang, Z.; Xia, W.; Qu, C.; Qiu, B.; Tabassum, H.; Gao, S.; Zou, R. Edge-abundant porous fe3o4 nanoparticles docking in nitrogen-rich graphene aerogel as efficient and durable electrocatalyst for oxygen reduction. *Chemelectrochem* **2017**, *4*, 2442–2447. [CrossRef]
43. Wang, N.; Li, L.; Zhao, D.; Kang, X.; Tang, Z.; Chen, S. Graphene composites with cobalt sulfide: Efficient trifunctional electrocatalysts for oxygen reversible catalysis and hydrogen production in the same electrolyte. *Small* **2017**, *13*, 1701025. [CrossRef]
44. Chen, L.; Liu, X.; Zheng, L.; Li, Y.; Guo, X.; Wan, X.; Liu, Q.; Shang, J.; Shui, J. Insights into the role of active site density in the fuel cell performance of Co-N-C catalysts. *Appl. Catal. B Environ.* **2019**, *256*, 117849. [CrossRef]
45. Zhang, J.; Chen, G.; Müllen, K.; Feng, X. Carbon-rich nanomaterials: Fascinating hydrogen and oxygen electrocatalysts. *Adv. Mater.* **2018**, *30*, 1800528. [CrossRef]

Article

A Theoretical Study of Fe Adsorbed on Pure and Nonmetal (N, F, P, S, Cl)-Doped $Ti_3C_2O_2$ for Electrocatalytic Nitrogen Reduction

Heng Luo [1,2], Xiaoxu Wang [3], Chubin Wan [1,*], Lu Xie [2], Minhui Song [2] and Ping Qian [1,2]

[1] Department of Physics, University of Science and Technology Beijing, Beijing 100083, China; s20190798@xs.ustb.edu.cn (H.L.); qianping@ustb.edu.cn (P.Q.)
[2] Beijing Advanced Innovation Center for Materials Genome Engineering, University of Science and Technology Beijing, Beijing 100083, China; b20170555@xs.ustb.edu.cn (L.X.); s20191385@xs.ustb.edu.cn (M.S.)
[3] DP Technology, Beijing 100083, China; wangxx@dp.tech
* Correspondence: cbwan@ustb.edu.cn

Abstract: The possibility of using transition metal (TM)/MXene as a catalyst for the nitrogen reduction reaction (NRR) was studied by density functional theory, in which TM is an Fe atom, and MXene is pure $Ti_3C_2O_2$ or $Ti_3C_2O_{2-x}$ doped with N/F/P/S/Cl. The adsorption energy and Gibbs free energy were calculated to describe the limiting potentials of N_2 activation and reduction, respectively. N_2 activation was spontaneous, and the reduction potential-limiting step may be the hydrogenation of N_2 to *NNH and the desorption of *NH_3 to NH_3. The charge transfer of the adsorbed Fe atoms to N_2 molecules weakened the interaction of N≡N, which indicates that Fe/MXene is a potential catalytic material for the NRR. In particular, doping with nonmetals F and S reduced the limiting potential of the two potential-limiting steps in the reduction reaction, compared with the undoped pure structure. Thus, Fe/MXenes doped with these nonmetals are the best candidates among these structures.

Keywords: DFT; MXene; nitrogen reduction; electrocatalysis; Gibbs free energy

Citation: Luo, H.; Wang, X.; Wan, C.; Xie, L.; Song, M.; Qian, P. A Theoretical Study of Fe Adsorbed on Pure and Nonmetal (N, F, P, S, Cl)-Doped $Ti_3C_2O_2$ for Electrocatalytic Nitrogen Reduction. *Nanomaterials* **2022**, *12*, 1081. https://doi.org/10.3390/nano12071081

Academic Editor: Frederik Tielens

Received: 3 March 2022
Accepted: 23 March 2022
Published: 25 March 2022

Publisher's Note: MDPI stays neutral with regard to jurisdictional claims in published maps and institutional affiliations.

Copyright: © 2022 by the authors. Licensee MDPI, Basel, Switzerland. This article is an open access article distributed under the terms and conditions of the Creative Commons Attribution (CC BY) license (https://creativecommons.org/licenses/by/4.0/).

1. Introduction

Ammonia is a raw material for the production of various fertilizers and is a potential energy source that is easy to store and transport, environmentally friendly, and relatively safe. Ammonia synthesis is important in agricultural production and energy development. However, most ammonia synthesis still relies on the Hubble–Bosch method proposed in the 20th century, which requires harsh reaction conditions (400–600 °C and 20–40 MPa) [1–3]. This method consumes a large amount of energy and causes significant greenhouse gas emissions [4]. In addition, other negative effects, such as adverse effects on the equipment under high-temperature and high-pressure conditions, need to be considered. Therefore, the development of environmentally friendly and less energy-demanding methodologies for NH_3 synthesis is urgently needed. Electrocatalytic ammonia synthesis has attracted increasing attention owing to its high efficiency and environmental friendliness. The introduction of electrical energy has a remarkable influence on N_2 activation and changes the reaction pathways [5], which is beneficial for the development of new stable and efficient catalysts.

New catalysts can be developed from unique structures, such as core–shell Ni–Au nanoparticles for CO_2 hydrogenation [6], or from new materials. The excellent physical, electronic, and chemical properties of two-dimensional (2D) materials have attracted extensive scientific research [7–13]. In addition, 2D materials, such as molybdenum disulfide, graphene, and metal–organic frameworks (MOFs) [14–16], have emerged as potential candidates for electrochemical nitrogen reduction reactions (NRRs). Notably, MXene, a new member of the 2D material family that joined in 2011 [17], has developed rapidly in the past nine years [17–19]. The general formula of MXene is $M_{n+1}X_nT_x$, where M represents early transition metals (TMs), X represents carbon or nitrogen, T_x represents the surface functional groups O, OH, or F, and n = 1, 2, 3. MXenes are synthesized by the chemical etching of A layers in the MAX ($M_{n+1}AX_n$) phase. Although a variety of 2D MXenes have been theoretically predicted [20], only a few have been synthesized. MXenes are applied in a wide range of fields, including electrocatalysis [21], hydrogen storage [22,23], lithium-ion batteries [24,25], and supercapacitors [26]. MXene is a potential candidate for electrochemical NRRs (e-NRRs) because of its large specific surface area, adjustable structure, and excellent stability [27–29].

MXene-based electrocatalysts for the e-NRR can be divided into two categories: pure MXene and MXene-based hybrid electrocatalysts [30]. Pure MXene is a potential candidate for the e-NRR. For example, Azofra et al. [31] found that M_3C_2 exhibited good N_2 capture and activation behavior. However, bare-metal atoms on the surface of M_3C_2 are considered active sites [31,32], which tend to bind to functional groups such as oxygen groups; thus, the electrical conductivity is decreased, and the active sites are inactivated. Pure MXene still faces challenges as a catalyst for the e-NRR; therefore, MXene hybrids have been designed. Li et al. [33] loaded nanosized Au particles onto Ti_3C_2 nanosheets (Au/Ti_3C_2) for the e-NRR. Their research indicated that the hybrid is conducive to N_2 chemisorption and decreases the activation energy barrier. Au/Ti_3C_2 shows excellent catalytic performance. MnO_2-decorated $Ti_3C_2T_x$ (MnO_2-$Ti_3C_2T_x$) has also been studied as an efficient electrocatalyst for ammonia synthesis under environmental conditions [34]. MnO_2 and $Ti_3C_2T_x$ synergistically promote electrocatalytic activity to achieve superior catalytic activity. In addition, single-atom catalysts (SACs) have been widely studied because of their low cost, superior performance, and full use of metal atoms. Gao et al. [5] studied the reaction pathways and overpotentials of $Ti_3C_2O_2$-supported TM (Fe, Co, Ru, Rh) SACs. These MXene hybrids, including noble metal–MXene, TM oxide–MXene, and MXene-based SACs, have effectively changed the catalytic performance, providing more possibilities for the screening of new efficient and stable catalysts.

In this study, a 2D MXene, $Ti_3C_2O_2$, was modified with nonmetals (N, F, P, S, and Cl) and adsorbed TM (Fe atom, Fe/$Ti_3C_2O_{2-x}$) to study the catalytic performance of the e-NRR. Gibbs free energy (ΔG) was used to analyze the reaction pathway and limit the potential of each catalyst, and the main potential-limiting steps of the reaction were determined as *N_2 + H → *NHH and *NH_3 → NH_3.

2. Computational Methods

Density functional theory (DFT) calculations were performed using the Vienna ab initio simulation package v. 5.4.4. (University of Vienna, Vienna, Austria) [35,36]. The generalized gradient approximation with Perdew–Burke–Ernzerhof was used as an exchange-correlation function [37]. The projector-augmented wave method was adopted to describe the effect of the core electrons on the valence electron density [38]. The cut-off energy was set to 600 eV. The convergence criteria for the energy and force were 10^{-5} eV and 10^{-2} eV/Å, respectively. The thickness of the vacuum layer was more than 20 Å to avoid interactions in the z-direction, and the x-and y-directions were set as periodic boundary conditions. A 3 × 3 × 1 supercell was used for all the structures. For geometric optimization, the Brillouin zones were sampled with 4 × 4 × 1 Monkhorst–Pack meshes [39], and DFT-D3 was used to accurately describe Van der Waals interactions [40]. Charge transfer was computed by Bader charge population analysis [41,42] and the electron localization function (ELF) was analyzed using the VESTA code [43].

The substitution energies (ΔE_{sub}) of doping different nonmetallic elements (N/F/P/S/Cl) on the surface of $Ti_3C_2O_2$ can be expressed as

$$\Delta E_{sub} = E_{NM-Ti_3C_2O_{2-x}} - E_{Ti_3C_2O_2} + E_O - E_{NM} \qquad (1)$$

where E_O and E_{NM} represent the energies of a single O atom and nonmetallic elements (N, F, P, S, Cl), respectively, and were calculated using H_2 [44], H_2O [45], NH_3 [46], HF [47], H_3PO_4 [48], H_2S [49], and HCl [50] from the Open Quantum Materials Database (OQMD) [51,52].

The adsorption energy (ΔE_{ads}) of Fe anchored on NM-Ti$_3$C$_2$O$_{2-x}$ (NM represents the surface nonmetals, O, N, F, P, S, and Cl) was calculated using the following formula:

$$\Delta E_{ads} = E_{Fe/NM-Ti_3C_2O_{2-x}} - E_{NM-Ti_3C_2O_{2-x}} - E_{Fe} \quad (2)$$

ΔG was calculated as described by Nørskov et al. [53]. Under standard reaction conditions, the chemical potential of a proton and electron pair ($\mu[H^+ + e^-]$) is equal to half that of gaseous hydrogen ($\mu[H_2]$). ΔG was calculated using the following formula:

$$\Delta G = \Delta E_{DFT} + \Delta ZPE - T\Delta S - neU + \Delta G_{pH} \quad (3)$$

where ΔE is the potential energy change calculated by DFT, ΔZPE is the zero-point energy correction, and it is calculated by calculating the frequency of the adsorbed species. $T\Delta S$ is the entropy correction, which is usually available from some database, where $T = 298$ K; ΔG_{pH} and neU are the contributions from the pH and electrode potential (U), respectively; n is the number of electrons transferred; U is the applied bias. ΔG_{pH} is defined as

$$\Delta G_{pH} = -k_B T \ln[H^+] = pH \times K_B T \ln 10 \quad (4)$$

where k_B is Boltzmann's constant. For all the calculations, the pH was set to zero. The ΔE_{ads} values of different adsorbates were calculated as follows:

$$\Delta E_{ads} = E_{cat-mol} - E_{cat} - E_{mol} \quad (5)$$

where $\Delta E_{cat-mol}$ is the energy of the entire adsorption structure, E_{cat} is the energy of the catalyst, and E_{mol} is the energy of the adsorbate molecules such as N$_2$ and N$_x$H$_y$.

3. Results and Discussion

3.1. Geometric Structure

Bare Ti$_3$C$_2$ is a hexagonal lattice with P$\bar{3}m$1 group symmetry, five atomic layers of Ti–C–Ti–C–Ti, two exposed Ti layers, and an experimental lattice constant of 3.057 Å [54]. After structural optimization, $a = b = 3.020$ Å, which was in good agreement with the experimental values. Bare MXenes are unstable under relevant NRR operating conditions [55], and they are always functionalized by electronegative functional groups [56], as they are chemically exfoliated from the bulk MAX phase by HF [17,57]. O-terminated Ti$_3$C$_2$ was used for further experiments. There are different possibilities for the adsorption of O on Ti$_3$C$_2$. According to previous studies [5], the most stable structure is O adsorbed at the hollow sites of the contralateral surface Ti atoms, as shown in Figure 1a,b. Nonmetallic elements (N/F/P/S/Cl) were used to modify the Ti$_3$C$_2$O$_2$ surface. ΔE_{sub} indicates the stability of a surface before and after doping with nonmetallic elements. The ΔE_{sub} values for N, F, P, S, and Cl were 1.79, −1.04, 0.81, −0.27, and −1.01 eV, respectively. The structure became more stable after doping with F, S, and Cl when $\Delta E_{sub} < 0$ and became more unstable after doping with N and P when $\Delta E_{sub} > 0$. Among these doping situations, doping with F had the best stability, compared with doping with other nonmetallic elements.

Pure Ti$_3$C$_2$O$_2$ and Ti$_3$C$_2$O$_2$ modified with nonmetallic elements (Figure S1) were used to support single Fe atoms. Two different hollow sites (H1 and H2) and an O-top site on the surface were considered, as shown in Figure 1a. The O-top was unstable, and the E_{ads} values of Fe adsorbed on H1 and H2 are listed in Table 1. Except for the F-doped structures, the Fe atoms preferred to adsorb on the H1 site, as the E_{ads} was smaller. Notably, in the F-doped structure, the Fe atom was adsorbed on the next-nearest H1 site (Figure 1e). As shown in Table 1, the doping of N, F, P, and S facilitates the adsorption of Fe, while it is more difficult for Fe to adsorb on the Cl-doped structure. Figure 1c–h show the most stable adsorption positions for the different catalysts.

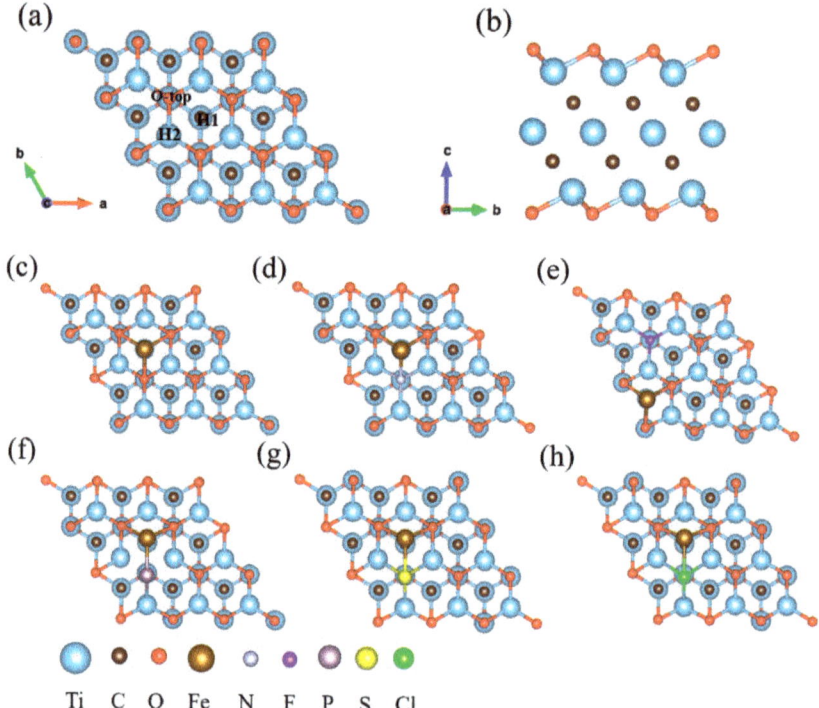

Figure 1. (a) Top view and different adsorption sites on $Ti_3C_2O_2$ and (b) side view of $Ti_3C_2O_2$. The most stable structure of Fe adsorbed on (c) $Ti_3C_2O_2$, (d) N-doped $Ti_3C_2O_2$, (e) F-doped $Ti_3C_2O_2$, (f) P-doped $Ti_3C_2O_2$, (g) S-doped $Ti_3C_2O_2$, and (h) Cl-doped $Ti_3C_2O_2$.

Table 1. Adsorption energies of Fe adsorbed on different sites and N_2 adsorbed on different catalysts, the charge on N_2, and the charge transferred after N_2 adsorption.

Species	E_{ads} of Fe (eV)		E_{ads} of N_2 (eV)	Charge Transferred on N_2 (e)
	H1	H2		
Fe/$Ti_3C_2O_2$	−3.57	−3.30	−0.92	0.19
Fe/N-$Ti_3C_2O_{2-x}$	−4.32	−3.90	−0.77	0.15
Fe/F-$Ti_3C_2O_{2-x}$	−3.61	−3.60	−0.78	0.18
Fe/P-$Ti_3C_2O_{2-x}$	−5.12	−4.68	−0.55	0.13
Fe/S-$Ti_3C_2O_{2-x}$	−4.33	−4.02	−0.59	0.16
Fe/Cl-$Ti_3C_2O_{2-x}$	−3.39	−3.11	−0.85	0.21

3.2. N_2 Adsorption

Based on the Fe/NM-$Ti_3C_2O_{2-x}$ structure, N_2 adsorption was calculated using E_{ads}. There are two different positions for N_2 adsorption, and advanced research has shown that N_2 adsorption is closer end to end than side to side [5]. Figure 2a–f show the most stable structure of N_2 adsorbed on different catalysts from end to end, and Figure 2g–l show the ELF of these structures. E_{ads} ranged from −0.55 eV to −0.92 eV, which indicates that the N_2 adsorption has strong spontaneity, and the absolute value of E_{ads} from small to large was in the order: Fe/P-$Ti_3C_2O_{2-x}$ < Fe/S-$Ti_3C_2O_{2-x}$ < Fe/N-$Ti_3C_2O_{2-x}$ < Fe/F-$Ti_3C_2O_{2-x}$ < Fe/Cl-$Ti_3C_2O_{2-x}$ < Fe/$Ti_3C_2O_2$ (Table 1). After N_2 adsorption, the N≡N bond lengths in Fe/$Ti_3C_2O_2$, Fe/N-$Ti_3C_2O_{2-x}$, Fe/F-$Ti_3C_2O_{2-x}$, Fe/P-$Ti_3C_2O_{2-x}$, Fe/S-$Ti_3C_2O_{2-x}$, and Fe/Cl-$Ti_3C_2O_{2-x}$ are 1.128, 1.125, 1.129, 1.123, 1.126, and 1.130 Å, respectively. Compared with the N≡N bond length in the gas phase (1.11 Å), all of them became longer. The calculation of charge transfer is shown in Table 1. The results show that N_2 gains electrons in all these catalysts and the translated charges increase with an increase in the number of valence electrons from N to O or from P to S and Cl in the same period. However, doping with F did not obey this rule, which may be due to the special adsorption site of Fe. Fe was adsorbed on the first nearest H1 site and followed the trend from N to O and F. These findings were consistent with those of Wang et al. [58]. A strong positive correlation exists between the electron gains of N_2 and the change in bond length: N_2 on Fe/Cl-$Ti_3C_2O_{2-x}$ gained the most electrons and had the largest increase in bond length relative to the gas phase, whereas N_2 on Fe/P-$Ti_3C_2O_{2-x}$ gained the least electrons and had the smallest increment in bond length relative to the gas phase.

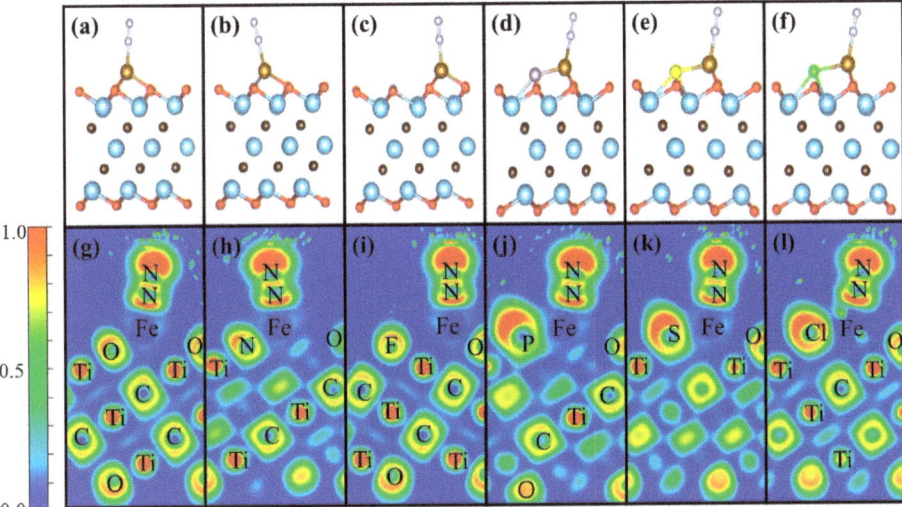

Figure 2. Most stable structures of N_2 adsorbed on (**a**) Fe/$Ti_3C_2O_2$, (**b**) Fe/N-$Ti_3C_2O_{2-x}$, (**c**) Fe/F-$Ti_3C_2O_{2-x}$, (**d**) Fe/P-$Ti_3C_2O_{2-x}$, (**e**) Fe/S-$Ti_3C_2O_{2-x}$, and (**f**) Fe/Cl-$Ti_3C_2O_{2-x}$ and ELFs of N_2 adsorbed on (**g**) Fe/$Ti_3C_2O_2$, (**h**) Fe/N-$Ti_3C_2O_{2-x}$, (**i**) Fe/F-$Ti_3C_2O_{2-x}$, (**j**) Fe/P-$Ti_3C_2O_{2-x}$, (**k**) Fe/S-$Ti_3C_2O_{2-x}$, and (**l**) Fe/Cl-$Ti_3C_2O_{2-x}$.

The partial density of states of N_2 adsorbed on Fe/$Ti_3C_2O_2$ or Fe/NM-$Ti_3C_2O_{2-x}$ (Figure 3) shows spin-up and spin-down of the d orbital of the Fe atom and the p orbital of the N atom. At the Fermi level, almost no spin-up was observed, whereas the spin-down was more obvious, and the d orbital of Fe effectively overlapped with the P orbital of N near the Fermi level. The electrons in the occupied d orbital of Fe/NM-$Ti_3C_2O_{2-x}$ transferred to the antibonding orbitals of N_2, as shown in Table 1, and the adsorbed N_2 on different catalysts gained electrons from 0.13 e to 0.21 e, thus lowering the bond energy of N_2.

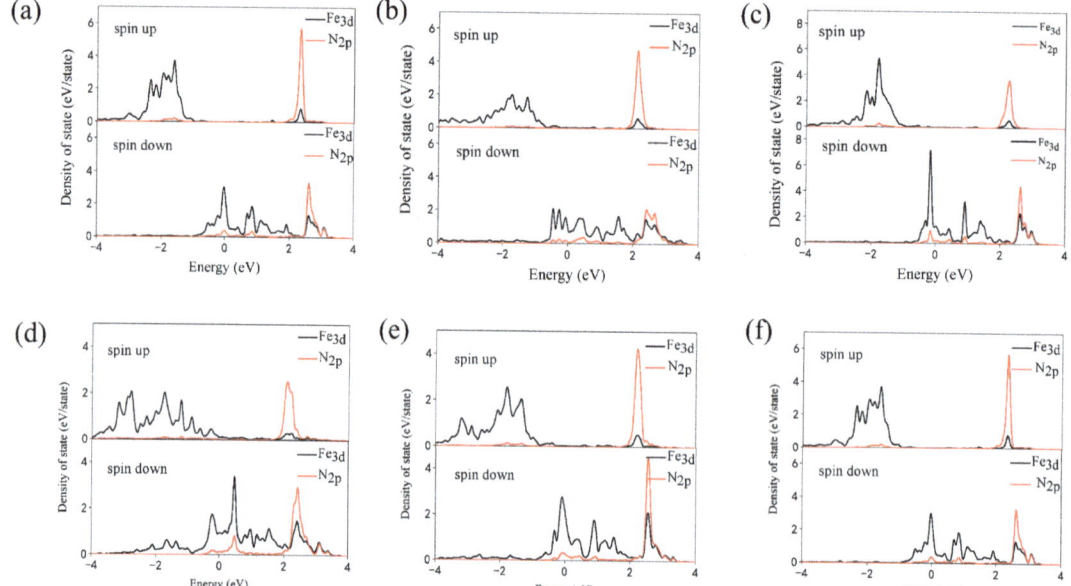

Figure 3. Partial density of states of N_2 adsorbed on (**a**) Fe/Ti$_3$C$_2$O$_2$, (**b**) Fe/N-Ti$_3$C$_2$O$_{2-x}$, (**c**) Fe/F-Ti$_3$C$_2$O$_{2-x}$, (**d**) Fe/P-Ti$_3$C$_2$O$_{2-x}$, (**e**) Fe/S-Ti$_3$C$_2$O$_{2-x}$, and (**f**) Fe/Cl-Ti$_3$C$_2$O$_{2-x}$.

3.3. N_2 Reduction Mechanism

The overall e-NRR reaction on the cathode is

$$N_2(g) + 6(H^+ + e^-) \rightarrow 2NH_3(g) \tag{6}$$

and the anode reactions provide protons and electrons. Liu et al. [59] summarized the mechanism of the e-NRR. The e-NRR is divided into dissociation and association mechanisms by different hydrogenation (protonation and reduction) sequences and the breaking of the N≡N triple bond. In the dissociation mechanism, the N≡N bond is broken during the adsorption process (* denotes the adsorption site).

$$2* + N_2 \rightarrow 2*N \tag{7}$$

Then, two separated N atoms on the surface of the catalysts receive protons and electrons, and ammonia is formed in the last hydrogenation step:

$$*N + H^+ + e^- \rightarrow *NH \tag{8}$$

$$*NH + H^+ + e^- \rightarrow *NH_2 \tag{9}$$

$$*NH_2 + H^+ + e^- \rightarrow *NH_3 \tag{10}$$

$$*NH_3 \rightarrow NH_3 \tag{11}$$

In the association mechanism, the N≡N bond breaks at a certain hydrogenation step. According to the hydrogenation sequence, it can be further classified into distal, alternating, and enzymatic pathways. The hydrogenation step in the enzymatic pathway is similar to that in the alternating pathway; the difference is that N_2 adsorbs side to side in the enzymatic pathway, but ends in the distal and alternating pathways. For the distal and alternating pathways, the first two steps are

$$* + N_2 \rightarrow *N_2 \qquad (12)$$

$$*N_2 + H^+ + e^- \rightarrow *N_2H \qquad (13)$$

In the distal pathway, the N atom moves away from the catalytically gained protons and electrons, releasing the first NH$_3$ molecule, as follows:

$$*N_2H + H^+ + e^- \rightarrow *NNH_2 \qquad (14)$$

$$*NNH_2 + H^+ + e^- \rightarrow *N + NH_3 \qquad (15)$$

Hydrogenation then occurs on the remaining N atom and releases the second NH$_3$ molecule according to Reactions (8)–(11). In the alternating pathway, hydrogenation occurs on two newton atoms alternatively, and NH$_3$ is formed until the N≡N bond is completely broken.

$$*N_2H + H^+ + e^- \rightarrow *NHNH \qquad (16)$$

$$*NHNH + H^+ + e^- \rightarrow *NHNH_2 \qquad (17)$$

$$*NHNH_2 + H^+ + e^- \rightarrow *NH_2NH_2 \qquad (18)$$

$$*NH_2NH_2 + H^+ + e^- \rightarrow *NH_2 + NH_3 \qquad (19)$$

After the first NH$_3$ is released, the remaining *NH$_2$ obtains protons and electrons and releases the second ammonia according to Reactions (10) and (11). Figure 4 shows the other mixed pathways that follow neither the distal nor alternating pathways but a combination of two paths. Optimized structures of all the possible elementary steps in NRR is showed in Figure S2.

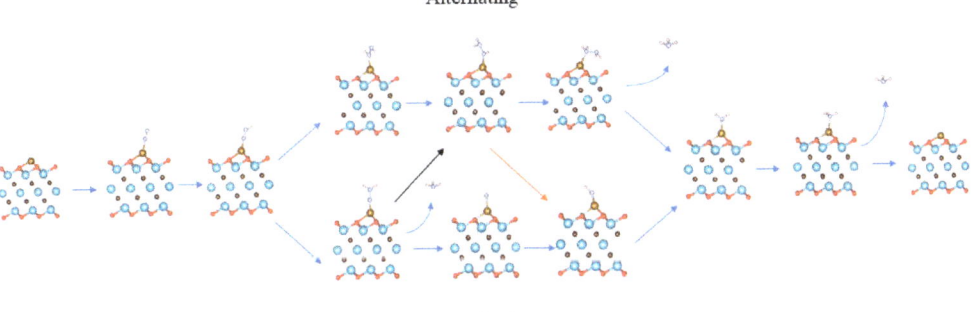

Figure 4. Possible pathway and reaction intermediates for NRR with the associated mechanism. Dark brown, blue, red, brown, light blue, and light pink represent C, Ti, O, Fe, N, and H, respectively.

The ΔG values calculated by DFT calculations considered all correction terms, including the zero-point energy, temperature, and entropy corrections. Table 2 illustrates the E_{ZPE} and entropy corrections (*TS*) of different reaction intermediates on Fe/Ti$_3$C$_2$O$_2$ using the *TS* values obtained from the National Institute of Standards and Technology [60] at T = 298 K. The catalyst as a substrate is immobilized, although the surface is different, we compared the zero-point energy with the study of Ling [61]; the difference is marginal, as N$_2$ reduction also occurred on the transition metal atoms in Ling's research, and only the E_{ZPE} of NH$_3$ was significantly different. NH$_3$ is a gas phase, not an adsorbent, so other research was also compared [5]. The calculated E_{ZPE} and *TS* of H$_2$ are 0.27 and 0.4 eV [60], respectively.

Table 2. E_{ZPE} and TS of different reaction intermediates on Fe/Ti$_3$C$_2$O$_2$, T = 298 K.

Adsorption Species	E_{ZPE} (eV)	E'_{ZPE} (eV)	E_{ZPE} Difference (eV)	TS [60] (eV)
N$_2$	0.15	0.15 [61]	0	0.59
*N≡N	0.19	0.20 [61]	0.01	0.23
*N=NH	0.47	0.49 [61]	0.02	0.20
*N−NH$_2$	0.78	0.82 [61]	0.04	0.25
*N	0.09	0.08 [61]	0.01	0.06
*NH	0.31	0.35 [61]	0.04	0.14
*NH$_2$	0.63	0.65 [61]	0.02	0.18
*NH$_3$	1.00	1.02 [61]	0.02	0.23
*NH=NH	0.81	0.80 [61]	0.01	0.25
*NH−NH$_2$	1.11	1.13 [61]	0.02	0.31
*NH$_2$−NH$_2$	1.50	1.49 [61]	0.01	0.27
NH$_3$	0.92	0.96 [5]	0.04	0.60

As shown in Figure 5a–f, for all structures, the first protonation was likely to generate *NNH species; the ΔG values for Fe/Ti$_3$C$_2$O$_2$, Fe/N-Ti$_3$C$_2$O$_{2-x}$, Fe/F-Ti$_3$C$_2$O$_{2-x}$, Fe/P-Ti$_3$C$_2$O$_{2-x}$, Fe/S-Ti$_3$C$_2$O$_{2-x}$, and Fe/Cl-Ti$_3$C$_2$O$_{2-x}$ increased to 0.90, 1.04, 0.85, 0.99, 0.88, and 1.01 eV, respectively. The second step is more likely to form *NNH$_2$ instead of the *NHNH species in the alternate path, as the energy requirements are higher, and the increments in ΔG for Fe/Ti$_3$C$_2$O$_2$, Fe/N-Ti$_3$C$_2$O$_{2-x}$, Fe/F-Ti$_3$C$_2$O$_{2-x}$, Fe/P-Ti$_3$C$_2$O$_{2-x}$, Fe/S-Ti$_3$C$_2$O$_{2-x}$, and Fe/Cl-Ti$_3$C$_2$O$_{2-x}$ were 0.1, 0.06, 0.12, −0.05, 0.12, and 0.07 eV to form *NNH$_2$, respectively. In the subsequent hydrogenation steps, the intermediate configuration in the alternating pathway was easier to form than the first NH$_3$ molecule desorption in the distal pathway. The first NH$_3$ is not desorbed until the fifth proton is added, and adsorptive *NH$_3$ is formed when the sixth proton is added. The reaction *NNH$_2$ → *NHNH$_2$ → *NH$_2$NH$_2$ → *NH$_2$ → *NH$_3$ is exothermic, and larger energy input is required until the adsorptive *NH$_3$ is desorbed to form the second NH$_3$ molecule. The ΔG values of Fe/Ti$_3$C$_2$O$_2$, Fe/N-Ti$_3$C$_2$O$_{2-x}$, Fe/F-Ti$_3$C$_2$O$_{2-x}$, Fe/P-Ti$_3$C$_2$O$_{2-x}$, Fe/S-Ti$_3$C$_2$O$_{2-x}$, and Fe/Cl-Ti$_3$C$_2$O$_{2-x}$ were 1.95, 1.11, 0.97, 1.07, 1.09, 0.99 eV, respectively. However, it was reported that the use of an acidic electrolyte can promote NH$_3$ desorption, as the protonation of adsorbed NH$_3$ to form NH$_4^+$ can easily proceed [62,63], so the actual energy barrier is even smaller. For all these structures, the two potential limiting steps were the first hydrogenation of N$_2$ to form the *NNH species and the last process of NH$_3$ desorption to form the second NH$_3$ molecule. Compared with the original structure, nonmetallic doping was beneficial for the desorption of the last NH$_3$ molecule, but only the doping of F and S was beneficial for the formation of *NNH and NH$_3$.

Figure 6 shows the most possible reaction pathway for different catalysts. All these structures are likely to follow the mixed pathway: N$_2$ → *N$_2$ → *NNH → *NNH$_2$ → *NHNH$_2$ → *NH$_2$NH$_2$ → *NH$_2$ → *NH$_3$ → NH$_3$. In addition, the doping of nonmetals has a remarkable effect on NRR. For N$_2$ adsorption, E_{ads} is reduced, compared with the nondoped structure, which may be the reason why NH$_3$ desorption is easier in the last step. In the hydrogenation process, the doping of different nonmetals also makes each step of the hydrogenation easier or harder. The doping of N, P, and Cl makes it difficult for *N$_2$ to form *NNH, whereas F and S facilitate the formation of *NNH from *N$_2$. From *NNH to *NNH$_2$, only the doping of P shows an obvious impact and makes the transformation occur spontaneously. In comparison, the other doped nonmetals do not show a great effect. The doping of nonmetal also does not have much influence on *NNH$_2$ → *NHNH$_2$ → *NH$_2$NH$_2$ → *NH$_2$ → *NH$_3$, as these reactions are exothermic for all structures. Considering the stability of nonmetal doping, the best catalysts may be Fe/F-Ti$_3$C$_2$O$_{2-x}$ and Fe/S-Ti$_3$C$_2$O$_{2-x}$.

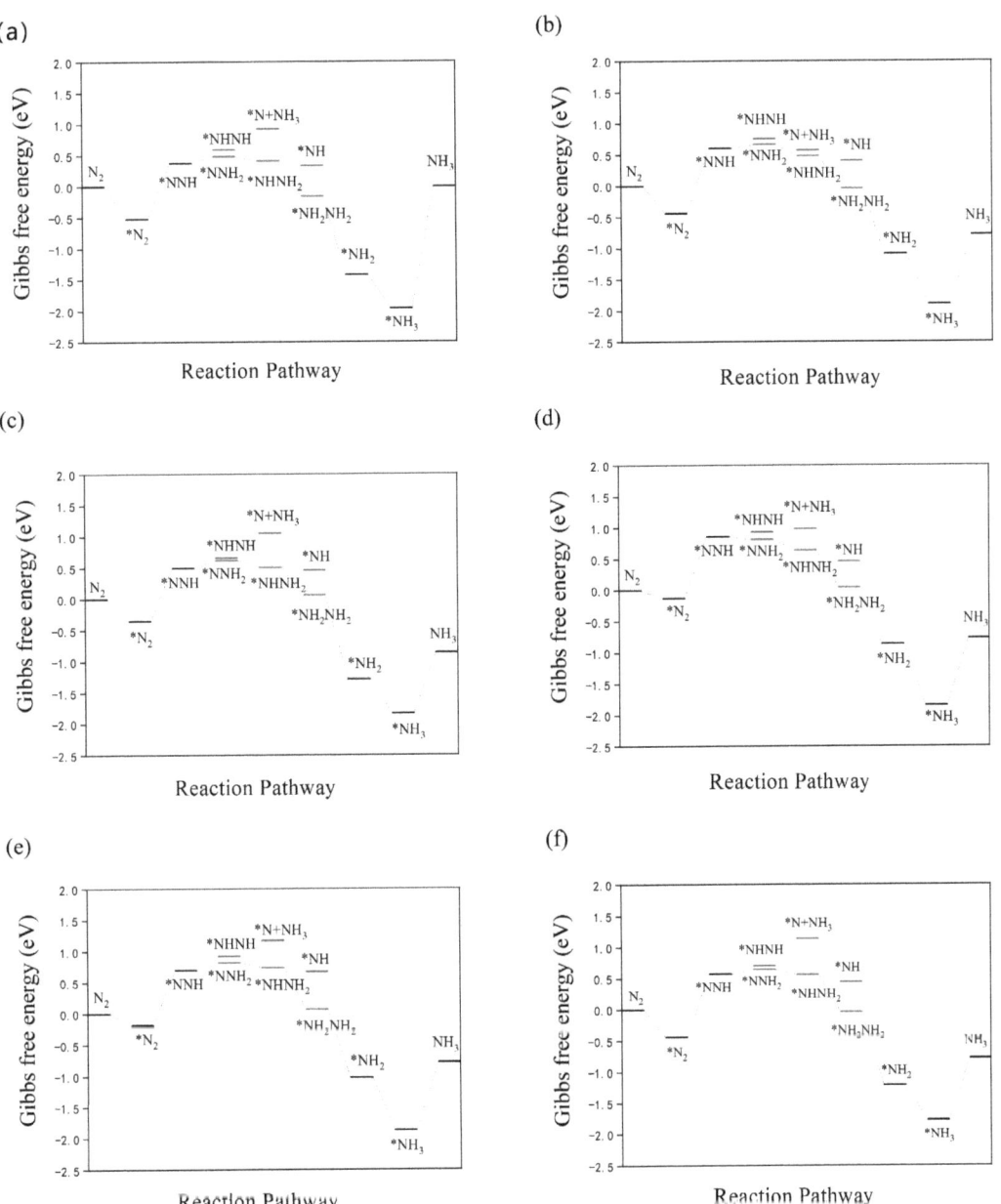

Figure 5. Gibbs free energy diagrams of (**a**) Fe/Ti$_3$C$_2$O$_2$, (**b**) Fe/N-Ti$_3$C$_2$O$_{2-x}$, (**c**) Fe/F-Ti$_3$C$_2$O$_{2-x}$, (**d**) Fe/P-Ti$_3$C$_2$O$_{2-x}$, (**e**) Fe/S-Ti$_3$C$_2$O$_{2-x}$, and (**f**) Fe/Cl-Ti$_3$C$_2$O$_{2-x}$.

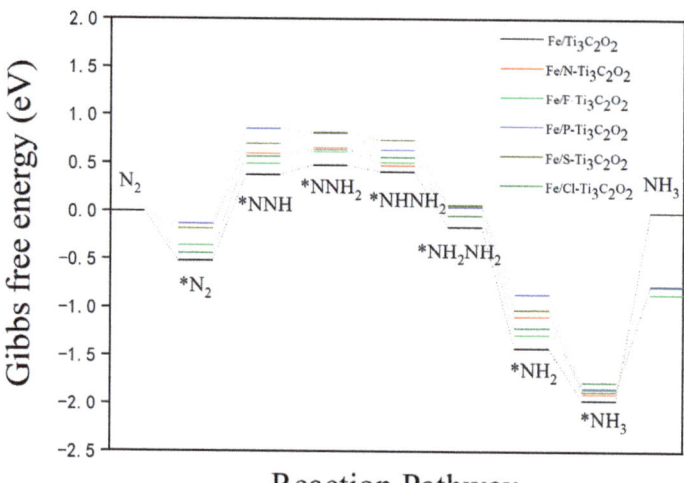

Figure 6. Nitrogen reduction reaction pathways for all structures.

4. Conclusions

The reaction pathway of the TM atom, Fe, adsorbed on pure $Ti_3C_2O_2$ and surface nonmetal (N/F/P/S/Cl)-doped $Ti_3C_2O_2$ as the N_2 reduction reaction catalyst was calculated using DFT. The main limiting steps of the reaction are $*N_2 + H \rightarrow *NNH$ and $*NH_3 \rightarrow NH_3$, and the limiting potentials of the two steps can reach 0.85–1.01 and 0.97–1.95 eV, respectively. Compared with pure $Ti_3C_2O_2$, nonmetal doping has an impact on catalytic performance. The doped nonmetal (N/F/P/S/Cl) reduces the energy barrier to form NH_3 in the last step, and only the doping of F and S is beneficial to the formation of $*NNH$ in the first step and the desorption of $*NH_3$ in the last step. Therefore, the materials doped with F and S are considered better candidate materials for NRR among the tested catalysts. Our research demonstrates a feasible way to search for new NRR catalysts by modifying the surface of MXenes and loading TM atoms as new catalysts.

Supplementary Materials: The following are available online at https://www.mdpi.com/article/10.3390/nano12071081/s1, Figure S1: Top and side views of $Ti_3C_2O_2$ and its nonmetal doped structure, Figure S2: Optimized structures of all the possible elementary steps in NRR, taking $Ti_3C_2O_2$ as an example. Other nonmetal-doped $Ti_3C_2O_2$ show similar geometric structure.

Author Contributions: The study was planned and designed by H.L., X.W., C.W. and P.Q. DFT calculations were performed by H.L. and X.W. The manuscript was prepared by H.L., L.X. and M.S. H.L., X.W., C.W., L.X., M.S. and P.Q. discussed the results and commented on the manuscript together. All authors have read and agreed to the published version of the manuscript.

Funding: This research was funded by the National Key Research and Development Program of China (Grant Nos. 2021YFB0700500 and 2018YFB0704300) and the National Natural Science Foundation of China (Grant No. 11975043).

Institutional Review Board Statement: Not applicable.

Informed Consent Statement: Not applicable.

Data Availability Statement: The datasets generated during and/or analyzed during the current study are available from the corresponding author.

Conflicts of Interest: The authors declare no conflict of interest.

References

1. Van der Ham, C.J.M.; Koper, M.T.M.; Hetterscheid, D.G.H. Challenges in reduction of dinitrogen by proton and electron transfer. *Chem. Soc. Rev.* **2014**, *43*, 5183–5191. [CrossRef] [PubMed]
2. Guo, C.; Ran, J.; Vasileff, A.; Qiao, S.Z. Rational design of electrocatalysts and photo(electro)catalysts for nitrogen reduction to ammonia (NH_3) under ambient conditions. *Energy Environ. Sci.* **2018**, *11*, 45–56. [CrossRef]
3. Zhao, Y.; Zhao, Y.; Shi, R.; Wang, B.; Waterhouse, G.I.N.; Wu, L.Z.; Tung, C.H.; Zhang, T. Tuning Oxygen Vacancies in Ultrathin TiO_2 Nanosheets to Boost Photocatalytic Nitrogen Fixation up to 700 nm. *Adv. Mater.* **2019**, *31*, 1806482. [CrossRef] [PubMed]
4. Smill, V.; Streatfeild, R.A. Enriching the earth: Fritz Haber, Carl Bosch, and the transformation of world food production. *Electron. Green J.* **2002**, *43*, 622–623. [CrossRef]
5. Gao, Y.; Zhuo, H.; Cao, Y.; Sun, X.; Zhuang, G.; Deng, S.; Zhong, X.; Wei, Z.; Wang, J. A theoretical study of electrocatalytic ammonia synthesis on single metal atom/MXene. *Cuihua Xuebao/Chin. J. Catal.* **2019**, *40*, 152–159. [CrossRef]
6. Wang, R. The dynamics of the peel. *Nat. Catal.* **2020**, *3*, 333–334. [CrossRef]
7. Sun, M.; Liu, H.; Qu, J.; Li, J. Earth-Rich Transition Metal Phosphide for Energy Conversion and Storage. *Adv. Energy Mater.* **2016**, *6*, 1600087. [CrossRef]
8. Cong, L.; Xie, H.; Li, J. Hierarchical Structures Based on Two-Dimensional Nanomaterials for Rechargeable Lithium Batteries. *Adv. Energy Mater.* **2017**, *7*, 1601906. [CrossRef]
9. Fu, Q.; Bao, X. Surface chemistry and catalysis confined under two-dimensional materials. *Chem. Soc. Rev.* **2017**, *46*, 1842–1874. [CrossRef]
10. Novoselov, K.S.; Mishchenko, A.; Carvalho, A.; Neto, A.H.C. 2D materials and van der Waals heterostructures. *Science* **2016**, *353*, aac9439. [CrossRef]
11. Shan, A.; Teng, X.; Zhang, Y.; Zhang, P.; Xu, Y.; Liu, C.; Li, H.; Ye, H.; Wang, R. Interfacial electronic structure modulation of Pt-MoS2 heterostructure for enhancing electrocatalytic hydrogen evolution reaction. *Nano Energy* **2022**, *94*, 106913. [CrossRef]
12. Su, Y.; Cao, S.; Shi, L.; Qian, P. Investigation of biaxial strain behavior and phonon-limited mobility for γ graphyne: First-principles calculation. *J. Appl. Phys.* **2021**, *130*, 195703. [CrossRef]
13. Chan, H.; Wang, H.; Song, K.; Zhong, M.; Shi, L.; Qian, P. Origin of phonon-limited mobility in two-dimensional metal dichalcogenides. *J. Phys. Condens. Mat.* **2022**, *34*, 013003. [CrossRef] [PubMed]
14. Suryanto, B.H.R.; Kang, C.S.M.; Wang, D.; Xiao, C.; Zhou, F.; Azofra, L.M.; Cavallo, L.; Zhang, X.; Macfarlane, D.R. Rational Electrode-Electrolyte Design for Efficient Ammonia Electrosynthesis under Ambient Conditions. *ACS Energy Lett.* **2018**, *3*, 1219–1224. [CrossRef]
15. Deng, J.; Liu, C. Boron-Doped Graphene Catalyzes Dinitrogen Fixation with Electricity. *Chem* **2018**, *4*, 1773–1774. [CrossRef]
16. Luo, S.; Li, X.; Zhang, B.; Luo, Z.; Luo, M. MOF-Derived Co_3O_4@NC with Core-Shell Structures for N_2 Electrochemical Reduction under Ambient Conditions. *ACS Appl. Mater. Interfaces* **2019**, *11*, 26891–26897. [CrossRef]
17. Naguib, M.; Kurtoglu, M.; Presser, V.; Lu, J.; Niu, J.; Heon, M.; Hultman, L.; Gogotsi, Y.; Barsoum, M.W. Two-dimensional nanocrystals produced by exfoliation of Ti_3AlC_2. *Adv. Mater.* **2011**, *23*, 4248–4253. [CrossRef]
18. Zhang, P.; Wang, D.; Zhu, Q.; Sun, N.; Fu, F.; Xu, B. Plate-to-Layer Bi_2MoO_6/MXene-Heterostructured Anode for Lithium-Ion Batteries. *Nano-Micro Lett.* **2019**, *11*, 81. [CrossRef]
19. Liu, Y.T.; Zhang, P.; Sun, N.; Anasori, B.; Zhu, Q.Z.; Liu, H.; Gogotsi, Y.; Xu, B. Self-Assembly of Transition Metal Oxide Nanostructures on MXene Nanosheets for Fast and Stable Lithium Storage. *Adv. Mater.* **2018**, *30*, 1707334. [CrossRef]
20. Tang, Q.; Zhou, Z.; Chen, Z. Innovation and discovery of graphene-like materials via density-functional theory computations. *Wiley Interdiscip. Rev. Comput. Mol. Sci.* **2015**, *5*, 360–379. [CrossRef]
21. Liu, A.; Liang, X.; Ren, X.; Guan, W.; Gao, M.; Yang, Y.; Yang, Q.; Gao, L.; Li, Y.; Ma, T. Recent Progress in MXene-Based Materials: Potential High-Performance Electrocatalysts. *Adv. Funct. Mater.* **2020**, *30*, 2003437. [CrossRef]
22. Hu, Q.; Sun, D.; Wu, Q.; Wang, H.; Wang, L.; Liu, B.; Zhou, A.; He, J. MXene: A new family of promising hydrogen storage medium. *J. Phys. Chem. A* **2013**, *117*, 14253–14260. [CrossRef] [PubMed]
23. Wu, X.; Wang, Z.; Yu, M.; Xiu, L.; Qiu, J. Stabilizing the MXenes by Carbon Nanoplating for Developing Hierarchical Nanohybrids with Efficient Lithium Storage and Hydrogen Evolution Capability. *Adv. Mater.* **2017**, *29*, 1607017. [CrossRef] [PubMed]
24. Naguib, M.; Come, J.; Dyatkin, B.; Presser, V.; Taberna, P.L.; Simon, P.; Barsoum, M.W.; Gogotsi, Y. MXene: A promising transition metal carbide anode for lithium-ion batteries. *Electrochem. Commun.* **2012**, *16*, 61–64. [CrossRef]
25. Luo, J.; Tao, X.; Zhang, J.; Xia, Y.; Huang, H.; Zhang, L.; Gan, Y.; Liang, C.; Zhang, W. Sn^{4+} Ion Decorated Highly Conductive Ti_3C_2 MXene: Promising Lithium-Ion Anodes with Enhanced Volumetric Capacity and Cyclic Performance. *ACS Nano* **2016**, *10*, 2491–2499. [CrossRef]
26. Yan, J.; Ren, C.E.; Maleski, K.; Hatter, C.B.; Anasori, B.; Urbankowski, P.; Sarycheva, A.; Gogotsi, Y. Flexible MXene/Graphene Films for Ultrafast Supercapacitors with Outstanding Volumetric Capacitance. *Adv. Funct. Mater.* **2017**, *27*, 1701264. [CrossRef]
27. Zhong, Y.; Xia, X.H.; Shi, F.; Zhan, J.Y.; Tu, J.P.; Fan, H.J. Transition metal carbides and nitrides in energy storage and conversion. *Adv. Sci.* **2015**, *3*, 1500286. [CrossRef]
28. Li, Z.; Wu, Y. 2D Early Transition Metal Carbides (MXenes) for Catalysis. *Small* **2019**, *15*, 1804736. [CrossRef]
29. Zhao, Q.; Zhu, Q.; Miao, J.; Zhang, P.; Wan, P.; He, L.; Xu, B. Flexible 3D Porous MXene Foam for High-Performance Lithium-Ion Batteries. *Small* **2019**, *15*, 1904293. [CrossRef]

30. Sun, J.; Kong, W.; Jin, Z.; Han, Y.; Ma, L.; Ding, X.; Niu, Y.; Xu, Y. Recent advances of MXene as promising catalysts for electrochemical nitrogen reduction reaction. *Chin. Chem. Lett.* **2020**, *31*, 953–960. [CrossRef]
31. Azofra, L.M.; Li, N.; Macfarlane, D.R.; Sun, C. Promising prospects for 2D d^2-d^4 M_3C_2 transition metal carbides (MXenes) in N_2 capture and conversion into ammonia. *Energy Environ. Sci.* **2016**, *9*, 2545–2549. [CrossRef]
32. Li, N.; Chen, X.; Ong, W.J.; Macfarlane, D.R.; Zhao, X.; Cheetham, A.K.; Sun, C. Understanding of Electrochemical Mechanisms for CO_2 Capture and Conversion into Hydrocarbon Fuels in Transition-Metal Carbides (MXenes). *ACS Nano* **2017**, *11*, 10825–10833. [CrossRef] [PubMed]
33. Liu, D.; Zhang, G.; Ji, Q.; Zhang, Y.; Li, J. Synergistic Electrocatalytic Nitrogen Reduction Enabled by Confinement of Nanosized Au Particles onto a Two-Dimensional Ti_3C_2 Substrate. *ACS Appl. Mater. Interfaces* **2019**, *11*, 25758–25765. [CrossRef]
34. Kong, W.; Gong, F.; Zhou, Q.; Yu, G.; Ji, L.; Sun, X.; Asiri, A.M.; Wang, T.; Luo, Y.; Xu, Y. An MnO_2-Ti_3C_2T: X MXene nanohybrid: An efficient and durable electrocatalyst toward artificial N_2 fixation to NH_3 under ambient conditions. *J. Mater. Chem. A* **2019**, *7*, 18823–18827. [CrossRef]
35. Kresse, G.; Furthmüller, J. Efficiency of ab-initio total energy calculations for metals and semiconductors using a plane-wave basis set. *Comput. Mater. Sci.* **1996**, *6*, 15–50. [CrossRef]
36. Kresse, G.; Furthmüller, J. Efficient iterative schemes for ab initio total-energy calculations using a plane-wave basis set. *Phys. Rev. B Condens. Matter Mater. Phys.* **1996**, *54*, 11169. [CrossRef]
37. Perdew, J.P.; Burke, K.; Ernzerhof, M. Generalized gradient approximation made simple. *Phys. Rev. Lett.* **1996**, *77*, 3865–3868. [CrossRef]
38. Blöchl, P.E. Projector augmented-wave method. *Phys. Rev. B* **1994**, *50*, 17953–17979. [CrossRef]
39. Monkhorst, H.J.; Pack, J.D. Special points for Brillouin-zone integrations. *Phys. Rev. B* **1976**, *13*, 5188–5192. [CrossRef]
40. Grimme, S.; Antony, J.; Ehrlich, S.; Krieg, H. A consistent and accurate ab initio parametrization of density functional dispersion correction (DFT-D) for the 94 elements H-Pu. *J. Chem. Phys.* **2010**, *132*, 154104. [CrossRef]
41. Henkelman, G.; Arnaldsson, A.; Jónsson, H. A fast and robust algorithm for Bader decomposition of charge density. *Comput. Mater. Sci.* **2006**, *36*, 354–360. [CrossRef]
42. Sanville, E.; Kenny, S.D.; Smith, R.; Henkelman, G. Improved grid-based algorithm for Bader charge allocation. *J. Comput. Chem.* **2007**, *28*, 899–908. [CrossRef] [PubMed]
43. Momma, K.; Izumi, F. VESTA 3 for three-dimensional visualization of crystal, volumetric and morphology data. *J. Appl. Crystallogr.* **2011**, *44*, 1272–1276. [CrossRef]
44. Ozerov, R.; Kogan, V.; Zhdanov, G.; Kukhto, O. The crystalline structure of solid isotopes of hydroxy. *Sov. Phys. Cryst.* **1962**, *6*, 507–508.
45. Bernal, J.D.; Fowler, R.H. A theory of water and ionic solution, with particular reference to hydrogen and hydroxyl ions. *J. Chem. Phys.* **1933**, *1*, 515–548. [CrossRef]
46. Olovsson, I.; Templeton, D.H. X-ray study of ammonia and ammonia monohydrate. *Am. Crystallogr. Assoc. Progr. Abstr.* **1959**, *12*, 832–836.
47. Johnson, M.W.; Sándor, E.; Arzi, E. The crystal structure of deuterium fluoride. *Acta Crystallogr. Sect. B Struct. Crystallogr. Cryst. Chem.* **1975**, *31*, 1998–2003. [CrossRef]
48. Furberg, S.; Landmark, P.; Gardell, S.; Magnéli, A.; Magnéli, A.; Pestmalis, H.; Åsbrink, S. The Crystal Structure of Phosphorous Acid. *Acta Chem. Scand.* **1957**, *11*, 1505–1511. [CrossRef]
49. Cockcroft, J.K.; Fitch, A.N. The solid phases of deuterium sulphide by powder neutron diffraction. *Z. Krist. New Cryst. Struct.* **1990**, *193*, 1–19. [CrossRef]
50. Sándor, E.; Farrow, R.F.C. Crystal structure of solid hydrogen chloride and deuterium chloride. *Nature* **1967**, *213*, 171–172. [CrossRef]
51. Saal, J.E.; Kirklin, S.; Aykol, M.; Meredig, B.; Wolverton, C. Materials design and discovery with high-throughput density functional theory: The open quantum materials database (OQMD). *JOM* **2013**, *65*, 1501–1509. [CrossRef]
52. Kirklin, S.; Saal, J.E.; Meredig, B.; Thompson, A.; Doak, J.W.; Aykol, M.; Rühl, S.; Wolverton, C. The Open Quantum Materials Database (OQMD): Assessing the accuracy of DFT formation energies. *NPJ Comput. Mater.* **2015**, *1*, 15010. [CrossRef]
53. Rossmeisl, J.; Logadottir, A.; Nørskov, J.K. Electrolysis of water on (oxidized) metal surfaces. *Chem. Phys.* **2005**, *319*, 178–184. [CrossRef]
54. Mashtalir, O.; Naguib, M.; Mochalin, V.N.; Dall'Agnese, Y.; Heon, M.; Barsoum, M.W.; Gogotsi, Y. Intercalation and delamination of layered carbides and carbonitrides. *Nat. Commun.* **2013**, *4*, 1716. [CrossRef] [PubMed]
55. Johnson, L.R.; Sridhar, S.; Zhang, L.; Fredrickson, K.D.; Raman, A.S.; Jang, J.; Leach, C.; Padmanabhan, A.; Price, C.C.; Frey, N.C.; et al. MXene Materials for the Electrochemical Nitrogen Reduction-Functionalized or Not? *ACS Catal.* **2020**, *10*, 253–264. [CrossRef]
56. Gao, G.; O'Mullane, A.P.; Du, A. 2D MXenes: A New Family of Promising Catalysts for the Hydrogen Evolution Reaction. *ACS Catal.* **2017**, *7*, 494–500. [CrossRef]
57. Tong, Y.; He, M.; Zhou, Y.; Zhong, X.; Fan, L.; Huang, T.; Liao, Q.; Wang, Y. Electromagnetic wave absorption properties in the centimetre-band of $Ti_3C_2T_x$ MXenes with diverse etching time. *J. Mater. Sci. Mater. Electron.* **2018**, *29*, 8078–8088. [CrossRef]
58. Wang, X.; Su, Y.; Song, M.; Song, K.; Chen, M.; Qian, P. Design single nonmetal atom doped 2D Ti_2CO_2 electrocatalyst for hydrogen evolution reaction by coupling electronic descriptor. *Appl. Surf. Sci.* **2021**, *556*, 149778. [CrossRef]

59. Liu, D.; Chen, M.; Du, X.; Ai, H.; Lo, K.H.; Wang, S.; Chen, S.; Xing, G.; Wang, X.; Pan, H. Development of Electrocatalysts for Efficient Nitrogen Reduction Reaction under Ambient Condition. *Adv. Funct. Mater.* **2021**, *31*, 2008983. [CrossRef]
60. National Institute of Standards and Technology. Available online: https://janaf.nist.gov/ (accessed on 30 December 2021).
61. Ling, C.; Ouyang, Y.; Li, Q.; Bai, X.; Mao, X.; Du, A.; Wang, J. A General Two-Step Strategy–Based High-Throughput Screening of Single Atom Catalysts for Nitrogen Fixation. *Small Methods* **2019**, *3*, 1800376. [CrossRef]
62. Chun, H.J.; Apaja, V.; Clayborne, A.; Honkala, K.; Greeley, J. Atomistic Insights into Nitrogen-Cycle Electrochemistry: A Combined DFT and Kinetic Monte Carlo Analysis of NO Electrochemical Reduction on Pt(100). *ACS Catal.* **2017**, *7*, 3869–3882. [CrossRef]
63. Clayborne, A.; Chun, H.-J.; Rankin, R.B.; Greeley, J. Elucidation of Pathways for NO Electroreduction on Pt(111) from First Principles. *Angew. Chem.* **2015**, *127*, 8373–8376. [CrossRef]

Article

Improving the Energetic Stability and Electrocatalytic Performance of Au/WSSe Single-Atom Catalyst with Tensile Strain

Shutao Zhao [1], Xiao Tang [2], Jingli Li [3], Jing Zhang [3], Di Yuan [3], Dongwei Ma [4,*] and Lin Ju [3,*]

[1] Key Laboratory of Functional Materials and Devices for Informatics of Anhui Higher Education Institutes, School of Physics and Electronic Science, Fuyang Normal University, Fuyang 236037, China
[2] College of Science, Institute of Materials Physics and Chemistry, Nanjing Forestry University, Nanjing 210037, China
[3] School of Physics and Electrical Engineering, Anyang Normal University, Anyang 455000, China
[4] Key Laboratory for Special Functional Materials of Ministry of Education, School of Materials Science and Engineering, Henan University, Kaifeng 475004, China
* Correspondence: madw@henu.edu.cn (D.M.); julin@aynu.edu.cn (L.J.)

Abstract: In the areas of catalysis and renewable energy conversion, the development of active and stable electrocatalysts continues to be a highly desirable and crucial aim. Single-atom catalysts (SACs) provide isolated active sites, high selectivity, and ease of separation from reaction systems, becoming a rapidly evolving research field. Unfortunately, the real roles and key factors of the supports that govern the catalytic properties of SACs remain uncertain. Herein, by means of the density functional theory calculations, in the Au/WSSe SAC, built by filling the single Au atom at the S vacancy site in WSSe monolayer, we find that the powerful binding between the single Au atom and the support is induced by the Au d and W d orbital hybridization, which is caused by the electron transfer between them. The extra tensile strain could further stabilize the Au/WSSe by raising the transfer electron and enhancing the orbital hybridization. Moreover, by dint of regulating the antibonding strength between the single Au atom and H atom, the extra tensile strain is capable of changing the electric-catalytic hydrogen evolution reaction (HER) performance of Au/WSSe as well. Remarkably, under the 1% tensile strain, the reaction barrier (0.06 eV) is only one third of that of free state. This theoretical work not only reveals the bonding between atomic sites and supports, but also opens an avenue to improve the electric-catalytic performance of SACs by adjusting the bonding with outer factors.

Keywords: single-atom catalyst; Au/WSSe; electrocatalysis; tensile strain

1. Introduction

As a new family of catalysts, single-atom catalysts (SACs) typically offer isolated active sites, great selectivity, and simplicity in separation from reaction systems. They have recently sparked a lot of attention throughout the globe due to their distinctive structural characteristics, which include optimized metal utilization, customized active sites, and astonishing catalytic activities [1–6]. Nevertheless, due to their high surface energy, single-atom sites are prone to sintering and aggregating into thermodynamically stable nanoclusters [7,8]. Sintering must be averted by adding the proper supports to enhance the local coordination environment, electrical characteristics, and strong metal-support interactions.

The recent emergence of Janus two-dimensional (2D) transition metal dichalcogenides (TMDs), which refer to layers with different surfaces (e.g., MoSSe and WSSe), have piqued intense research interest due to their distinctive characteristics and potential energy conversion applications [9–12]. The intrinsic dipole in Janus 2D materials induced by the out-of-plane asymmetric structure could strengthen the adsorption of molecules or atoms on the surface [13], which might result in a better production environment for SACs [14]. Transition-metal adatoms, in turn, could efficiently adjust the Janus TMDs' dipole moments [15]. It has been claimed that, by increasing the inherent dipole, the adsorption of appropriate transition-metal adatoms can cause the interactions between H_2O and MoSSe to transform from weak electrostatic van der Waals (vdW) to powerful chemical bonding, considerably enhancing the adsorption of H_2O molecules and laying the groundwork for photocatalytic water-splitting processes [14].

Additionally, although a range of potential SACs with inexpensive supports are emerging as appealing candidates for heterogeneous catalysis [16,17], it is unclear which precise functionalities and important roles of the supports really impact the catalytic capacities of these SACs. For the purpose of investigating the bonding between atomic sites and supports and the effect of bonding on improving the catalytic performance, in this work, we chose single Au atom and Janus WSSe monolayer to construct SAC, and study the interaction between them by analyzing interfacial transfer electron and electronic orbital coupling. Moreover, we applied external tensile strain to further stabilize the SAC and adjust the electric-catalytic performance for hydrogen evolution reaction (HER).

2. Calculation Method

In this study, we performed the DFT calculations for both geometrical relaxations and electronic structures using the Vienna Ab initio Simulation Package (VASP) program (Version 5.3, Hanger Group, University of Vienna) [18,19]. To represent the electron–ion interaction, we employed the projector augmented wave (PAW) pseudo potentials [20,21]. As the exchange-correlation functional, we selected generalized gradient approximations of Perdew–Burke–Ernzerhof (GGA-PBE) [22]. We placed a 20 Å vertical vacuum interval between each sample and the nearby mirror images to prevent interactions. We used Grimme's DFT-D3 method to deal with the vdW force [12,23]. A $3 \times 3 \times 1$ gamma-pack k-mesh regulated the Brillouin zone. The convergence conditions for the force and energy were 10^{-2} eV/Å and 10^{-5} eV, respectively, with the cutoff energy set at 500 eV. Despite the fact that tungsten is a heavy metal, since the influence of spin-orbit coupling (SOC) on the band gap of the WSSe monolayer has been shown to be minimal [12], we did not use the SOC correction in this instance to conserve computational resources. Additionally, the computational hydrogen electrode (CHE) model was used to perform the Gibbs free energy calculation [24]; meanwhile, the implicit solvent model included in VASPsol was used to account for the solvent effect [25,26]. In the Supporting Information, more Gibbs free energy simulation specifics (as listed in Table S1) are provided.

3. Results and Discussion

Introducing Au single-atom or cluster into electro- and photo-catalysts by doping or adsorbing has been demonstrated as an efficient approach to improve the catalytic performance. For the electrocatalysis, the accession of Au not only increases the conductivity of the catalyst, but also causes a strong electronic interaction at their interface, boosting the electrocatalytic performance of many compounds, such as Au/Ni_3S_2 [27] and Au/TiO_2 [28,29]. As to the photocatalysis, due to the prevention of charge recombination in the area of the space charge, together with the prolonged light absorption caused by the surface plasmon resonance (SPR) effect, loading Au could greatly improve visible light catalytic activity in many systems, such as 2% Au loaded $SnO_2/g\text{-}C_3N_4$ [30] and Au decorated $WO_3/g\text{-}C_3N_4$ [31].

In our work, for the SAC constructed by adding single Au atom on WSSe monolayer, generally, there may be two main configurations. Specifically, one is that the single Au atom adsorbs on the surface of WSSe monolayer (case 1); the other is that the single Au atom replaces one of the atoms on the surface of WSSe monolayer (case 2). Hence, first of all, we confirmed the more favorable configuration by comparing the adsorption energy (E_{ad}) of case 1 with the formation energy (E_{for}) of case 2, respectively. Lower value indicates more stable. After that, we studied the interaction between them through analyzing interfacial transfer electron and electronic orbital coupling. Moreover, we applied external tensile strain with the purpose of further stabilizing the SAC and tuning the electric-catalytic HER performance.

3.1. Single Au Atom Adsorbed Janus WSSe Monolayer

To pursue the energetically stable configuration of SAC built by single Au atom adsorbed Janus WSSe monolayer, as shown in Figure S1, we took two absorption cases into consideration, namely the single Au atom adsorbing on the Se and S sides, respectively. For each absorption case, we examined four possible adsorption sites, namely centre, bond, S (Se), and W sites labeled as C_s, B_s, S_s and W_s for S layer, and C_{se}, B_{se}, Se_{se} and W_{se} for Se layer, respectively. The optimized structures for all these samples are shown in the inset of Figure 1. Based on the total energy of these systems, we found that, on the S layer, the single Au atom tended to stay on the S_s site; meanwhile on the Se layer, it was likely to locate at the C_{se} site. Furthermore, we computed the E_{ad} to estimate the adsorption strength, which is defined as follows,

$$E_{ad} = E_{ad\_sys} - E_{sub} - E_{Au} \quad (1)$$

where E_{ad_sys} and E_{sub} are the total energy of WSSe monolayer with and without single Au atom adsorption. E_{Au} equals to -3.274 eV, which is the total energy of per Au atom in stable Au solid, obtained from the Materials Project database. The E_{ad} of single Au atom adsorbed on Ss and Cse sites are 0.763 eV and 0.834 eV. On the basis of the definition of E_{ad}, the smaller value demonstrates a higher stability. That is to say, the single Au atom prefers to stay at the S layer, which may be explained by the higher electronegativity of S element (5.85 eV) than the one of Se element (5.76 eV) [32]. As shown in Figure 1a, the single Au atom could be fixed by its near S atoms through chemical bonds. However, the positive values of E_{ad} indicate that all these adsorption models are not stable. The single Au atom adsorbed on the surface of WSSe probably aggregates into cluster, which means that single Au atom adsorption is not a feasible way to make up SAC.

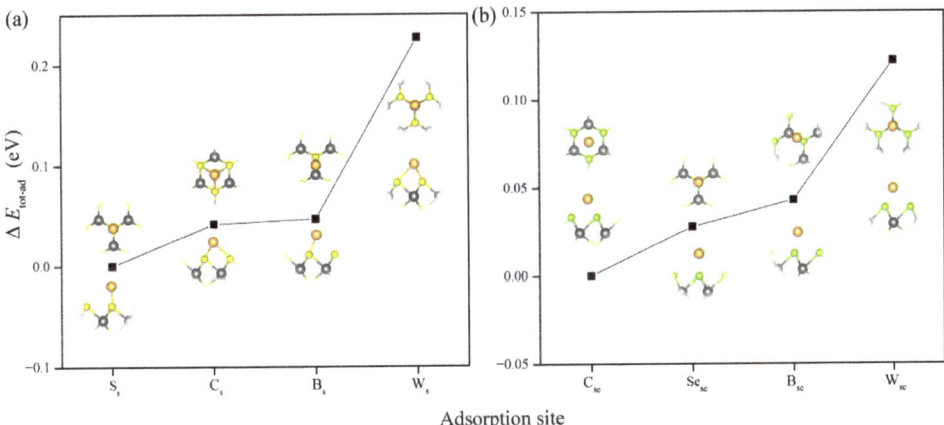

Figure 1. Relative total energy (ΔE) of single Au atom adsorbed WSSe monolayer with the four adsorption sites on the (**a**) S and (**b**) Se layers, respectively. For each case, the energy of the system with the lowest total energy is taken as the reference value. The insets show the top view (above) and the side view (below) of the optimized structures of these systems. The yellow, green, grey, and golden balls represent S, Se, W and Au atoms, respectively.

3.2. Single Au Atom Doped Janus WSSe Monolayer (Au/WSSe)

In the WSSe monolayer, there are three kinds of atoms, namely W, S, and Se atoms, which means that the single Au atom has three choices to replace one atom of the WSSe monolayer. Before this substitution process, a W/S/Se vacancy should first be formed [33].

According to our previous work [34], the formation energy of S vacancy (0.49 eV/atom) is much lower than the ones of Se (3.77 eV/atom) and W (4.97 eV/atom) vacancies. Therefore, herein, we only consider the case that the single Au atom fills in the S vacancy. The E_{for} for this situation is gained by the following equation,

$$E_{for} = E_{doped\_sys} - E_{vac-s} - E_{Au} \qquad (2)$$

where E_{doped_sys} and E_{vac-s} are the total energy of Au/WSSe and WSSe monolayer with one S vacancy, respectively. The calculated E_{for} is −0.55 eV. This negative value means the S vacancy could work as an anchored site to capture a single Au atom, which is similar to the case of Pt single atoms on the nanosized onion-like carbon (Pt$_1$/OLC) [35]. Moreover, by running ab initio molecular dynamics simulations (AIMD) with the Nos'e thermostat model as implemented in VASP, the thermal stability of Au/WSSe was evaluated. The outcome of the AIMD simulations on Au/WSSe is depicted in Figure S2, where the negligible geometric reconstructions and small total energy variations suggest that Au/WSSe may be thermostable at ambient temperature. Besides the Au element, we also considered the single atom of its congeners (i.e., Ag and Cu) to build SACs. However, as shown in Figure S3, both their formation energies were positive, indicating they are not as stable as Au/WSSe. Therefore, we do not discuss the cases of Ag/WSSe and Cu/WSSe further.

To further enhance the stability of Au/WSSe, we applied uniaxial tensile strain to it, which is widely used to tune the morphologic and electric structures of two dimensional materials [10,12,36,37]. As displayed in Figure 2, the calculated results stated that the E_{for} became lower with higher tensile strain. Moreover, as shown in Figure 3a, the single Au atom came closer to the surface of the WSSe monolayer as the tensile strain rose, which could be verified by the lower height (h) of the single Au atom from the W plane (see Figure 3b). Furthermore, based on the calculated results obtained with the climbing image nudged elastic band (CI-NEB) method (see Figure S4), we found that the extra tensile strain could make the diffusion barriers of single Au atom increase (from 2.75 eV at free state to 3.10 eV under 5% tensile strain), which also revealed that introducing uniaxial tensile strain could increase the stability of Au/WSSe.

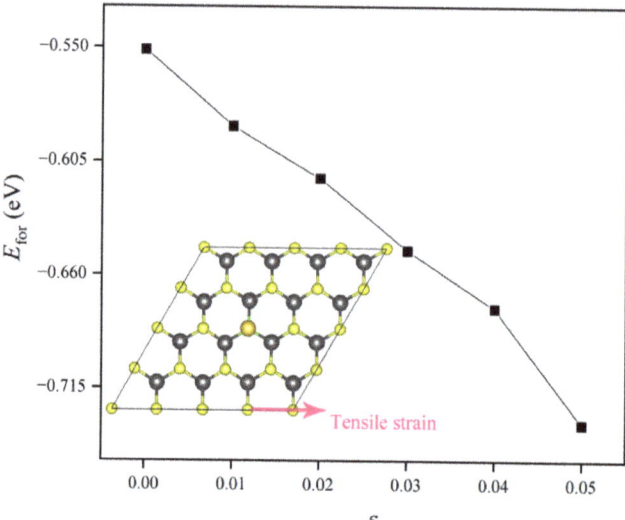

Figure 2. Formation energy of Au/WSSe under different tensile strain.

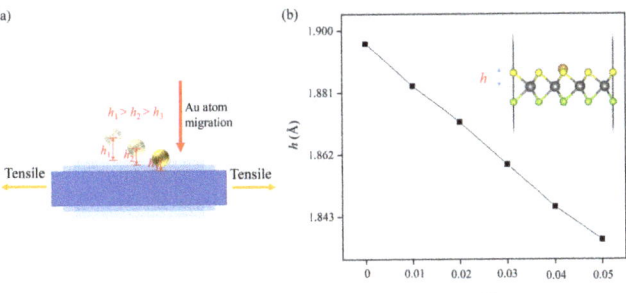

Figure 3. (a) The diagram of single Au atom position change over external tensile stress. (b) The height of the single Au atom from the W plane in Au/WSSe under different tensile strain.

3.3. Interfacial Interaction between Single Au Atom and WSSe Monolayer

As plotted in Figure S5, compared with the case of the pristine WSSe monolayer, the results of total density of states (TDOS) demonstrated that both the adsorption and dopant of a single Au atom could induce obvious spin-splits in the gap states at/near the Fermi level, which caused magnetic moments of 1.00 and 0.94 μB, respectively. As mentioned earlier, the doped system is more stable than the adsorption one; therefore, we focused on the Au/WSSe and explored the origin of its tensile strain-dependent stability by researching the interfacial interaction between the single Au atom and WSSe monolayer, which could be reflected from the electronic orbitals coupling. As plotted in Figure S5c, the TDOS shows that there was one obvious orbital hybridization peak near the Fermi level in each spin direction. In addition, we assessed the accurate electronic structure of Au/WSSe using the Heyd–Scuseria–Ernzerhof (HSE06) hybrid functional [38] to prevent the underestimation of the band gap calculated within PBE. As shown in Figure S6, the gap state near the Fermi level was dominated by the spin-up states, which agreed with the PBE result.

Based on the analysis of PDOS, we found that this orbital hybridization was mainly contributed by the coupling of the d orbitals of the single Au atom and its three nearest neighboring W atoms (see Figure 4a). Numerous additional SACs have also been identified to exhibit the strong contact between atomic species and the nearby atoms on the support [39–41]. Notably, when the external tensile strain was (ε = 5%) applied, as shown in Figure 4b, the orbital hybridization peak near the Fermi level split into two peaks, broadening the coupling energy range, which surely strengthened the interfacial interaction between the single Au atom and WSSe monolayer. Furthermore, the expansion of gap-states indicates an enhanced electric conductivity, which could be confirmed by experimentally measuring the I-V curve.

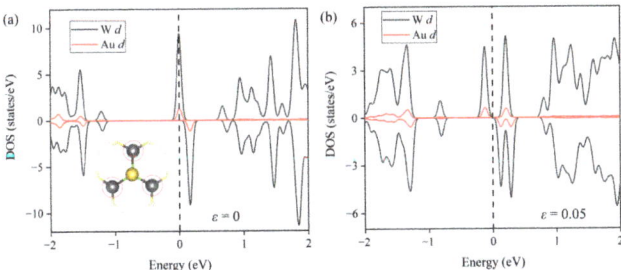

Figure 4. Partial density of states (PDOS) of the d orbitals of the single Au atom and its three nearest neighboring W atoms (circled with the red lines shown in the inset) in Au/WSSe (**a**) without and (**b**) with external tensile strain (ε = 5%).

Generally, orbital hybridization arises with electron transfer. Therefore, in the following work, we studied the electron gained by the single Au atom in the Au/WSSe on the basis of charge density difference ($\Delta\rho$) and Bader charge analysis. The $\Delta\rho$ is defined as follows [42],

$$\Delta\rho = \rho_{Au/WSSe} - \rho_{Au} - \rho_{vac-s}$$

where $\rho_{Au/WSSe}$, ρ_{Au}, and ρ_{vac-s} are the charge densities of Au/WSSe, single Au atom, and the WSSe monolayer with one S vacancy, respectively. From the electron redistribution shown in Figure 5a, it could be easily seen that there was some electron accumulation (pink area) around the single Au atom. The Bader charge analysis demonstrated that the single Au atom received 0.284 e from the support. The electron transfer process was able to separately make the single Au atom and the support become negatively and positively charged. Then, the binding strength between them was reinforced by the electrostatic force originated from the opposite polar components. For the case of Au/WSSe under extra tensile strain (ε = 5%), as illustrated in Figure 5b, the electron accumulation (pink area) around the single Au atom became larger, and more electrons migrated to the single Au atom from the substrate (see Figure 5c), which strengthened the bonding between the single Au atom and the substrate by raising the opposite polarization. Therefore, the Au/WSSe became more stable under larger tensile strain, in line with the results of formation energy shown in Figure 2.

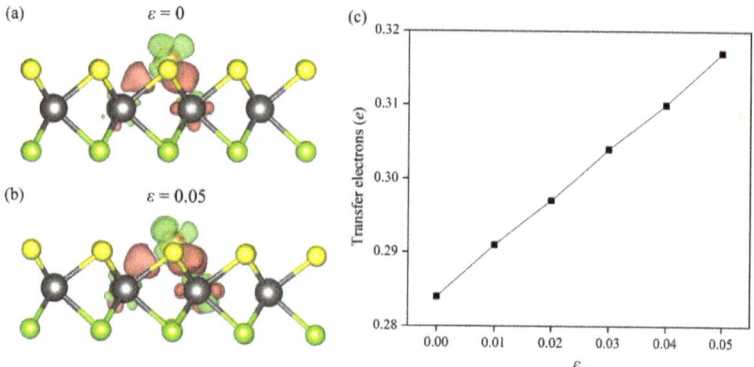

Figure 5. The 3D differential charge density plots of Au/WSSe (**a**) without and (**b**) with external tensile strain. The green and pink regions stand for electron depletion and accumulation, where the isosurfaces are set to 0.004 e/Å^3. (**c**) The charge gained by the single Au atom in Au/WSSe under different tensile strain.

3.4. Tunable Electric-Catalytic Performance of Au/WSSe for HER with Tensile Strain

Due to the high hydrogen adsorption free energy (ΔG_{H*} = 1.82 eV) [43], the pristine Janus WSSe monolayer is inert to the electric-catalytic HER, which is similar to other layered transition metal dichalcogenides [44,45]. Nevertheless, for the Au/WSSe, as shown in Figure 6a, the ΔG_{H*} dropped down to -0.042 eV, due to the high activity of the single Au atom. Unfortunately, the second hydrogenation step (H* → H$_2$*, see Figure 6b) proceeded a bit laboriously (ΔG_{H_2*} = 0.19 eV), because of the over-powerful binding strength of the intermediate H*. Many previous works reported that the electric-catalytic performance of SAC was sensitive to the electron distribution of the single metal atom [1,3,4]. In addition, as stated earlier, the gained electron of the single Au atom in Au/WSSe could be adjusted by the extra tensile strain. Hereby, we calculated the ΔG_{H*} on Au/WSSe under different additional tensile strain. As shown in Figure 6a, the ΔG_{H*} rises as the extra tensile strain becomes heavier.

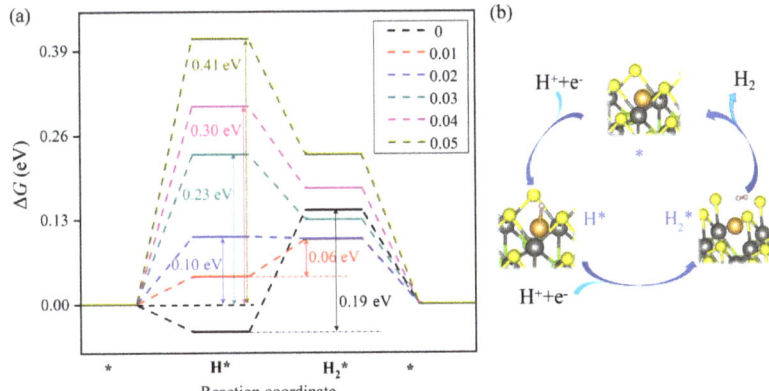

Figure 6. (a) The free-energy profile for electrochemical HER on Au/WSSe under different tensile strain. (b) The optimized configurations of the intermediates in the HER process. * stands for the adsorption site.

One possible explanation for this phenomena mentioned above could be as illustrated in Figure 7. The intrinsic 6s electron of the single Au atom combines with the 1s electron of the H atom to form the bonding orbital, while the gained electron of the single Au atom from the support has to fill the empty antibonding orbital. As the external tensile strain increases, the antibonding orbital is enhanced with more gained electrons of the single Au atom filled, making the adsorption capacity of Au/WSSe for H atom decline. Furthermore, in view of the whole HER, the reaction barriers on Au/WSSe under relative smaller tensile strain (ε = 1% and 2%) are lower than those on the free Au/WSSe. Especially, in the case of ε = 1%, the reaction barrier (0.06 eV) is only one third of the corresponding one on free Au/WSSe, indicating the application of appropriate tensile strain could improve the electric-catalytic performance of Au/WSSe.

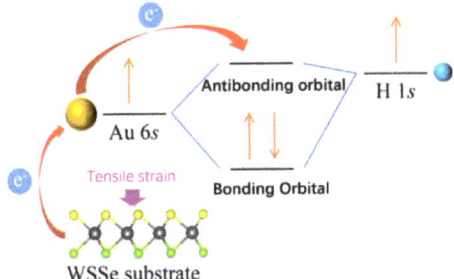

Figure 7. The schematic diagram of the mechanism of external tensile strain affecting adsorption capacity of Au/WSSe for H atom.

4. Conclusions

SACs optimize the usage of metal atoms, which is crucial for supported noble metal catalysts, in particular. Furthermore, SACs have considerable promise for obtaining high activity and selectivity thanks to their homogeneous and well-defined single-atom dispersion. In this work, on the basis of energetic stability, we found that a single Au atom that adsorbs on the surface of perfect WSSe monolayer tends to aggregate; meanwhile, filling the single Au atom at the site of the S vacancy in the WSSe monolayer to build Au/WSSe could keep the single Au atom dispersed, predicting a potential path for fabricating SAC. The powerful binding between the single Au atom and the support in Au/WSSe is induced by the Au d and W d orbital hybridization, which is caused by the electron transfer between them.

The extra tensile strain could further stabilize the Au/WSSe by raising the transfer electron and enhancing the orbital hybridization. Moreover, the extra tensile strain also is able to tune the electric-catalytic HER performance of Au/WSSe by changing the antibonding strength between the single Au atom and H atom. According to the Sebastian principles, the suitable application of extra tensile strain could accelerate the HER rate by reducing the reaction barriers. Especially, with the application of 1% tensile strain, the reaction barrier of HER is merely 0.06 eV, which is smaller than most common electrocatalysts. To sum up, for the first time, our work not only reveals the coupling between the atomic sites and supports in the SAC of Au/WSSe, but also proposes an effective path to improve its electric-catalytic HER performance by tuning the coupling with appropriate tensile strain.

Supplementary Materials: The following supporting information can be downloaded at: https://www.mdpi.com/article/10.3390/nano12162793/s1, Figure S1: The blue dashed circle indicates the adsorption sites considered in our work at the S and Se layers; Figure S2: The *ab initio* molecular dynamics (AIMD) simulations of Au/WSSe for 5 ps with a time step of 1 fs at 300 K; Figure S3: The optimized configurations of Cu/WSSe and Ag/WSSe, as well as the formation energy (E_{for}) for each system; Figure S4: Atomic configurations for the diffusion of the Au single atoms from one favourable doping site to its neighbouring one at the defective WSSe surface at free state and under 5% tensile strain, including initial state, transition state and final state; Figure S5: Calculated projected density of states at PBE level for pristine WSSe monolayer, single Au atom adsorbed WSSe monolayer, and free Au/WSSe; Figure S6: The band structure of Au/WSSe obtained with HSE06 functional; Table S1: Zero-pint Energy Correction, Entropy Contribution, Total Energy, and the Gibbs Free Energy of H* and H$_2$* Adsorbates on Au/WSSe under different tensile strain; Detailed calculation information on the free energy differences of hydrogenation steps for HER [24].

Author Contributions: Supervision, L.J.; project administration, L.J. and D.M.; Software, L.J.; data curation, S.Z.; formal analysis, S.Z. and J.Z.; funding acquisition, L.J., D.Y. and S.Z.; investigation, S.Z., X.T. and J.L.; Writing—original draft, S.Z., D.Y., D.M. and L.J. All authors have read and agreed to the published version of the manuscript.

Funding: This work is supported by Henan Scientific Research Fund for Returned Scholars, the Foundation for University Youth Key Teacher by the Henan Province (Grant No. 2019GGJS190), the Natural Science Research Project and Cultivation Funding for Excellent Youth Scholars of Anhui Province (Grant Nos. KJ2020A0544 and gxgnfx2022030), Open Project of Key Laboratory of Functional Materials and Devices for Informatics of Anhui Higher Education Institutes (Grant No. FSKFKT002), Fuyang Normal University Horizontal Project (Grant Nos. SXHZ202011 and HX2022036), and College Students Innovation Fund of Anyang Normal University (Grant No. 202210479073).

Data Availability Statement: The data presented in this study are available in Supplementary Materials.

Conflicts of Interest: The authors declare no conflict of interest.

References

1. Ju, L.; Tan, X.; Mao, X.; Gu, Y.; Smith, S.; Du, A.; Chen, Z.; Chen, C.; Kou, L. Controllable CO_2 electrocatalytic reduction via ferroelectric switching on single atom anchored In$_2$Se$_3$ monolayer. *Nat. Commun.* **2021**, *12*, 5128. [CrossRef] [PubMed]
2. Hannagan, R.T.; Giannakakis, G.; Flytzani-Stephanopoulos, M.; Sykes, E.C.H. Single-Atom Alloy Catalysis. *Chem. Rev.* **2020**, *120*, 12044. [CrossRef] [PubMed]
3. Zhang, Y.; Yang, J.; Ge, R.; Zhang, J.; Cairney, J.M.; Li, Y.; Zhu, M.; Li, S.; Li, W. The effect of coordination environment on the activity and selectivity of single-atom catalysts. *Coordin. Chem. Rev.* **2022**, *461*, 214493. [CrossRef]
4. Xu, H.; Zhao, Y.; Wang, Q.; He, G.; Chen, H. Supports promote single-atom catalysts toward advanced electrocatalysis. *Coordin. Chem. Rev.* **2022**, *451*, 214261. [CrossRef]
5. Li, Z.; Ji, S.; Liu, Y.; Cao, X.; Tian, S.; Chen, Y.; Niu, Z.; Li, Y. Well-Defined Materials for Heterogeneous Catalysis: From Nanoparticles to Isolated Single-Atom Sites. *Chem. Rev.* **2020**, *120*, 623–682. [CrossRef] [PubMed]
6. Wang, A.; Li, J.; Zhang, T. Heterogeneous single-atom catalysis. *Nat. Rev. Chem.* **2018**, *2*, 65–81. [CrossRef]
7. Su, J.; Ge, R.; Dong, Y.; Hao, F.; Chen, L. Recent progress in single-atom electrocatalysts: Concept, synthesis, and applications in clean energy conversion. *J. Mater. Chem. A* **2018**, *6*, 14025–14042. [CrossRef]
8. Quan, Z.; Wang, Y.; Fang, J. High-Index Faceted Noble Metal Nanocrystals. *Accounts Chem. Res.* **2013**, *46*, 191–202. [CrossRef] [PubMed]

9. Lu, A.Y.; Zhu, H.; Xiao, J.; Chuu, C.P.; Han, Y.; Chiu, M.H.; Cheng, C.C.; Yang, C.W.; Wei, K.H.; Yang, Y.; et al. Janus monolayers of transition metal dichalcogenides. *Nat. Nanotechnol.* **2017**, *12*, 744–749. [CrossRef]
10. Ju, L.; Bie, M.; Zhang, X.; Chen, X.; Kou, L. Two-dimensional Janus van der Waals heterojunctions: A review of recent research progresses. *Front. Phys.* **2021**, *16*, 13201. [CrossRef]
11. Ju, L.; Bie, M.; Shang, J.; Tang, X.; Kou, L. Janus transition metal dichalcogenides: A superior platform for photocatalytic water splitting. *J. Phys. Mater.* **2020**, *3*, 022004. [CrossRef]
12. Ju, L.; Bie, M.; Tang, X.; Shang, J.; Kou, L. Janus WSSe Monolayer: An Excellent Photocatalyst for Overall Water Splitting. *ACS Appl. Mater. Interfaces* **2020**, *12*, 29335–29343. [CrossRef]
13. Ma, X.; Wu, X.; Wang, H.; Wang, Y. A Janus MoSSe monolayer: A potential wide solar-spectrum water-splitting photocatalyst with a low carrier recombination rate. *J. Mater. Chem. A* **2018**, *6*, 2295–2301. [CrossRef]
14. Ma, X.; Yong, X.; Jian, C.-C.; Zhang, J. Transition Metal-Functionalized Janus MoSSe Monolayer: A Magnetic and Efficient Single-Atom Photocatalyst for Water-Splitting Applications. *J. Phys. Chem. C* **2019**, *123*, 18347–18354. [CrossRef]
15. Tao, S.; Xu, B.; Shi, J.; Zhong, S.; Lei, X.; Liu, G.; Wu, M. Tunable Dipole Moment in Janus Single-Layer MoSSe via Transition-Metal Atom Adsorption. *J. Phys. Chem. C* **2019**, *123*, 9059–9065. [CrossRef]
16. Xiong, Y.; Dong, J.; Huang, Z.Q.; Xin, P.; Chen, W.; Wang, Y.; Li, Z.; Jin, Z.; Xing, W.; Zhuang, Z.; et al. Single-atom Rh/N-doped carbon electrocatalyst for formic acid oxidation. *Nat. Nanotechnol.* **2020**, *15*, 390–397. [CrossRef]
17. Chen, Y.; Ji, S.; Chen, C.; Peng, Q.; Wang, D.; Li, Y. Single-Atom Catalysts: Synthetic Strategies and Electrochemical Applications. *Joule* **2018**, *2*, 1242–1264. [CrossRef]
18. Kresse, G.; Furthmüller, J. Efficient iterative schemes for ab initio total-energy calculations using a plane-wave basis set. *Phys. Rev. B* **1996**, *54*, 11169–11186. [CrossRef] [PubMed]
19. Kresse, G.; Furthmüller, J. Efficiency of ab-initio total energy calculations for metals and semiconductors using a plane-wave basis set. *Comput. Mater. Sci.* **1996**, *6*, 15–50. [CrossRef]
20. Blöchl, P.E. Projector augmented-wave method. *Phys. Rev. B* **1994**, *50*, 17953–17979. [CrossRef]
21. Kresse, G.; Joubert, D. From ultrasoft pseudopotentials to the projector augmented-wave method. *Phys. Rev. B* **1999**, *59*, 1758–1775. [CrossRef]
22. Perdew, J.P.; Burke, K.; Ernzerhof, M. Generalized Gradient Approximation Made Simple. *Phys. Rev. Lett.* **1996**, *77*, 3865–3868. [CrossRef]
23. Grimme, S. Semiempirical GGA-type density functional constructed with a long-range dispersion correction. *J. Comput. Chem.* **2006**, *27*, 1787–1799. [CrossRef]
24. Nørskov, J.K.; Rossmeisl, J.; Logadottir, A.; Lindqvist, L.; Kitchin, J.R.; Bligaard, T.; Jónsson, H. Origin of the Overpotential for Oxygen Reduction at a Fuel-Cell Cathode. *J. Phys. Chem. B* **2004**, *108*, 17886–17892. [CrossRef]
25. Ju, L.; Shang, J.; Tang, X.; Kou, L. Tunable Photocatalytic Water Splitting by the Ferroelectric Switch in a 2D AgBiP$_2$Se$_6$ Monolayer. *J. Am. Chem. Soc.* **2020**, *142*, 1492–1500. [CrossRef]
26. Mao, X.; Kour, G.; Zhang, L.; He, T.; Wang, S.; Yan, C.; Zhu, Z.; Du, A. Silicon-doped graphene edges: An efficient metal-free catalyst for the reduction of CO_2 into methanol and ethanol. *Catal. Sci. Technol.* **2019**, *9*, 6800–6807. [CrossRef]
27. Liu, H.; Cheng, J.; He, W.; Li, Y.; Mao, J.; Zheng, X.; Chen, C.; Cui, C.; Hao, Q. Interfacial electronic modulation of Ni_3S_2 nanosheet arrays decorated with Au nanoparticles boosts overall water splitting. *Appl. Catal. B-Environ.* **2022**, *304*, 120935. [CrossRef]
28. Guo, M.; Du, J. First-principles study of electronic structures and optical properties of Cu, Ag, and Au-doped anatase TiO_2. *Phys. B* **2012**, *407*, 1003–1007. [CrossRef]
29. Yang, Z.; Wu, R.; Goodman, D.W. Structural and electronic properties of Au on TiO_2 (110). *Phys. Rev. B* **2000**, *61*, 14066–14071. [CrossRef]
30. Zada, A.; Humayun, M.; Raziq, F.; Zhang, X.; Qu, Y.; Bai, L.; Qin, C.; Jing, L.; Fu, H. Exceptional Visible-Light-Driven Cocatalyst-Free Photocatalytic Activity of g-C_3N_4 by Well Designed Nanocomposites with Plasmonic Au and SnO_2. *Adv. Energy Mater.* **2016**, *6*, 1601190. [CrossRef]
31. Humayun, M.; Ullah, H.; Cao, J.; Pi, W.; Yuan, Y.; Ali, S.; Tahir, A.A.; Yue, P.; Khan, A.; Zheng, Z.; et al. Experimental and DFT Studies of Au Deposition Over WO_3/g-C_3N_4 Z-Scheme Heterojunction. *Nanomicro Lett.* **2019**, *12*, 7. [CrossRef]
32. Ghosh, D.C.; Chakraborty, T. Gordy's electrostatic scale of electronegativity revisited. *J. Mol. Struc. THEOCHEM* **2009**, *906*, 87–93. [CrossRef]
33. Ma, D.; Ju, W.; Li, T.; Zhang, X.; He, C.; Ma, B.; Lu, Z.; Yang, Z. The adsorption of CO and NO on the MoS_2 monolayer doped with Au, Pt, Pd, or Ni: A first-principles study. *Appl. Surf. Sci.* **2016**, *383*, 98–105. [CrossRef]
34. Ju, L.; Liu, P.; Yang, Y.; Shi, L.; Yang, G.; Sun, L. Tuning the photocatalytic water-splitting performance with the adjustment of diameter in an armchair WSSe nanotube. *J. Energy Chem.* **2021**, *61*, 228–235. [CrossRef]
35. Liu, D.; Li, X.; Chen, S.; Yan, H.; Wang, C.; Wu, C.; Haleem, Y.A.; Duan, S.; Lu, J.; Ge, B.; et al. Atomically dispersed platinum supported on curved carbon supports for efficient electrocatalytic hydrogen evolution. *Nat. Energy* **2019**, *4*, 512–518. [CrossRef]
36. Jin, C.; Tang, X.; Tan, X.; Smith, S.C.; Dai, Y.; Kou, L. A Janus MoSSe monolayer: A superior and strain-sensitive gas sensing material. *J. Mater. Chem. A* **2019**, *7*, 1099–1106. [CrossRef]
37. Liu, S.; Yin, H.; Singh, D.J.; Liu, P.-F. Ta_4SiTe_4: A possible one-dimensional topological insulator. *Phys. Rev. B* **2022**, *105*, 195419. [CrossRef]

38. Heyd, J.; Scuseria, G.E.; Ernzerhof, M. Hybrid functionals based on a screened Coulomb potential. *J. Chem. Phys.* **2003**, *118*, 8207–8215. [CrossRef]
39. Wan, G.; Lin, X.-M.; Wen, J.; Zhao, W.; Pan, L.; Tian, J.; Li, T.; Chen, H.; Shi, J. Tuning the Performance of Single-Atom Electrocatalysts: Support-Induced Structural Reconstruction. *Chem. Mater.* **2018**, *30*, 7494–7502. [CrossRef]
40. Yuan, S.; Pu, Z.; Zhou, H.; Yu, J.; Amiinu, I.S.; Zhu, J.; Liang, Q.; Yang, J.; He, D.; Hu, Z.; et al. A universal synthesis strategy for single atom dispersed cobalt/metal clusters heterostructure boosting hydrogen evolution catalysis at all pH values. *Nano Energy* **2019**, *59*, 472–480. [CrossRef]
41. O'Connor, N.J.; Jonayat, A.S.M.; Janik, M.J.; Senftle, T.P. Interaction trends between single metal atoms and oxide supports identified with density functional theory and statistical learning. *Nat. Catal.* **2018**, *1*, 531–539. [CrossRef]
42. Ju, L.; Dai, Y.; Wei, W.; Li, M.; Huang, B. DFT investigation on two-dimensional GeS/WS$_2$ van der Waals heterostructure for direct Z-scheme photocatalytic overall water splitting. *Appl. Surf. Sci.* **2018**, *434*, 365–374. [CrossRef]
43. Er, D.; Ye, H.; Frey, N.C.; Kumar, H.; Lou, J.; Shenoy, V.B. Prediction of Enhanced Catalytic Activity for Hydrogen Evolution Reaction in Janus Transition Metal Dichalcogenides. *Nano Lett.* **2018**, *18*, 3943–3949. [CrossRef]
44. Zhang, J.; Jia, S.; Kholmanov, I.; Dong, L.; Er, D.; Chen, W.; Guo, H.; Jin, Z.; Shenoy, V.B.; Shi, L.; et al. Janus Monolayer Transition-Metal Dichalcogenides. *ACS Nano* **2017**, *11*, 8192–8198. [CrossRef]
45. Li, H.; Tsai, C.; Koh, A.L.; Cai, L.; Contryman, A.W.; Fragapane, A.H.; Zhao, J.; Han, H.S.; Manoharan, H.C.; Abild-Pedersen, F.; et al. Corrigendum: Activating and optimizing MoS$_2$ basal planes for hydrogen evolution through the formation of strained sulphur vacancies. *Nat. Mater.* **2016**, *15*, 364. [CrossRef]

Article

A Stretchable and Self-Healing Hybrid Nano-Generator for Human Motion Monitoring

Yongsheng Zhu [1], Fengxin Sun [1], Changjun Jia [1], Tianming Zhao [2,*] and Yupeng Mao [1,*]

[1] Physical Education Department, Northeastern University, Shenyang 110819, China; 2001276@stu.neu.edu.cn (Y.Z.); 2171435@stu.neu.edu.cn (F.S.); 2071367@stu.neu.edu.cn (C.J.)
[2] College of Sciences, Northeastern University, Shenyang 110819, China
* Correspondence: zhaotm@stumail.neu.edu.cn (T.Z.); maoyupeng@pe.neu.edu.cn (Y.M.)

Abstract: Transparent stretchable wearable hybrid nano-generators present great opportunities in motion sensing, motion monitoring, and human-computer interaction. Herein, we report a piezoelectric-triboelectric sport sensor (PTSS) which is composed of TENG, PENG, and a flexible transparent stretchable self-healing hydrogel electrode. The piezoelectric effect and the triboelectric effect are coupled by a contact separation mode. According to this effect, the PTSS shows a wide monitoring range. It can be used to monitor human multi-dimensional motions such as bend, twist, and rotate motions, including the screw pull motion of table tennis and the 301C skill of diving. In addition, the flexible transparent stretchable self-healing hydrogel is used as the electrode, which can meet most of the motion and sensing requirements and presents the characteristics of high flexibility, high transparency, high stretchability, and self-healing behavior. The whole sensing system can transmit signals through Bluetooth devices. The flexible, transparent, and stretchable wearable hybrid nanogenerator can be used as a wearable motion monitoring sensor, which provides a new strategy for the sports field, motion monitoring, and human-computer interaction.

Keywords: self-powered; sports monitoring; hydrogel; hybrid nano-generator

Citation: Zhu, Y.; Sun, F.; Jia, C.; Zhao, T.; Mao, Y. A Stretchable and Self-Healing Hybrid Nano-Generator for Human Motion Monitoring. *Nanomaterials* **2022**, *12*, 104. https://doi.org/10.3390/nano12010104

Academic Editors: Rongming Wang, Shuhui Sun and Jung Woo Lee

Received: 17 November 2021
Accepted: 28 December 2021
Published: 29 December 2021

Publisher's Note: MDPI stays neutral with regard to jurisdictional claims in published maps and institutional affiliations.

Copyright: © 2021 by the authors. Licensee MDPI, Basel, Switzerland. This article is an open access article distributed under the terms and conditions of the Creative Commons Attribution (CC BY) license (https://creativecommons.org/licenses/by/4.0/).

1. Introduction

The Chinese diving team won gold (silver) medals in the men's and women's ten-meter platform events in the Tokyo Olympic Games in 2020. This excellent performance is related to "Rip entry" technology. Athletes' entry into the water is a fluid-solid coupling problem [1–4]. This technology helps athletes to reduce the impact force when their hands enter into the water in a square shape. On the contrary, when athletes' hands enter the water in a wedge shape, the water shows the characteristic of escaping in the direction of lower pressure. The larger the wedge angle, the higher the wave height of the free liquid surface and the greater the splash [5,6]. Therefore, athletes must change their hands into a square shape immediately before entering the water [7]. In other words, athletes need to complete this technology according to the direction of rotation, the folding speed of the palms, and the orientation of the palms. "Rip entry" motion is not a simple process of moving the palms inward. In the process of training, this technology should be scientifically monitored by coaches and athletes, in order to learn high-quality movements. At present, some researchers have been studying "Rip entry" technology in sports biomechanics theory [8–10]. However, the monitoring of this technology is still an isolated field. First of all, the palms and finger joints are small joints, and a single high-speed camera cannot capture the motion at all [11–13]. In addition, athletes' palms need to remain in a square shape before entering the water. Recent studies show that most wearable sensors can only bend in the longitudinal axis [14,15]. It is necessary to improve the flexibility and stretchability of these sensors for twist motions. Meanwhile, capacitor/resistance sensors present a risk of electricity leaking when the sensors are in the water [16–21]. Therefore, a soft, flexible, and self-powered sensor needs to be developed which can be used to monitor "Rip entry" technology. Moreover, this would also be practical for the scientific training of diving events and would provide some new ideas for other sports technology monitoring such as backhand twisting and pulling action in table tennis, shot put, and so on.

In recent years, more and more attention has been paid to health and sports. With the development of IOTs and big data, a large number of flexible intelligent motion monitoring devices have been integrated. Wang's group created TENG (triboelectric nano-generator) [22–24] and PENG (piezoelectric nano-generator) [25,26]. These devices possess excellent working performance, a simple structure, long-life advantages, and are low cost [27–35]. Some researchers have coupled TENG and PENG together [36–42]. This can transfer the pressure, frequency, and acceleration of irregular and low frequency body motions signal to electronic equipment to achieve wireless transmission. Meanwhile, researchers have considered how TENG and PENG work underwater and the studies have shown that they have a good design structure and function [43–47]. However, the devices are big for body motion monitoring and they have rigid structures.

In this study, a self-powered sensor is prepared which is coupled with PENG and TENG. PVDF is used as the sensitive layer of PENG. TENG adopts the single-electrode method, transparent PTFE is the negative electrode layer of TENG, and the skin is the positive electrode layer. The transparent PTFE and PVDF use the common electrode and the PTFE is not only the negative layer of TENG, but also the protective layer of PENG. Through this combined mode, power generation performance is improved largely. PTSSs can convert mechanical energy into electrical energy, which provides more possibilities for carbon neutralization and peak emission of carbon dioxide. It is noted that hydrogel is used as the electrode of the sensor. As a flexible electrode, the hydrogel has the following characteristics: stretchability, lightweight, small size, high transparency, good biocompatibility, simple production, and low cost. Considering the comfortability of sport monitoring and complex body motion characteristics, the sports monitoring sensor which is composed of a coupling device of TENG, PENG, and stretchable hydrogel electrodes, can be used for body sport training and health monitoring to develop more intelligent and comfortable applications.

2. Materials and Methods

2.1. Materials

Poly (vinylidene fluoride) (PVDF) powder was bought from Qinshang plastic Co., Ltd. (Suzhou, China). N, N-Dimethylformamide (DMF), deionized water, Acrylamide (AM), Lithium chloride (LiCl), N′,N′-methylene diacrylamide (MBA), Ammonium persulphate (APS), and N′,N′,N′,N′-Tetramethylethylenediamine (TMLD) were bought from Jintong letai chemical industry products Co., Ltd. (Beijing, China). Dow Corning 3140 RTV and Svlgard 184 were bought from Xinheng trading Co., Ltd. (Tianjin, China). The transmittance PTFE was bought from Taizhou huafu plastic industry Co., Ltd. (Taizhou, China). The latex was bought from Ji'nan Chuangyuan Chemical Co., Ltd. (Jinan, China).

2.2. Preparation

Fabrication of hydrogel: AM as monomer, MBA as cross-linking agent, APS as initiator, and TMED as catalyst. Firstly, 12 g acrylamide powder and 14 g lithium chloride particles were added in 40 mL pure water to stir using a magnetic stirrer, the magnetic stirrer was set to 600 revolutions per minute. Furthermore, the 4.23 mol/L and 8.24 mol/L of AM and LiCl mixture solution were prepared, respectively. Secondly, after stirring for 5 min, the 0.04 mol/L APS and 0.03 mol/L MBA were added to the mixture solution to be stirred together for 5 min, following this it should stand for 5 min to get the pre solution. Thirdly, three drops of TMEDA were dropped in the pre solution and stirred for 2 min to accelerate commissure for the hydrogel. Finally, the mixture solution was dropped on the culture plate to gain PAAM-LiCl hydrogel (Figure 1c).

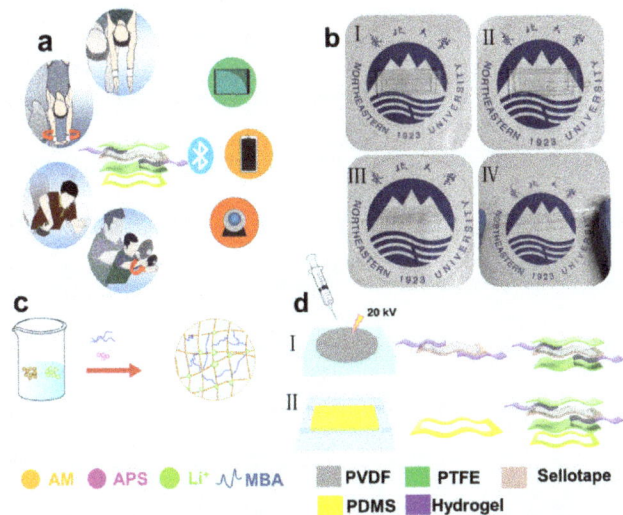

Figure 1. The drawing of the PTSS and its applications. (**a**) Scene of how the PTSS is used for sports. (**b**) Optical image of the PTSS (**I**), TENG (**II**), PENG (**III**), and hydrogel (**IV**). (**c**) Manufacturing step of the hydrogel. (**d**) Manufacturing step of the PVDF film and PDMS substrate and schematic diagram of the PTSS combination.

Fabrication of the PVDF film: Firstly, the 15 wt% P(VDF-TrFE) polymer powder and 85 wt% DMF solution were stirred in a water bath at 60 °C for 4 h. For removing bubbles, semitransparent mixture solution was set in a vacuum drying oven for 30 min. Secondly, the mixed solution was dropped on the bottom silicon rotating surface of the spin coater, which was rotated at 400 rpm for the 20 s. Then, the PVDF/MBA mixture was dried at 80 °C for 15 min. After the second step was repeated three times, the mixture solution was dried at 120 °C for 12 h, and the multilayer film was prepared (Figure 1(dI)).

Fabrication of the PDMS substrate: Firstly, the PDMS solution and solidifying agent were mixed together with a 10:1 weight ratio for 3 min. Secondly, the mixture solution was stirred by an ultrasonic stirrer for 5 min. Thirdly, after being stirred for 5 min, we dropped it on the mold and we put it in the heating furnace at 80 °C for 10 min to get the PDMS substrate film. Finally, the PDMS was cut according to our needs (Figure 1(dII)).

Fabrication of latex film: The latex solution was poured into the glass mold and dried at room temperature for 24 h to get latex film with a thickness of 2 mm.

Fabrication of the whole sensor: Firstly, the PVDF edges were sealed with insulating tape to avoid short-circuiting. This is because when the hydrogel is applied on the PVDF film, the edges of the film are very thin, and if there is no anti-short circuit treatment, the hydrogel contacts with double sides and it can cause short-circuiting. Secondly, double hydrogel electrodes were covered on both sides of the PVDF and then both sides were tightly fixed by transmittance PTFE films. The PENG was packaged by Dow Corning 3140 RTV. Therefore, the hydrogel could not dehydrate in a short time, as this improves the stability and extends its service life. Thirdly, the PMDS was hollowed out at middle position to let the skin contact the PTFE. It was noted that the PDMS should be fixed to the negative dipole side of the PVDF to allow the TENG and PENG to couple together. In the basic test, the latex film was attached to the PMDS substate to replace the skin. In the practical test, the latex was removed and skin contacted the PEFE directly.

2.3. Characterization and Measurement

The PTSS was fixed on the stepping motor to simulate joint movement. The different amplitudes and frequencies were used to test PTSS characteristics. Sensing signals were generated by the PTSS and collected by oscilloscopes (sto1102c, Shenzhen, China). The stress-strain test was carried out by an optical microscope (Scientific compass Co., Ltd., Cmt6103 manufacturer MTS Mets industrial system, Shenyang, China).

3. Results

Diving belongs to the category of difficult and beautiful events The motions of diving are fast and subtle; these motions cannot be captured and analyzed by sports monitoring equipment easily. However, the PTSS can monitor these kinds of motions. The PTSS can be applied to body joint surfaces easily to monitor the subtle changes of athletes. It can meet the requirements of many sports events with difficult skills monitoring. In addition, it can capture the motion signals of diving athletes and transmit them without a power supply in real-time. Meanwhile, for monitoring the violent motions of athletes, the PTSS can be combined with two transparent and stretchable hydrogel electrodes. The electrodes only conduct electricity, but also have excellent stretchability. Figure 1a shows a scene of athletes wearing a PTSS to monitor the motions and sense of signal transmission. In this process, the athletes wear PTSS to monitor angle and frequency changes. Then, the sensing signal generated by the PTSS is transmitted to other electrical equipment. At the same time, the PTSS can convert mechanical energy into electrical energy, which provides more possibilities for carbon neutralization and peak emission of carbon dioxide in the sports monitoring field. Figure 1b shows the optical image of PTSS. The PTSS (I) is made up of TENG (II), PENG (III), and hydrogel electrodes (IV). This sensor has the characteristics of high transparency and high stretchability. The characteristic of high transparency can ensure that the coach can observe detailed motions in order to analyze athletes' motions by combining the sensing signals. It is noted that excellent stretchability is the core of this work. In order to cope with strenuous exercise, we use a soft and stretchable hydrogel electrode to represent the traditional rigid electrode. It can conduct electricity and respond to enormous deformation, ensuring the smooth transmission of sensing signals. Figure 1c shows the manufacturing steps of the hydrogel. AM, MBA, APS, and TMED are mixed in a certain proportion to form a hydrogel electrode. Figure 1d shows the manufacturing steps of the PTSS. The PTSS is composed of PENG and TENG. Step I shows the manufacturing steps of PENG. In this process, PVDF powder is manufactured into PVDF film. Then, the PVDF film is covered with the anti-short circuit treatment. Finally, the double hydrogel electrodes are fixed on the PVDF by the PTFE. Step II shows the manufacturing steps of the PDMS. The PDMS solution and curing agent are mixed according to the weight ratio of 10:1, then bubbles are removed, and the mixed solution is heated and molded. Then, the PDMS film is cut into a hollow structure and attached to PENG so that the PTFE substrate of PENG can rub against the skin.

Figure 2 shows the coupling mechanism of TENG and PENG. TENG and PENG use a common electrode. The TENG system adopts a single electrode mode. The coupling mechanism can be divided into four sections. When the athlete is in a static state, the PTSS is in a free state with no external force and no electrical output (State I and Figure 2(bI)). When an external force is applied to the PTSS, the deformation happens on the PENG firstly, and the piezoelectric signal is produced (State II and Figure 2(bII)) with the external force increasing continuously. The skin/latex contacts the PTFE, the skin/latex tends to lose electrons, while the PTFE tends to acquire electrons, electrons transfer from the skin to the PTFE. The negative side of the PVDF and negative layer of TENG share the same electrode. The contact electrification and piezoelectricity occur simultaneously. Therefore, the electrons of the PENG and TENG move in the opposite direction on the circuit (State III and Figure 2(bIII)). When the external force disappears, the electrons flow back. The electrons of the PTFE flow to the ground. Then, the electrons of PENG and TENG move on the circuit in opposite directions again (State IV and Figure 2(bIV)). This unique working mechanism can be well applied to sports. Many sports are periodic, thus, many motion structures can be directly coupled with this mechanism in sports. Therefore, PTSS can be applied to human motion monitoring perfectly. Figure 2c shows the SEM (HITACHI S-4800) image of PENG. The Figure 2(cI) shows the side-view SEM image of the whole device. The PTSS is packaged by the PTFE film. Figure 2(cII–cIV) are the enlarged view of Figure 2(cI). There is no gap between the hydrogel and the PVDF layer. Hence, the PENG unit can hardly be influenced by the triboelectric signal.

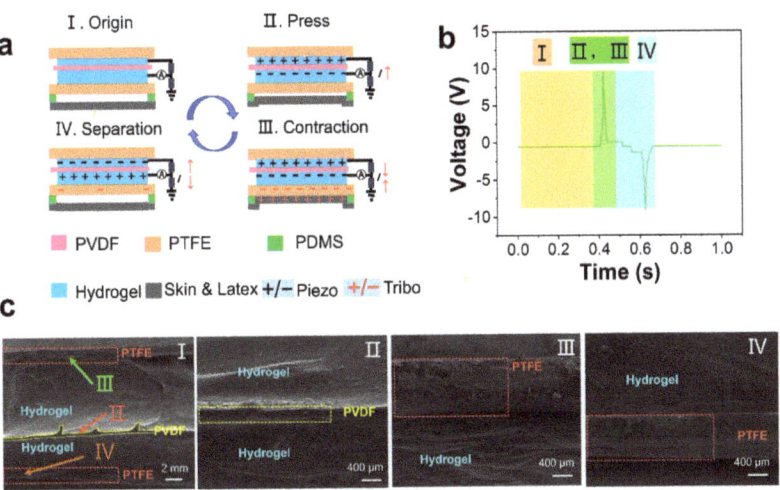

Figure 2. The mechanism of PTSS and SEM of PENG. (**a**) Coupling mechanism of TENG and PENG. (**b**) Corresponding signal. (**c**) The SEM image of PENG.

In this study, the outputting voltage is the major factor in the monitoring process. We injected water mist into the airtight box and the humidity achieved 100%. Figure S1 shows the cross-sectional microscopic image of the PTSS and three views of the device. It can be clearly seen that there is no air gap in the PENG unit, which can avoid the influence of the triboelectric effect between the hydrogel and the PVDF. The device is packaged by Dow Corning 3140 RTV. Therefore, the hydrogel cannot dehydrate in a short time, as this improves the stability and extends its service life. A stepper motor was used to test the outputting voltage of PENG, TENG, and PTSS at the 100% RH environment (Figure S2) and air, respectively. The average outputting voltage of PENG, TENG, and PTSS was 3.96 V, 3.08 V, and 8.6 V, respectively in Figure 3a–c. Figure S3 shows the data measured by PENG, TENG, and PTSS under the air condition and tested force, no deformation happened to them. We applied a tiny force to them; the applied force was 15 N. It is evident that PENG and TENG can be coupled together. Compared with PENG and TENG, the PTSS has an excellent sensing performance. Some movements cannot be detected by a single device. TENG cannot monitor the twist movement clearly. The outputting and response of PENG are low. The PTSS combines the merits of TENG and PENG. It can ensure a high output and accurate detection under complicated conditions such as twisting and bending. In addition, PTSS can increase the range of movement monitoring. We tested the 360° wrist rotation by PENG, TENG and PTSS in the Figure S4. Compared to single PENG and TENG, the PTSS can monitor more detailed signals. It is a device that can monitor various movement states. It can help the coach to better analyze the signal. The coach can observe clearer signals in the monitoring process. Figure 3d–f shows the power and outputting voltage of the PENG unit, the TENG unit, and the PTSS at different load resistances. As the resistance increases, the outputting voltage also increases. When the load resistances are 2.5 MΩ, 6 MΩ, and

9 MΩ, the power of PTSS, TENG, and PENG reaches the maximum. Therefore, the inherent resistance of PTSS, TENG, and PENG are 2 MΩ, 6 MΩ, and 9 MΩ, respectively.

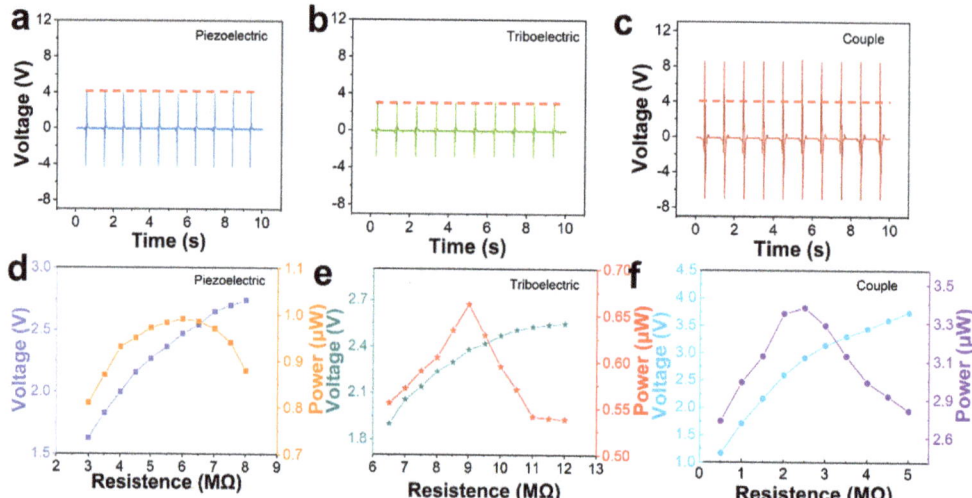

Figure 3. Electrical properties of PENG, TENG, and PTSS in a 100% RH environment. (**a–c**) Comparison of output voltages of PENG, TENG and PTSS. (**d–f**) Power and outputting voltage of PENG, TENG, and PTSS at different load resistances.

Hydrogel is used as an electrode; hydrogel has excellent characteristics and can be well used for motion monitoring. We tested the resistance of hydrogel and found that with the stretching of hydrogel, its resistance increases. When the hydrogel is stretched from 2 cm to 12 cm, the resistance changes from 54.5 kΩ to 96.8 kΩ (Figure S5). Even if the resistance of the hydrogel changes with stretching, the inherent resistance of the PTSS is much larger than that of the hydrogel. Therefore, the change in hydrogel resistance does not affect the outputting voltage. Compared with a metal electrode, the hydrogel electrode has a similar sensing signal. The hydrogel has good stretchability, biocompatibility, and comfort properties and these are better than those of the metal electrode (Figure S5). Figure 4a shows the transparency of the PTSS and each part of it. The average transmittance of hydrogel for the PDMS, PTFE, PVDF, and PTSS are 96.06%, 93.55%, 92.92%, 90.79%, and 76%, respectively. The above data demonstrate that the PTSS has high transmittance, thus a coach can observe detailed motions conveniently. Figure 4b shows the stress-strain curve of different APS ratios of the hydrogel. We found that although the proportion of APS is very small, the content of APS is an important index that affects hydrogel stretchability and solidification time. The stress-strain curves were tested by changing the APS content to 0.01 mol/L, 0.03 mol/L, 0.05 mol/L, and 0.07 mol/L, respectively, while the concentration of other materials remained constant. We found that the hydrogel with a concentration of 0.01 mol/L APS cannot be cured (Figure S6). According to the results, the 0.03 mol/L APS hydrogel has the best tensile property which can be stretched to 2317.53% at least. This is because the maximum range of the machine is 2317.53%. The 0.05 mol/L and 0.07 mol/L APS hydrogel are 1815.21% and 1400.93%, respectively. The optical images of the tensile property are shown in Figure S7. The excellent tensile properties can prevent the electrode from being damaged. It ensures the continuous operation of the motion monitor. Figure 4c,d show that hydrogels can be stretched in many structures. In sport monitoring, in order to meet the needs of sports, hydrogels can be equipped with different structures to reduce the impact on athletes.

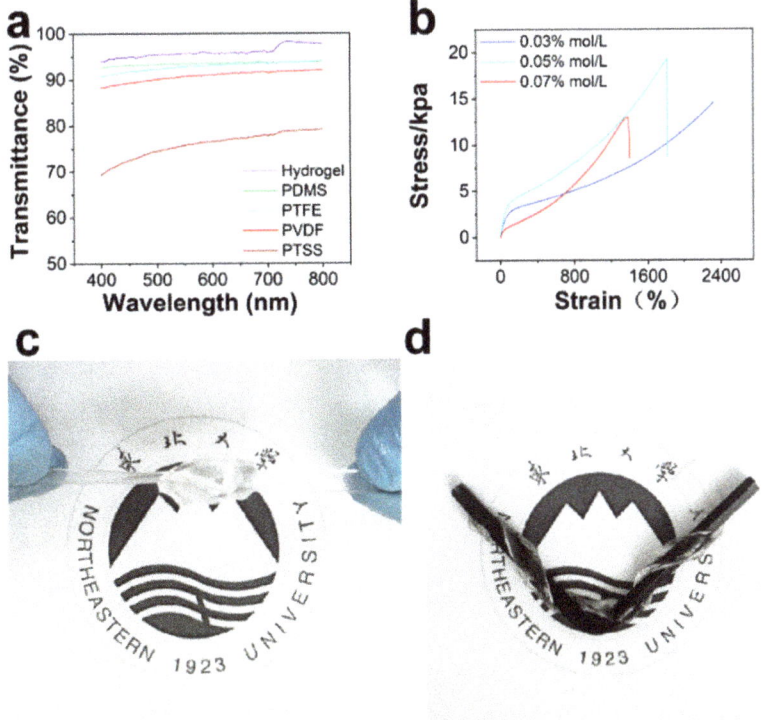

Figure 4. Physical properties of hydrogel (**a**) Transparency of each part of the PTSS. (**b**) Stress-strain curve of different APS ratios of hydrogel. (**c**,**d**) Scene of stretching hydrogel in any shapes.

As a motion monitoring sensor, its sensing accuracy and size, including bend angle, frequency, and twist angle should be taken into account. Usually, water sports are monitored under high humidity conditions. We did the above experiments in 100% RH (Figure S2), all of which were tested using a stepping motor. Figure 5a shows the PTSS outputting voltage at different bending angles. When the bend angle is 7.53°, 12.63°, 17.84° and 23.19°, the outputting voltage of PTSS in 100% RH is 8.6 V, 9.9 V, 11 V, and 11.6 V. The data measured under air condition is shown in Figure S8. Figure 5b shows the linear relationship between angles and voltages of the PTSS in the 100% RH environment. The Pearson coefficient is 0.98606, which indicates that it has a good linear relationship. The formula is:

$$V \approx 7.32 + 0.19 \times \text{degree} \tag{1}$$

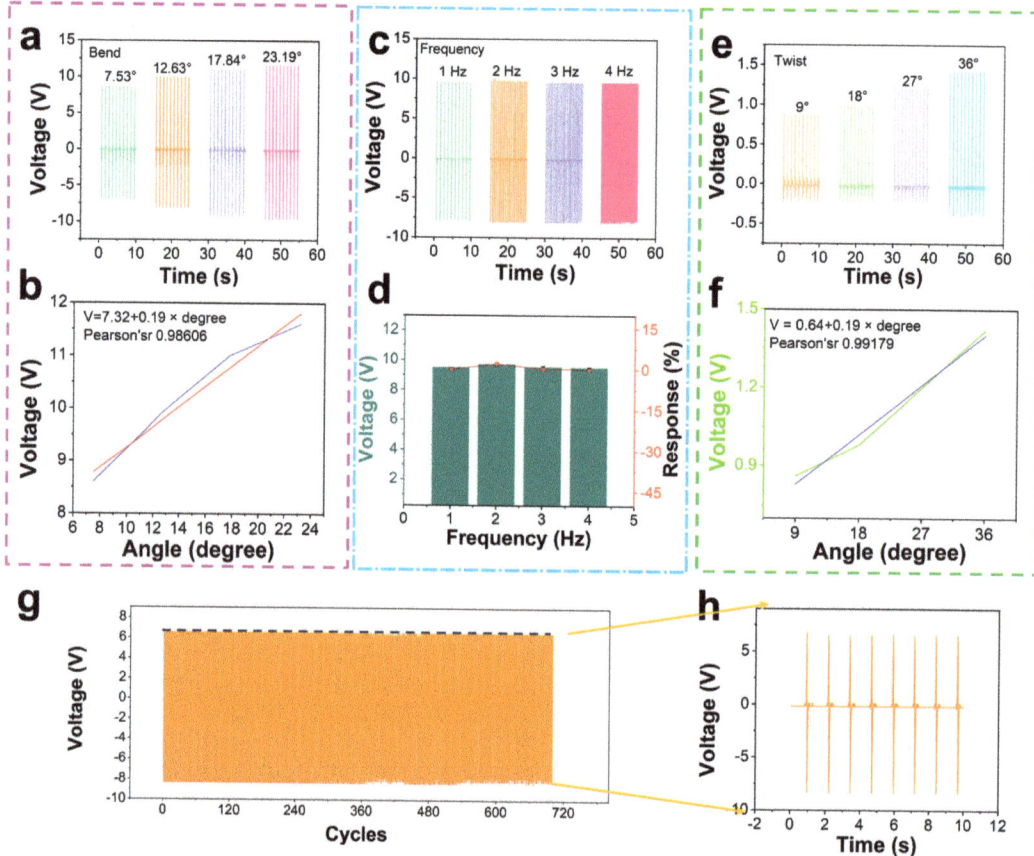

Figure 5. The performance test of PTSS. (**a**) Outputting voltage of PTSS at different bend angles in 100% RH. (**b**) The linear relation of angles and voltages. (**c,d**) Outputting voltage and response of PTSS at different frequencies in 100% RH. (**e**) Outputting voltage of PTSS at different twist angles in 100% RH. (**f**) Linear relation of twist angles and voltages. (**g**) Durability test of PTSS. (**h**) The detail of durability.

To study the relation of bend angle, frequency, and twist angel with output voltage. The response of the PTSS can be calculated from the following equation:

$$R\% = \left| \frac{V_0 - V_i}{V_i} \right| \times 100\%, \quad (2)$$

where V_0 and V_i are the outputting voltage of 7.53° (first data) and other angles. When the bend angle is 7.53°, 2.63°, 7.84°, and 23.19°, the response of the PTSS is 0, 13.13%, 21.82%, and 25.86% (Figure S9). Figure 5c shows the outputting voltage of the PTSS at the same angle and at different frequencies. When the frequency is 1 Hz, 2 Hz, 3 Hz, and 4 Hz, the outputting voltage is 9.5 V, 9.7 V, 9.52 V, and 9.51 V. The data measured under air condition has been shown in Figure S8. Figure 5d shows the responses of the PTSS at the same angle and different frequencies. The two dielectric plates, with thicknesses of d1 and d2 and the relative dielectric constants ε_{r1} and ε_{r2}, respectively, are stacked face to face as two triboelectric layers. At the outer surface of the PTFE dielectric, a hydrogel layer is deposited as an electrode. The distance (x) between the two triboelectric layers can be varied under the agitation of mechanical force. After being forced to get in contact with each other, the inner surfaces of the two triboelectric layers will have opposite static charges (tribo-charges) with equal density of σ, as a result of contact electrification. For insulators, as discussed, it is reasonable to assume that the tribo-charges are uniformly distributed along the two surfaces with negligible decay. When the two triboelectric layers start to separate from each other, with increased x, a potential difference (V) between the two electrodes will be induced. The amount of transferred charges at the electrode, as driven by the induced potential, is defined as Q which also represents the instantaneous amount of charges on the electrode. With the above model, the V–Q–x relationship of such contact-mode TENG can be derived based on electrodynamics. Since the area size (S) of the PTFE and skin is several orders of magnitude larger than their separation distance ($d_1 + d_2 + x$) in the experimental case, it is reasonable to assume that the two electrodes are infinitely large. Under this assumption, the charges will uniformly distribute on the inner surfaces of the PTFE and skin. Inside the dielectrics and the air gap, the electric field only has the component in the direction perpendicular to the surface, with the positive value pointing to hydrogel. From the Gauss theorem, the electric field strength at each region is given by:

$$\text{Inside Dielectric skin}: E_1 = -\frac{Q}{S\varepsilon_0\varepsilon_{r1}} \tag{3}$$

$$\text{Inside the air gap}: E_{air} = \frac{-\frac{Q}{S} + \sigma(t)}{\varepsilon_0} \tag{4}$$

$$\text{Inside Dielectric PTFE}: E_2 = -\frac{Q}{S\varepsilon_0\varepsilon_{r2}} \tag{5}$$

The voltage between the two electrodes can be given by:

$$V = E_1 d_1 + E_2 d_2 + E_{air} x \tag{6}$$

Therefore, the outputting voltage is dependent on the surface charge density and the motion frequency depends on the voltage frequency [48,49]. When the frequency is 1 Hz, 2 Hz, 3 Hz, and 4 Hz, the response is 0, 2%, 0.2%, and 0.05%, respectively (Movie S1). The response demonstrates that when the bend angle is fixed, the motion frequency changes and the outputting voltage is stable. The frequency of the voltage peak occurrence is the same as that of motion. Therefore, PTSS can monitor the motion frequency with excellent performance. For example, in the short race, step frequency and step length are the absolute factors of the competition. Monitoring the step frequency at the starting stage, running stage, and sprint stage is a vital measure of an athlete's performance. The PTSS can monitor the frequency of every stage to provide visual data. Figure 5e shows the outputting voltage of the PTSS at different twist angles. When the twist angles are 9°, 18°, 27°, and 36°, the average voltages of the PTSS are 0.86 V, 0.98 V, 1.2 V, and 1.42 V, respectively. The data measured under air condition is shown in Figure S8. Figure 5f shows the linear relation of twist angle and voltage. The Pearson coefficient is 0.99179, which demonstrates that there is a good linear relationship. The formula is:

$$V \approx 0.64 + 0.19 \times \text{degree} \tag{7}$$

When the twisted angel is 9°, 18°, 27°, and 36°, the response is 0, 12.24%, 28.33%, and 39.43% (Figure S9). Figure 5g shows the durability test for PTSS. PTSS still has excellent outputting performance after working at a big bend angle for 720 cycles. It demonstrates that PTSS can work in violent motions. Meanwhile, we pressed the hydrogel for 3400 cycles, and it also had an excellent working performance (Movie S2). Figure 5h shows the details of durability. It shows that the device can maintain a stable sensing characteristic.

To meet the extreme monitoring conditions, we tested the performance of PTSS in the water (Movie S3). Figure 6a shows the outputting voltage of the wrist bending before entering the water. Before entering the water (Figure 6b), the PENG and TENG of the PTSS worked together, with an average voltage of 2.45 V, after entering the water, the outputting voltage dropped, with an average voltage of 1.25 V (Figure 6c). This is because the friction layer of the PTSS directly rubs against the skin, and the water makes electrons flow to the ground with the water. Therefore, TENG does not work at this time. However, the PENG is tightly fixed by the PTFE. The PENG has the excellent property of being waterproof. Figure 6d–f shows the biocompatibility of the PTSS. The PTSS was attached to the skin for 6 h, and there was no rejection reaction on the athlete's wrist. This demonstrates that PTSS has excellent biocompatibility and it can monitor the motion of athletes for a long time. Figure 6g–i shows the self-healing property of hydrogel. In the violent motion, the electrode may be damaged. The hydrogel can meet this problem because it has a good tensile ability and self-healing properties. Figure 6g shows that hydrogel can be used as a wire way to light the LED, and it has good stability and self-healing properties. After the hydrogel is cut, it can also be connected to conduct electricity (Figure 6h). At the same time, it also has a good stretching ability to the motion display (Movie S4), and hydrogels have self-healing properties and can be stretched 70.11% after repair (Figure S10). Due to the self-healing function of hydrogel, the hydrogel electrode can recover by itself even though the damage occurs at a high strength impact. It improves the stability and extends the service life [50]. It is noted that stretchability like this cannot meet the intense exercise demand, but it can be applied to many static motions such as weightlifting.

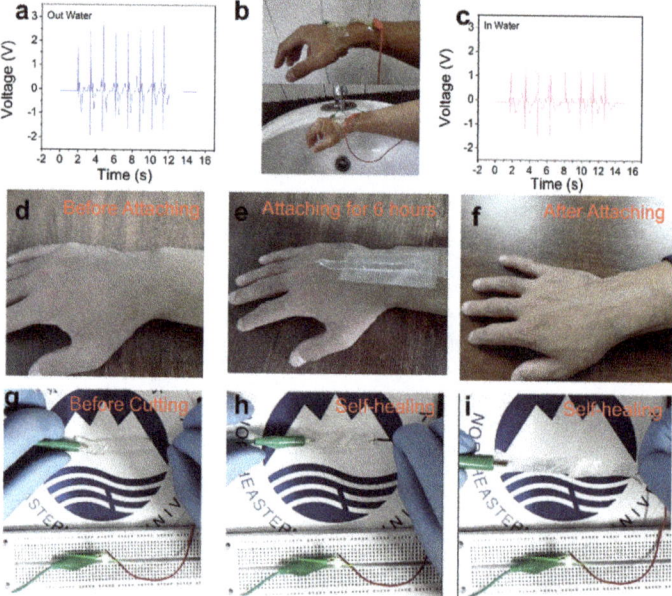

Figure 6. Scenes of waterproof, biocompatibility, and self-healing. (**a**–**c**) Test of waterproof ability. (**d**–**f**) Test of biocompatibility. (**g**–**i**) Self-healing property of hydrogel.

The technical motions of diving are difficult and highly ornamental. The referees score for the athletes' performance, including their handstands or upright preparation postures, aerial skills, the number of somersaults, and the spray size at the moment of entering the water. At present, the water rip entry motion of the world's top athletes is to cross their hands to form a square shape, and the water spray formed by this technical motion is small. In order to better monitor the athletes' technique of rip entry motion, we must monitor their wrist motions firstly, which are wrist bending, twisting, and rotation. The voltage generated by a series of motions is shown in Figure 7a,b. Figure 7a,b shows the outputting voltage and details of wrist bending motion, and its specific motions are shown in Movie S5. The bending of the athlete's wrist produces an upper wave peak I, and the straightening of the athlete's wrist produces a lower wave peak II. The reason why wave peak I is significantly greater than wave peak II is that the speed and strength of the bending motion are greater than those of the straightening motion. This kind of motion monitoring is also suitable for shotput wrist motion monitoring. Figure 7c,d shows twist motions and details of the twist, its specific motions are shown in Movie S6. The rotation motion of the wrist includes internal rotation and external rotation. In Figure 7d, part I is the internal rotation motion and part II is the external rotation motion. As shown in Figure 7d, wave peak I is significantly larger than wave peak II because the speed and strength of the internal rotation motion are greater than those of the external rotation motion. This kind of motion monitoring is also applicable to the wrist-twisting and pulling motion monitoring of table tennis. We believe that it will be an important research direction to apply this monitoring method to the stability monitoring of athletes' rotation technical motion. Figure 7e,f shows the rotation motion and details of the wrist. Part I of Figure 7f is the internal rotation motion and part II is the external rotation motion. Wave peak I is significantly greater than wave peak II, because the speed and strength of internal rotation motion are greater than those of external rotation motion (Movie S7). This kind of motion monitoring is also applicable to the wrist motion monitoring of diving athletes. Based on the limitations of this research, our group cannot quantitatively analyze its speed and strength, which is also a direction for our next study. Figure 7g–j show the motions and details of two athletes simulating 301C diving on land (Movies S8 and S9). Figure 7h,j show signal waveforms of two athletes in one motion cycle. Overall, the waveforms of the two athletes are similar, but there are also some differences. Figure 7j shows a 301C motion with a knee hugging reverse somersault for half a cycle and a difficulty coefficient of 1.8. 301C motion is divided into four stages. Stage I is the round body motion of holding the knees with both hands, and a certain angle of the wrist. As shown in Figure 7j, Athlete 2 has a large and complete knee hugging motion in stage I. Stage II is a tight state in which the arms are quickly extended from the knee hugging state to both sides of the body. At this time, the wrist experiences bending from holding the knee to straightening. Stage III is the water entry stage. The arms quickly move from both sides of the body to the top of the head, and the hands form a square shape. At this time, the wrist needs to rotate inward from the straight state to the square state. After entering the water, the wrist naturally returns to the normal state. The operation signal at this time and Figure 7f are the voltage generated by the same operation. Stage IV is the stage of re-embracing the knee, and the motion is the same as that in stage I. In addition, we have also monitored the twisting and pulling motion of table tennis (Movie S10), and the sensing signal is shown in Figure S11. It shows that PTSS can monitor multiple motions. Figure 7k,l are a schematic diagram of the wireless Bluetooth transmission system. The sensor is connected to a signal transmitting module. When the transmitting module receives the signal, it transmits the signal to the receiving module, and the LEDs are controlled by it. When the athlete is in a static state, there is no signal and Bluetooth only lights up one light, corresponding to the red circle state in Figure 7h. In contrast, when the athlete is in a dynamic state, the signal is produced and Bluetooth lights up three lights, corresponding to the black circle state in Figure 7h. The encapsulated device connects with the Bluetooth module. In the Bluetooth communication test, we only put the PTSS into the water. In this study, the main aim was to design a sensor which can be used to monitor the human movement underwater. The experiment showed that PTSS can monitor the bending and twist signal underwater. We have encapsulated the device and Bluetooth together. It will be introduced in subsequent work. Figure 7m shows the real-time wireless Bluetooth waveform display system (Movie S11). In the process of training, if we can collect information through mobile phones, it will be conducive to more direct monitoring of athletes' training status. Therefore, we built a wireless monitoring system composed of a hybrid nano-generator, digital multimeter with Bluetooth module and collection to demonstrate the possibility of human motion monitoring (Figure 7m). Wearable mixed-mode sensors can collect and process human motion data through a digital multimeter, and a mobile app can monitor the voltage in real-time. We have proved that it can be applied to human motion monitoring through a simple impact test. Meanwhile, it can also charge capacitor. Figure S12 shows curves of the PTSS charges capacitors. The PTSS charges are 0.1 µF, 0.22 µF, and 0.47 µF and the charging voltages are 2.05 V, 1.79 V and 0.86 V, respectively. The PTSS can not only generate electricity by itself, but can also convert mechanical energy into electrical energy. This is of great significance to carbon neutralization and peak carbon dioxide emissions. This potential application provides more possibilities for the field of motion detection.

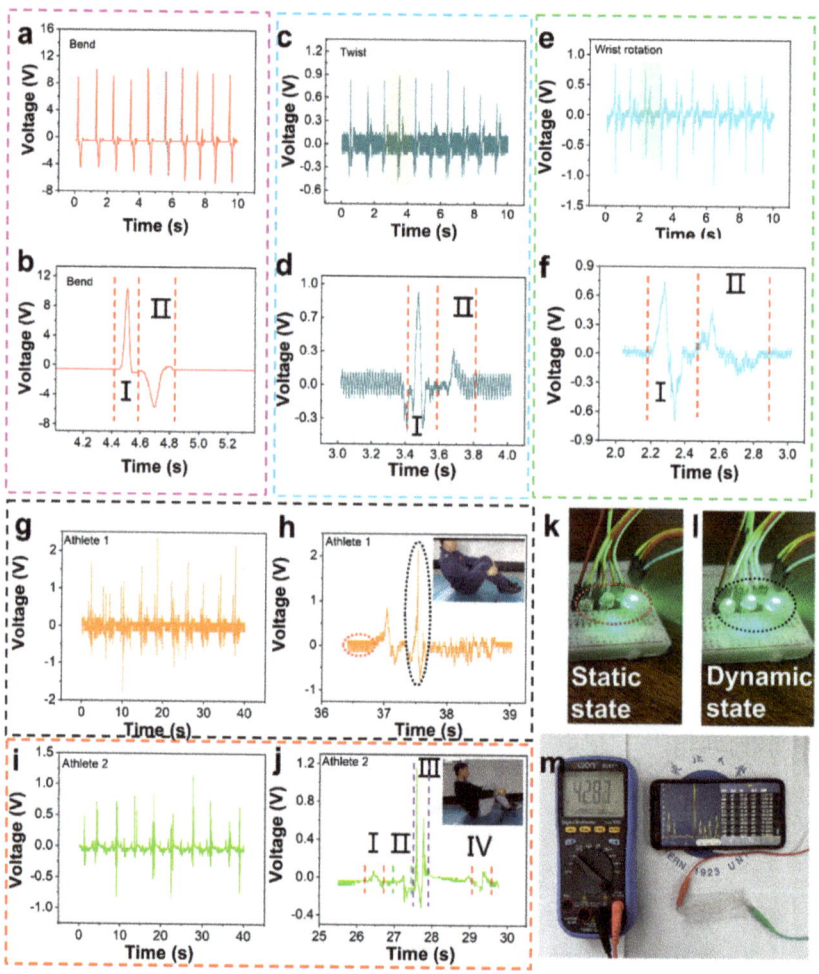

Figure 7. The actual test and wireless signal transmission system. (**a**,**b**) Wrist bending test and its response. (**c**,**d**) Wrist twist test and its response. (**e**,**f**) Wrist rotation test and its response. (**g**,**h**) Outputting voltage and details of Athlete 1's 301C diving motion. (**i**,**j**) Outputting voltage and details of Athlete 2's 301c diving motion. (**k**–**m**) Wireless Bluetooth transmission system.

4. Conclusions

To conclude, a new type of flexible stretchable self-healing composite nano-generator for human motion monitoring sensor has been developed. It is composed of PENG, TENG, and hydrogel electrodes. The contact-separation mode is adopted to realize the coupling of the piezoelectric effect and triboelectric effects, which further improve the sensitivity and measuring range. The hydrogel electrode has excellent stretchability and has a self-powered property which can meet the requirement of strenuous exercise. At the same time, PTSS can monitor different movements and events. It has a good monitoring effect on difficult and beautiful sports events. This research also solves the problem of wireless transmission and brings more opportunities for wireless big data and scientific sports.

Supplementary Materials: The following are available online at https://www.mdpi.com/article/10.3390/nano12010104/s1; Figure S1: (a) The cross-sectional microscopic image of PTSS (b–d) Three views of the device; Figure S2: The water mist airtight box; Figure S3: (a–c) The outputting voltage of PENG, TENG and PTSS in air test (d) Test force; Figure S4: The outputting voltage of 360° wrist rotation of PENG, TENG, and PTSS; Figure S5: (a,b) The conductivity of hydrogel (c) The outputting voltage of the metal electrode TENG; Figure S6: The image of 0.01 mol/L APS hydrogel; Figure S7: The tensile property of 0.03 mol/L, 0.05 mol/L, and 0.07 mol/L APS hydrogel; Figure S8: The PTSS outputting voltage at different bending angles, frequency, and twist in air; Figure S9: The responses of bending angel and twist angle in 100% RH environment; Figure S10: Stress-strain curve of self-healing hydrogel; Figure S11: The output voltage of twisting and pulling motion of table tennis; Figure S12: Charging characteristic curve; Movie S1: The frequency test of PTSS in 100% RH environment; Movie S2: The hydrogel being pressed for 3400 cycles; Movie S3: The performance of PTSS in water; Movie S4: The self-healing and stretch properties of hydrogel; Movie S5: The wrist bend motion and outputting voltage; Movie S6: The twist motion and outputting voltage; Movie S7: The rotation motion outputting voltage; Movie S8: The motions and details of Athlete 1 simulating 301C diving on land; Movie S9: The motions and details of Athlete 2 simulating 301C diving on land; Movie S10: The twisting and pulling motion of table tennis; Movie S11: The real-time wireless Bluetooth waveform display system.

Author Contributions: Y.M., T.Z. and Y.Z. put forward to the concept of the study. The data was collected, sorted out and analyzed by, T.Z. and Y.Z. C.J. and F.S. made the visualization. Y.M. wrote the manuscript. Y.Z. finished proof. All authors have read and agreed to the published version of the manuscript.

Funding: This research was funded by Winter Sports Training monitoring technical services. Grant number: 2020021300002.

Institutional Review Board Statement: Not applicable.

Informed Consent Statement: Informed consent was obtained from all subjects involved in the study.

Data Availability Statement: The data presented in this study are available on request from the corresponding author. The authors would like to thank Jilong Gao from Shiyanjia Lab (www.shiyanjia.com) for the Stress and strain analysis.

Acknowledgments: We thank Xiang Wu at Shenyang University of Technology for his helpful discussion.

Conflicts of Interest: The authors declare no conflict of interest.

References

1. Juarez, G.; Gastopoulos, T.; Zhang, Y.B.; Siegel, M.L.; Arratia, P.E. Splash control of drop impacts with geometric targets. *Phys. Rev. E* **2012**, *85*, 026319. [CrossRef] [PubMed]
2. Chen, L.; Kang, Q.J.; Mu, Y.T.; He, Y.L.; Tao, W.Q. A critical review of the pseudopotential multiphase lattice Boltzmann model: Methods and applications. *Int. J. Heat Mass Transf.* **2014**, *76*, 210–236. [CrossRef]
3. Zhou, W.F.; Liao, S.M.; Men, Y.Q. A fluid-solid coupled modeling on water seepage through gasketed joint of segmented tunnels. *Tunn. Undergr. Space Technol.* **2021**, *114*, 104008. [CrossRef]
4. Zou, Q.L.; Lin, B.Q. Fluid-Solid Coupling Characteristics of Gas-Bearing Coal Subjected to Hydraulic Slotting: An Experimental Investigation. *Energy Fuels* **2018**, *32*, 1047–1060. [CrossRef]
5. Wang, K.; Ma, X.; Bai, W.; Lin, Z.B.; Li, Y.B. Numerical simulation of water entry of a symmetric/asymmetric wedge into waves using OpenFOAM. *Ocean Eng.* **2021**, *227*, 108923. [CrossRef]
6. Wang, J.B.; Lugni, C.; Faltinsen, O.M. Experimental and numerical investigation of a freefall wedge vertically entering the water surface. *Appl. Ocean Res.* **2015**, *51*, 181–203. [CrossRef]

7. Zhang, S.; Xiang, W.; Zou, G.Z. Research on Body Type Correction of FINA Diving Difficulty Coefficient Based on Rigid Body System. *Complexity* **2021**, *2021*, 5519947. [CrossRef]
8. Kim, T.; Kim, S.M. Analysis of muscle activity and kinematic variables according to jump type of 3m springboard diving 109C skill-case study. *Korean J. Sport Sci.* **2019**, *30*, 907–915.
9. Heinen, T.; Supej, M.; Cuk, I. Performing a forward dive with 5.5 somersaults in platform diving: Simulation of different technique variations. *Scand. J. Med. Sci. Spor.* **2017**, *27*, 1081–1089. [CrossRef]
10. Young, C.; Reinkensmeyer, D.J. Judging complex movement performances for excellence: A principal components analysis-based technique applied to competitive diving. *Hum. Mov. Sci.* **2014**, *36*, 107–122. [CrossRef]
11. Wang, Y.R.; Liu, J.F.; Liu, G.J.; Tang, X.L.; Liu, P. Observation and analysis of high-speed human motion with frequent occlusion in a large area. *Meas. Sci. Technol.* **2009**, *20*, 125101. [CrossRef]
12. Wang, P.; Zhang, Q.; Jin, Y.L.; Ru, F. Studies and simulations on the flight trajectories of spinning table tennis ball via high-speed camera vision tracking system. *Proc. Inst. Mech. Eng. Part P J. Sports Eng. Technol.* **2019**, *233*, 210–226. [CrossRef]
13. Theobalt, C.; Albrecht, I.; Haber, J.; Magnor, M.; Seidel, H.P. Pitching a baseball—Tracking high-speed motion with multi-exposure images. *ACM Trans. Graph.* **2004**, *23*, 540–547. [CrossRef]
14. Yan, B.B.; Liu, G.Q.; He, J.; Luo, Y.H.; Yang, L.W.; Qi, H.F.; Sang, X.Z.; Wang, K.R.; Yu, C.X.; Yuan, J.H.; et al. Simultaneous Vector Bend and Temperature Sensing Based on a Polymer and Silica Optical Fibre Grating Pair. *Sensors* **2018**, *18*, 3507. [CrossRef] [PubMed]
15. Lee, S.; Reuveny, A.; Reeder, J.; Lee, S.; Jin, H.; Liu, Q.H.; Yokota, T.; Sekitani, T.; Isoyama, T.; Abe, Y.; et al. A transparent bending-insensitive pressure sensor. *Nat. Nanotechnol.* **2016**, *11*, 472–478. [CrossRef] [PubMed]
16. Xu, Q.H.; Lu, Y.T.; Zhao, S.Y.; Hu, N.; Jiang, Y.W.; Li, H.; Wang, Y.; Gao, H.Q.; Li, Y.; Yuan, M.; et al. A wind vector detecting system based on triboelectric and photoelectric sensors for simultaneously monitoring wind speed and direction. *Nano Energy* **2021**, *89*, 106382. [CrossRef]
17. Xu, Q.H.; Fang, Y.S.; Jing, B.Q.S.; Hu, N.; Lin, K.; Pan, Y.F.; Xu, L.; Gao, H.Q.; Yuan, M.; Chu, L.; et al. A portable triboelectric spirometer for wireless pulmonary function monitoring. *Biosens. Bioelectron.* **2021**, *187*, 113329. [CrossRef] [PubMed]
18. Yuan, M.; Li, C.H.; Liu, H.M.; Xu, Q.H.; Xie, Y.N. A 3D-printed acoustic triboelectric nanogenerator for quarter-wavelength acoustic energy harvesting and self-powered edge sensing. *Nano Energy* **2021**, *85*, 105962. [CrossRef]
19. Lin, H.B.; Liu, Y.; Chen, S.L.; Xu, Q.H.; Wang, S.T.; Hu, T.; Pan, P.F.; Wang, Y.Z.; Zhang, Y.L.; Li, N.; et al. Seesaw structured triboelectric nanogenerator with enhanced output performance and its applications in self-powered motion sensing. *Nano Energy* **2019**, *65*, 103944. [CrossRef]
20. Lin, H.B.; He, M.H.; Jing, Q.S.; Yang, W.F.; Wang, S.T.; Liu, Y.; Zhang, Y.L.; Li, J.; Li, N.; Ma, Y.W.; et al. Angle-shaped triboelectric nanogenerator for harvesting environmental wind energy. *Nano Energy* **2019**, *56*, 269–276. [CrossRef]
21. Xie, Y.N.; Wang, S.H.; Niu, S.M.; Lin, L.; Jing, Q.S.; Yang, J.; Wu, Z.Y.; Wang, Z.L. Grating-Structured Freestanding Triboelectric-Layer Nanogenerator for Harvesting Mechanical Energy at 85% Total Conversion Efficiency. *Adv. Mater.* **2014**, *26*, 6599–6607. [CrossRef] [PubMed]
22. Yang, H.M.; Wang, M.F.; Deng, M.M.; Guo, H.Y.; Zhang, W.; Yang, H.K.; Xi, Y.; Li, X.G.; Hu, C.G.; Wang, Z.L. A full-packaged rolling triboelectric-electromagnetic hybrid nanogenerator for energy harvesting and building up self-powered wireless systems. *Nano Energy* **2019**, *56*, 300–306. [CrossRef]
23. Li, S.Y.; Fan, Y.; Chen, H.Q.; Nie, J.H.; Liang, Y.X.; Tao, X.L.; Zhang, J.; Chen, X.Y.; Fu, E.G.; Wang, Z.L. Manipulating the triboelectric surface charge density of polymers by low-energy helium ion irradiation/implantation. *Energy Environ. Sci.* **2020**, *13*, 896–907. [CrossRef]
24. Yin, R.; Yang, S.Y.; Li, Q.M.; Zhang, S.D.; Liu, H.; Han, J.; Liu, C.T.; Shen, C.Y. Flexible conductive Ag nanowire/cellulose nanofibril hybrid nanopaper for strain and temperature sensing applications. *Sci. Bull.* **2020**, *65*, 899–908. [CrossRef]
25. Huang, X.; Liu, W.; Zhang, A.H.; Zhang, Y.; Wang, Z.L. Ballistic transport in single-layer MoS_2 piezotronic transistors. *Nano Res.* **2016**, *9*, 282–290. [CrossRef]
26. Lin, P.; Zhu, L.P.; Li, D.; Xu, L.; Pan, C.F.; Wang, Z.L. Piezo-Phototronic Effect for Enhanced Flexible MoS_2/WSe_2 van der Waals Photodiodes. *Adv. Funct. Mater.* **2018**, *28*, 1802849. [CrossRef]
27. Nguyen, V.; Kelly, S.; Yang, R. Piezoelectric peptide-based nanogenerator enhanced by single-electrode triboelectric nanogenerator. *APL Mater.* **2017**, *5*, 074108. [CrossRef]
28. Wang, Y.M.; Zeng, Q.F.; He, L.L.; Yin, P.; Sun, Y.; Hu, W.; Yang, R.S. Fabrication and application of biocompatible nanogenerators. *Iscience* **2021**, *24*, 102274. [CrossRef] [PubMed]
29. Jenkins, K.; Kelly, S.; Nguyen, V.; Wu, Y.; Yang, R.S. Piezoelectric diphenylalanine peptide for greatly improved flexible nanogenerators. *Nano Energy* **2018**, *51*, 317–323. [CrossRef]
30. Nguyen, V.; Zhu, R.; Jenkins, K.; Yang, R.S. Self-assembly of diphenylalanine peptide with controlled polarization for power generation. *Nat Commun.* **2016**, *7*, 13566. [CrossRef] [PubMed]
31. Tao, Z.; Yuan, H.; Ding, S.; Wang, Y.; Hu, W.; Yang, R.S. Diphenylalanine-based degradable piezoelectric nanogenerators enabled by polylactic acid polymer-assisted transfer. *Nano Energy* **2021**, *88*, 106229. [CrossRef]
32. Xue, C.Y.; Li, J.Y.; Zhang, Q.; Zhang, Z.B.; Hai, Z.Y.; Gao, L.B.; Feng, R.T.; Tang, J.; Liu, J.; Zhang, W.D.; et al. A Novel Arch-Shape Nanogenerator Based on Piezoelectric and Triboelectric Mechanism for Mechanical Energy Harvesting. *Nanomaterials* **2015**, *5*, 36–46. [CrossRef]

33. Wang, S.T.; He, M.H.; Weng, B.J.; Gan, L.H.; Zhao, Y.R.; Li, N.; Xie, Y.N. Stretchable and Wearable Triboelectric Nanogenerator Based on Kinesio Tape for Self-Powered Human Motion Sensing. *Nanomaterials* **2018**, *8*, 657. [CrossRef]
34. Ji, S.H.; Yun, J.S. Fabrication and Characterization of Aligned Flexible Lead-Free Piezoelectric Nanofibers for Wearable Device Applications. *Nanomaterials* **2018**, *8*, 657. [CrossRef]
35. Mao, Y.P.; Zhang, W.L.H.; Wang, Y.B.; Guan, R.X.; Liu, B.; Wang, X.H.; Sun, Z.; Xing, L.L.; Chen, S.; Xue, X.Y. Self-Powered Wearable Athletics Monitoring Nanodevice Based on ZnO Nanowire Piezoelectric-Biosensing Unit Arrays. *Sci. Adv. Mater.* **2019**, *11*, 351–359. [CrossRef]
36. Han, M.D.; Chen, X.X.; Yu, B.C.; Zhang, H.X. Coupling of Piezoelectric and Triboelectric Effects: From Theoretical Analysis to Experimental Verification. *Adv. Electron. Mater.* **2015**, *1*, 1500187. [CrossRef]
37. Yu, J.B.; Hou, X.J.; Cui, M.; Zhang, S.N.; He, J.; Geng, W.P.; Mu, J.L.; Chou, X.J. Highly skin-conformal wearable tactile sensor based on piezoelectric-enhanced triboelectric nanogenerator. *Nano Energy* **2019**, *64*, 103923. [CrossRef]
38. Zeeshan; Panigrahi, B.K.; Ahmed, R.; Mehmood, M.U.; Park, J.C.; Kim, Y.; Chun, W. Operation of a low-temperature differential heat engine for power generation via hybrid nanogenerators. *Appl. Energy* **2021**, *285*, 116385. [CrossRef]
39. Ma, M.Y.; Zhang, Z.; Zhao, Z.N.; Liao, Q.L.; Kang, Z.; Gao, F.F.; Zhao, X.; Zhang, Y. Self-powered flexible antibacterial tactile sensor based on triboelectric-piezoelectric-pyroelectric multi-effect coupling mechanism. *Nano Energy* **2019**, *66*, 104105. [CrossRef]
40. Zhu, Y.B.; Yang, B.; Liu, J.Q.; Wang, X.Z.; Chen, X.; Yang, C.S. An Integrated Flexible Harvester Coupled Triboelectric and Piezoelectric Mechanisms Using PDMS/MWCNT and PVDF. *J. Microelectromech. Syst.* **2015**, *24*, 513–515. [CrossRef]
41. Zhu, J.X.; Zhu, Y.L.; Wang, X.H. A Hybrid Piezoelectric and Triboelectric Nanogenerator with PVDF Nanoparticles and Leaf-Shaped Microstructure PTFE Film for Scavenging Mechanical Energy. *Adv. Mater. Interfaces* **2018**, *5*, 1700750. [CrossRef]
42. Wu, C.S.; Liu, R.Y.; Wang, J.; Zi, Y.L.; Lin, L.; Wang, Z.L. A spring-based resonance coupling for hugely enhancing the performance of triboelectric nanogenerators for harvesting low-frequency vibration energy. *Nano Energy* **2017**, *32*, 287–293. [CrossRef]
43. Xi, Y.; Wang, J.; Zi, Y.L.; Li, X.G.; Han, C.B.; Cao, X.; Hu, C.G.; Wang, Z.L. High efficient harvesting of underwater ultrasonic wave energy by triboelectric nanogenerator. *Nano Energy* **2017**, *38*, 101–108. [CrossRef]
44. Zelenovskiy, P.; Yuzhakov, V.; Nuraeva, A.; Kornev, M.; Shur, V.Y.; Kopyl, S.; Kholkin, A.; Vasilev, S.; Tofail, S.A.M. The Effect of Water Molecules on Elastic and Piezoelectric Properties of Diphenylalanine Microtubes. *IEEE Trns. Dielectr. Electr. Insul.* **2020**, *27*, 1474–1477. [CrossRef]
45. Tang, N.; Zheng, Y.B.; Yuan, M.M.; Jin, K.; Haick, H. High-Performance Polyimide-Based Water-Solid Triboelectric Nanogenerator for Hydropower Harvesting. *ACS Appl. Mater. Interfaces* **2021**, *13*, 32106–32114. [CrossRef]
46. Khandelwal, G.; Raj, N.; Kim, S.J. Materials Beyond Conventional Triboelectric Series for Fabrication and Applications of Triboelectric Nanogenerators. *Adv. Energy Mater.* **2021**, *11*, 32. [CrossRef]
47. Raj, N.; Ks, A.; Khandelwal, G.; Alluri, N.R.; Kim, S.J. A lead-free ferroelectric $Bi_{0.5}Na_{0.5}TiO_3$ based flexible, lightweight nanogenerator for motion monitoring applications. *Sustain. Energ. Fuels* **2020**, *4*, 5636–5644.
48. Niu, S.M.; Wang, S.H.; Lin, L.; Liu, Y.; Zhou, Y.S.; Hu, Y.F.; Wang, Z.L. Theoretical study of contact-mode triboelectric nanogenerators as an effective power source. *Energy Environ. Sci.* **2013**, *6*, 3576–3583. [CrossRef]
49. Liu, Y.; Niu, S.M.; Wang, Z.L. Theory of Tribotronics. *Adv. Electron. Mater.* **2015**, *1*, 11. [CrossRef]
50. Taylor, D.L.; Panhuis, M.I.H. Self-healing Hydrogels. *Adv. Mater.* **2016**, *28*, 9060–9093. [CrossRef]

Article

Structure and Magnetic Properties of ErFe$_x$Mn$_{12-x}$ (7.0 ≤ x ≤ 9.0, Δx = 0.2)

Penglin Gao [1,2], Yuanhua Xia [2], Jian Gong [2] and Xin Ju [1,*]

[1] School of Mathematics and Physics, University of Science and Technology Beijing, Beijing 100083, China; gaoyingluo@163.com

[2] Key Laboratory of Neutron Physics and Institute of Nuclear Physics and Chemistry, China Academy of Engineering Physics, Mianyang 621900, China; xiayuanhua2009@126.com (Y.X.); gongjian@hotmail.com (J.G.)

* Correspondence: jux@ustb.edu.cn

Citation: Gao, P.; Xia, Y.; Gong, J.; Ju, X. Structure and Magnetic Properties of ErFe$_x$Mn$_{12-x}$ (7.0 ≤ x ≤ 9.0, Δx = 0.2). *Nanomaterials* **2022**, *12*, 1586. https://doi.org/10.3390/nano12091586

Academic Editor: Jean-Pierre Bucher

Received: 2 April 2022
Accepted: 5 May 2022
Published: 7 May 2022

Publisher's Note: MDPI stays neutral with regard to jurisdictional claims in published maps and institutional affiliations.

Copyright: © 2022 by the authors. Licensee MDPI, Basel, Switzerland. This article is an open access article distributed under the terms and conditions of the Creative Commons Attribution (CC BY) license (https://creativecommons.org/licenses/by/4.0/).

Abstract: The magnetic interactions of iron-rich manganese-based ThMn$_{12}$ type rare earth metal intermetallic compounds are extremely complex. The antiferromagnetic structure sublattice and the ferromagnetic structure sublattice had coexisted and competed with each other. Previous works are focus on studying magnetic properties of RFe$_x$Mn$_{12-x}$ (x = 0–9.0, Δx = 0.2). In this work, we obtained a detailed magnetic phase diagram for iron-rich ErFe$_x$Mn$_{12-x}$ series alloy samples with a fine composition increment (Δx = 0.2), and studied the exchange bias effect and magnetocaloric effect of samples. ErFe$_x$Mn$_{12-x}$ series (x = 7.0–9.0, Δx = 0.2) alloy samples were synthesized by arc melting, and the pure ThMn$_{12}$-type phase structure was confirmed by X-ray diffraction (XRD). The neutron diffraction test was used to confirm the Mn atom preferentially occupying the 8i position and to quantify the Mn. The magnetic properties of the materials were characterized by a comprehensive physical property measurement system (PPMS). Accurate magnetic phase diagrams of the samples in the composition range 7.0–9.0 were obtained. Along with temperature decrease, the samples experienced paramagnetic, ferromagnetic changes for samples with x < 7.4 and x > 8.4, and paramagnetic, antiferromagnetic and ferromagnetic or paramagnetic, ferromagnetic and antiferromagnetic changes for samples with 7.4 ≤ x ≤ 8.2. The tunable exchange bias effect was observed for sample with 7.4 ≤ x ≤ 8.2, which resulting from competing magnetic interacting among ferromagnetic and antiferromagnetic sublattices. The maximum magnetic entropy change in an ErFe$_{9.0}$Mn$_{3.0}$ specimen reached 1.92 J/kg/K around room temperature when the magnetic field change was 5 T. This study increases our understanding of exchange bias effects and allows us to better control them.

Keywords: neutron diffraction; exchange-bias; magnetocaloric effect

1. Introduction

Manganese (Mn) is the only 3d-series element that forms a stable ThMn$_{12}$-type structure with rare earth elements [1,2], and it is mainly ferromagnetic and antiferromagnetic [3]. However, the pure ThMn$_{12}$-type rare earth iron compound RFe$_{12}$ does not exist. In the early 1980s, Yang et al. [4] found that a stable ternary rare earth iron intermetallic compound R(Fe$_x$Mn$_{1-x}$)$_{12}$ could be formed by substitution, thus setting off a surge of research into iron-rich ThMn$_{12}$-type compounds [5]. Subsequent studies have found that a number of tertiary elements can stabilize the ThMn$_{12}$ phase; their molecular formulas can be written as RFe$_x$M$_{12-x}$ or R(Fe,M)$_{12}$, where R is a rare earth element and M = Mn, V, Cr, Mo, W, Ti, Si, Al, Nb or Ga [6–9].

In the RMn$_{12}$ alloy, the strong antiferromagnetic interaction between manganese atoms prohibits interaction between rare earth atoms and manganese atoms, so the RMn$_{12}$ alloy has two magnetic ordering temperatures: R-R ferromagnetic ordering temperature, and Mn-Mn antiferromagnetic ordering temperature [10]. Iron (Fe) can replace Mn in large quantities (up to 75%) without changing the crystal structure [11]. Researchers [12–22] have investigated the structure and magnetic transitions of RFe_xMn_{12-x}-series materials (x = 0–9.0, Δx = 1) using neutron diffraction, magnetic measurements and electrical measurements and have found that magnetic interaction in the alloy is extremely complex. As the proportion of Fe increases, the material undergoes an antiferromagnetic → antiferromagnetic + ferromagnetic → ferromagnetic transition. Among the iron-rich RFe_xMn_{12-x}-series (x = 6.0–9.0) samples, only materials with integer values of x have been studied. This composition range includes the magnetic transition stage in which antiferromagnetism and ferromagnetism coexist in the material and plane anisotropy and axis anisotropy compete with each other. Therefore, it is necessary to prepare iron-rich RFe_xMn_{12-x}-series (x = 6.0–9.0) alloy samples with a finer composition change to obtain more detailed and complete magnetic phase diagrams, and thus be able to develop new aspects of applications for the material. We first studied YFe_xMn_{12-x}-series (x = 6.0–9.0) samples to obtain more complete magnetic phase diagrams for the materials and observed very large exchange bias effects and zero field cooling (ZFC) exchange bias effects in the samples [23]. After the discovery of exchange bias effect in Co/CoO nanoparticles, investigations have been mainly focused on a large number of heterogeneous structures such as magnetic bilayers, core-shell nanoparticles, and ferromagnetic nanoparticles embedded in antiferromagnetic matrix compounds [24–26]. So, it is necessary to further study exchange bias for the bulk metallic materials with exchange interactions occurring among the bulk sublattice. Firstly, we study how the magnetic atoms affect the EB effect in $ThMn_{12}$-type compounds. The second-order Stevens factor $α_J$ for Er atoms is >0, but the second-order crystal field coefficient (A20) of the rare earth sublattice in the $ThMn_{12}$ structure is negative, so magnetocrystalline anisotropy tends to the easy axis. We prepared $ErFe_xMn_{12-x}$-series ($7.0 \leq x \leq 9.0$, Δx = 0.2) alloy specimens have been prepared by arc melting to enable us to investigate the structure and magnetism of the alloy.

2. Experimental Methods

$ErFe_xMn_{12-x}$-series ($7.0 \leq x \leq 9.0$, Δx = 0.2) alloys were prepared by arc melting. The raw material was melted 4–5 times in an argon gas atmosphere according to the stoichiometric ratio to produce the alloy ingot; 5% more rare earth and 13% more Mn were added to compensate for volatilization in the melting process. A smaller current of 150 A was applied twice for melting, followed by a 200 A current once or twice to control the against excessive Mn volatilization. Specimens from the master alloy ingots were placed in sealed quartz tubes filled with argon and cooled down to room temperature after heat treatment at 1173 K for 2 days.

Phase purity was confirmed by a Cu target X-ray powder diffractometer (PANalytical, Almelo, The Netherlands) at room temperature. The high-resolution neutron diffraction spectrometer ($λ$ = 0.18846 nm) of Mianyang Research Reactor (CMRR, Mianyang, China) was used to analyze the crystal structure, in particularly for the positions of Mn atoms. Powdered alloy was bonded into a small cylinder with epoxy resin or the alloy ingot was shattered, so that we could select a small piece of regular shape for magnetic measurement. The ZFC and field cooling (FC) thermomagnetic curves ($M-T$ curves) of the samples were recorded, and the magnetic hysteresis loops ($M-H$ loops) of the samples under different FC and temperature conditions were measured by the comprehensive physical property measurement system (PPMS, Quantum Design (San Diego, CA, USA)).

3. Experimental Results and Analysis

A phase of the $ThMn_{12}$-type structure was formed in the $ErFe_xMn_{12-x}$-series ($7.0 \leq x \leq 9.0$) ingots, and some samples contained a small quantity of the Er(Fe, Mn)$_2$ phase. Heterogeneous Er$_2$(Fe, Mn)$_{17}$ and Er(Fe, Mn)$_2$ phases are formed in $ErFe_xMn_{12-x}$-series ($7.0 \leq x \leq 9.0$) alloys after heat treatment above 1273 K, which differentiates them from YFe_xMn_{12-x}-series ($7.0 \leq x \leq 9.0$) alloys. Long duration high-temperature heat treatment is therefore not suitable for this series of materials; 1173 K heat treatment for 48 h will produce homogeneous alloy samples with good crystal shapes.

The X-ray diffraction (XRD) spectra of the samples were examined before and after heat treatment. FullProf software [27] was used to refine the structure of the samples after heat treatment, and the relationship between the lattice constant and the composition of the samples was determined, as shown in Figure 1. With the increasing proportion of Fe, the lattice constant a decreased linearly and c remained unchanged.

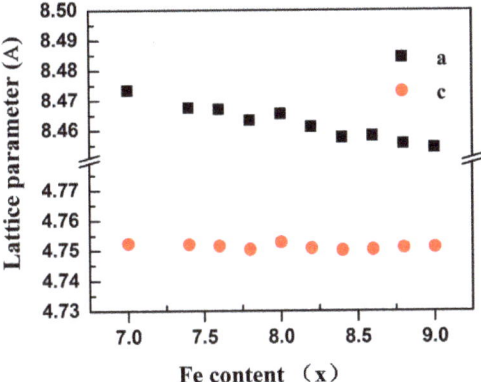

Figure 1. Variation of lattice constants a and c with Fe content of ErFe$_x$Mn$_{12-x}$ (7.0 ≤ x ≤ 9.0) series alloys after heat treatment.

The complete neutron diffraction spectra of some heat-treated samples were examined at room temperature, and the structure was refined using FullProf. The fitting spectrum is shown in Figure 2, and the crystal structure parameters are shown in Table 1. The samples formed a pure ThMn$_{12}$-type phase of space group I4/mmm (139), with rare earth Er atoms occupying the 2a position and Fe and Mn occupying three other unequal positions (8i, 8j, and 8f). Since the coherent neutron scattering lengths of Mn atoms ($b_{Mn} = -0.39$) and Fe atoms ($b_{Fe} = 0.95$) are significantly different, the relative proportions of Fe and Mn in the alloy samples can be obtained by fitting neutron diffraction data; the results are shown in Table 1. The Mn atom occupies the 8i position preferentially. The trend of change in the proportion of Mn in the materials was similar to that of the initial materials, although the proportion of Mn was slightly higher, which indicated that the proportion of compensated Mn in the initial materials was relatively high. The lattice constant a decreased as the proportion of Fe decreased, while the lattice constant c remained basically unchanged. This is because the Mn atom preferentially occupies the 8i position, and 8i–8i lies in the plane ab. Changes in the proportion of Mn therefore greatly influences the lattice constants a and b but has little effect on the lattice constant c.

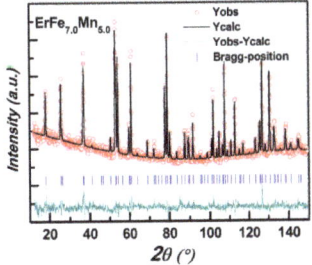 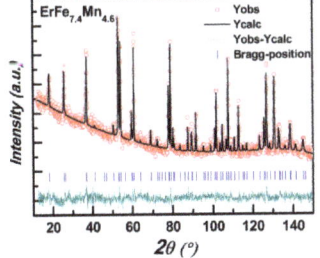 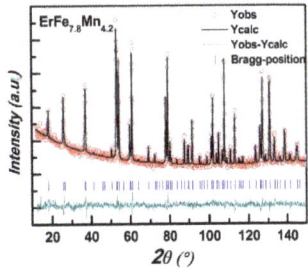

Figure 2. Cont.

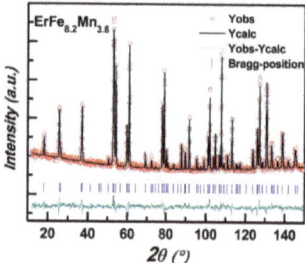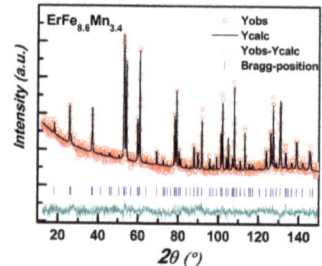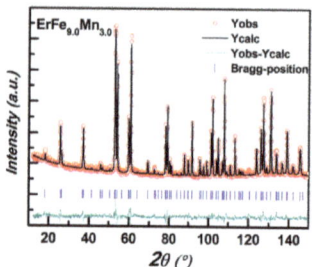

Figure 2. Refined neutron diffraction pattern of ErFe$_x$Mn$_{12-x}$ ($7.0 \leq x \leq 9.0$) series alloys (where red dots are experimental data, black curves are theoretical simulations, blue vertical bars are Bragg diffraction peak positions and the bottom green solid line is the difference curve).

Table 1. Information on crystal structure parameters of ErFe$_x$Mn$_{12-x}$ ($7.0 \leq x \leq 9.0$) series alloys.

ErFe$_x$Mn$_{12-x}$	a(Å)	c(Å)	occ, Fe, 8i	occ, Fe, 8j	occ, Fe, 8f	n, Fe	n, Mn	Rwp
ErFe$_{9.0}$Mn$_{3.0}$	8.45469(11)	4.75397(7)	0.476(0)	0.836(4)	0.928(8)	8.96	3.04	5.11
ErFe$_{8.6}$Mn$_{3.4}$	8.45777(24)	4.75346(16)	0.412(4)	0.788(12)	0.892(16)	8.368	3.632	4.54
ErFe$_{8.2}$Mn$_{3.8}$	8.46289(11)	4.75501(7)	0.344(0)	0.792(4)	0.908(8)	8.176	3.824	4.73
ErFe$_{7.8}$Mn$_{4.2}$	8.46758(16)	4.75572(11)	0.300(0)	0.704(8)	0.832(8)	7.344	4.656	3.88
ErFe$_{7.4}$Mn$_{4.6}$	8.47191(24)	4.75547(16)	0.284(0)	0.640(8)	0.764(8)	6.752	5.248	3.29
ErFe$_{7.0}$Mn$_{5.0}$	8.47767(19)	4.75605(13)	0.232(0)	0.636(4)	0.796(8)	6.656	5.344	7.28

Figure 3 shows the thermomagnetic curves of ErFe$_x$Mn$_{12-x}$-series ($7.0 \leq x \leq 9.0$) alloy samples in an external magnetic field of 50 Oe. T_C represents the Curie temperature, T_N is the Néel temperature, T_C and T_N is obtained by differentiating the $M-T$ curves under FC. T_f is the temperature corresponding to the bifurcation point in the ZFC and FC magnetization curves. As can be seen from the figure, the ZFC and FC $M-T$ curves of the samples both clearly bifurcated as the temperature decreased. T_f was slightly lower than the paramagnetic–ferromagnetic transition temperature of the samples due to the coexistence of Er-Er and Fe-Fe ferromagnetic exchanges interactions. Er-Fe, Er-Mn, Fe-Mn and Mn-Mn antiferromagnetic exchanges interactions, all interactions compete with each other, leading to spin frustration in the samples at low temperatures. For samples with $x > 7.2$, the FC $M-T$ curves initially increased to the maximum value and then decreased gradually as the temperature decreased. The curve steepened, and both the speed and amplitude of bending increased as the proportion of Fe decreased; it reached the maximum for $x = 7.8$ and then began to decrease and disappeared for $x = 7.2$. The magnetization curves for $x > 7.2$ samples were typical of ferrimagnetism magnetization curves. This was because light rare earth lattices and metal lattices are ferromagnetically arranged and heavy rare earth lattices and metal lattices are antiferromagnetically arranged in rare earth intermetallic compounds with a ThMn$_{12}$-type structure. Er is a heavy rare earth atom, so the samples had a ferrimagnetic structure in which the lattice magnetic moments of rare earth and transition metals were inversely arranged. As the temperature decreased, the magnetic moments of rare earth in the lattice increased rapidly and magnetic moments of transition metals increased slowly; the total magnetic moments of the samples initially increased to the maximum value and then decreased rapidly, and even showed a negative magnetic susceptibility. The $x = 7.2$ and $x = 7.0$ samples behave like pure ferro- or ferrimagnetic samples where high coercivity has developed already close to T_C. This causes the maximum in the ZFC curves very close to T_C.

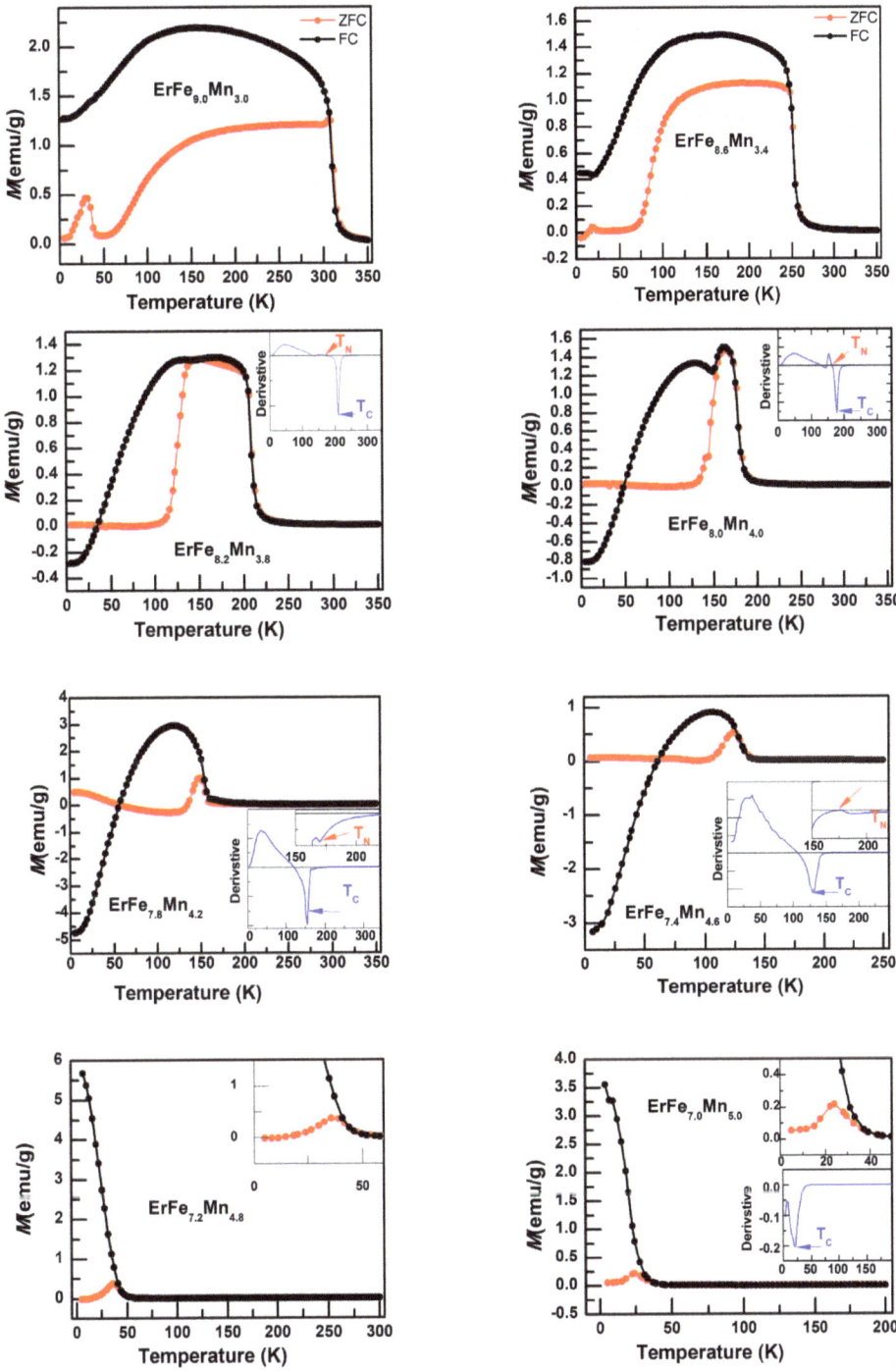

Figure 3. $M-T$ curves for ErFe$_x$Mn$_{12-x}$ ($7.0 \leq x \leq 9.0$) series alloys under zero field cooling (ZFC) and field cooling (FC) conditions, H = 50 Oe. (The inset shows the $M-T$ curves under FC after differentiation.)

In the YFe$_x$Mn$_{12-x}$-series (6.0 ≤ x ≤ 8.8) samples, as the proportion of Fe decreased, the T_C of the alloy rapidly decreased and the T_N slowly increased; the antiferromagnetic exchange magnetic ordering temperature of Mn-Mn was observed [23]. After rare earth Er atoms with magnetic moments replaced Y atoms without magnetic moments, the antiferromagnetic order of Mn-Mn was suppressed; the obvious antiferromagnetic order of Mn-Mn was only observed in the samples with the Fe proportion 7.4 ≤ x ≤ 8.2. The magnetic ordering temperature is shown in Table 2. Similar to YFe$_x$Mn$_{12-x}$, the ferromagnetic transition temperature of the alloy materials decreased rapidly as the proportion of Fe decreased.

Table 2. Magnetic ordering temperature, exchange bias field and coercive force field of ErFe$_x$Mn$_{12-x}$ (7.0 ≤ x ≤ 9.0) series alloys.

ErFe$_x$Mn$_{12-x}$	T_C (K)	T_f (K)	T_N (K)	H_E (kOe)	H_C (kOe)
Cooling Field	50 Oe	50 Oe	50 Oe	1000 Oe	1000 Oe
ErFe$_{9.0}$Mn$_{3.0}$	310	306			
ErFe$_{8.6}$Mn$_{3.4}$	250	248		−0.22	1.28
ErFe$_{8.2}$Mn$_{3.8}$	208	203	142	11.73	2.97
ErFe$_{8.0}$Mn$_{4.0}$	178	170	163		
ErFe$_{7.8}$Mn$_{4.2}$	154	160	169	6.615	9.54
ErFe$_{7.4}$Mn$_{4.6}$	128	126	176	11.08	4.52
ErFe$_{7.2}$Mn$_{4.8}$	22	44			
ErFe$_{7.0}$Mn$_{5.0}$	22	36		−1.27	28.11

Figure 4 shows the magnetic phase diagram of the ErFe$_x$Mn$_{12-x}$-series (7.0 ≤ x ≤ 9.0) alloys. The samples with x < 7.4 or x > 8.4 were mainly ferromagnetic. The samples with 7.4 ≤ x ≤ 8.2 were ferromagnetic and antiferromagnetic, and only the samples in this range of composition showed antiferromagnetic orders between different transition metal lattices. YFe$_x$Mn$_{12-x}$-series samples showed a clear exchange bias effect in the region where ferromagnetic interaction and antiferromagnetic interaction compete most intensely [23]. ErFe$_x$Mn$_{12-x}$-series samples may therefore similarly display exchange bias effects for 7.4 ≤ x ≤ 8.2. The FC M−H loops of some samples were measured, and the results are shown in Figure 5. The FC M−H loops of ErFe$_{8.2}$Mn$_{3.8}$, ErFe$_{7.8}$Mn$_{4.2}$ and ErFe$_{7.4}$Mn$_{4.6}$ samples all clearly had lateral shifts. The x = 7.4 and x = 7.8 samples had high coercivity, and the M−H loops were not completely closed when the applied field was 5T. The M−H loops were asymmetric, and lateral and vertical shifts occurred simultaneously. This indicates that the samples had very strong magnetocrystalline anisotropy at low temperatures, and that the antiferromagnetic interaction between the rare earth lattice and the transition metal lattice was the source of the anisotropy. When combined with the YFe$_x$Mn$_{12-x}$-series experimental results, we see that the exchange bias effect can be controlled by doping different rare earth elements in addition to altering the ratios of Fe and Mn.

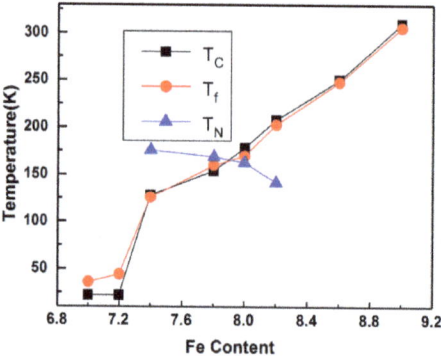

Figure 4. The magnetic phase diagram of ErFe$_x$Mn$_{12-x}$ (7.0 ≤ x ≤ 9.0) series alloys.

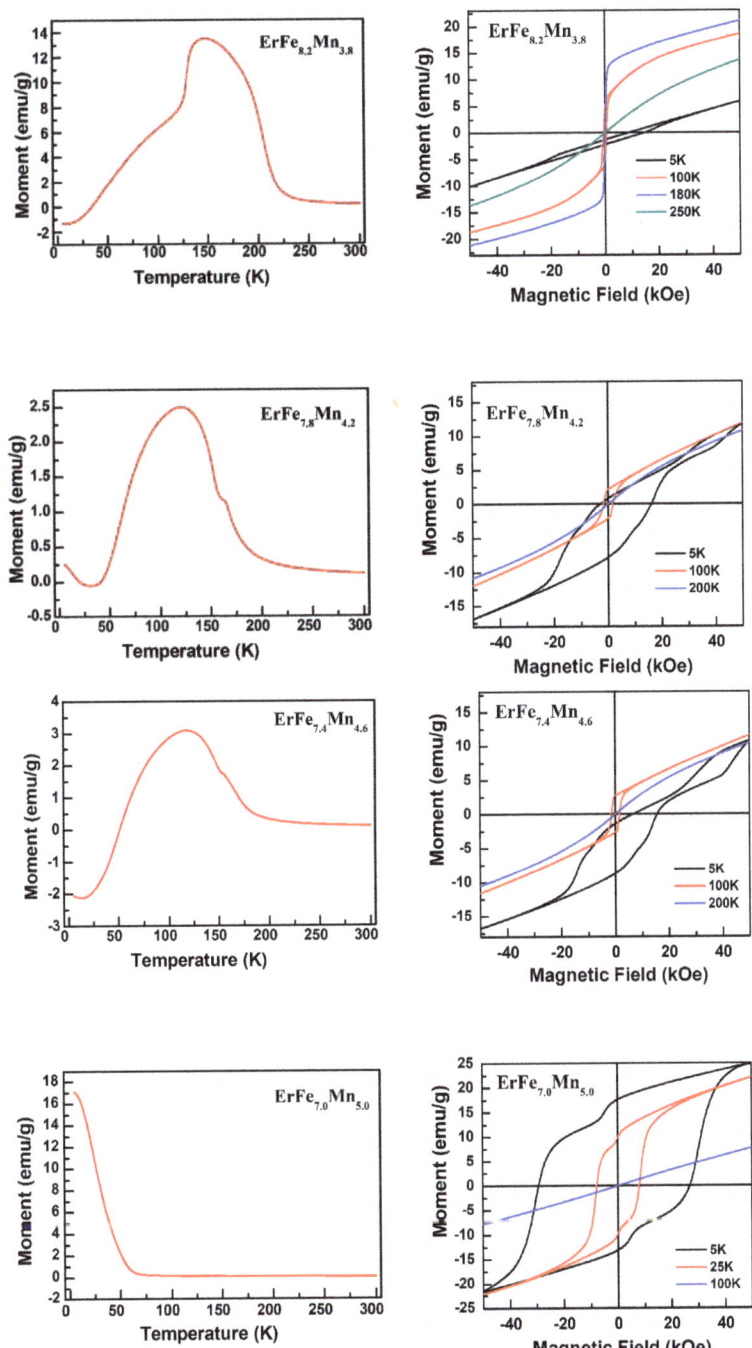

Figure 5. $M-T$ curve under field cooling condition (H = 1000 Oe) and $M-H$ curve after 1000 Oe field cooling of $ErFe_xMn_{12-x}$ (7.0 ≤ x ≤ 9.0) series alloys.

The ErFe$_{9.0}$Mn$_{3.0}$ compound had a Curie temperature of 310 K, and which is near the room temperature. The reverse magnetic moment of Er atom is decrease drastically as temperature increasing, so the samples may have had a considerable magnetocaloric effect near the Curie temperature. The isothermal magnetization curves in the temperature range 270–340 K were created, and are shown in Figure 6. The figure shows that as the temperature increased, magnetization intensity gradually decreased, and ferromagnetism was gradually transformed into paramagnetism. The isothermal magnetization curves were transformed to obtain the Arrott plot, as shown in Figure 7, in order to determine the type of phase transition occurring. There was no S-shaped curve in the Arrott plot, and no negative curve slope was observed, so the phase transition of the materials was also a second-order phase transition.

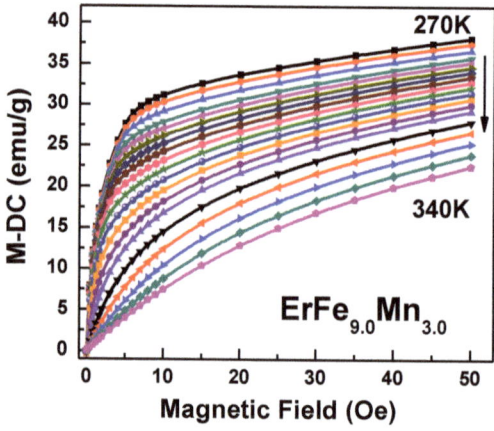

Figure 6. Isothermal magnetization curve of ErFe$_{9.0}$Mn$_{3.0}$.

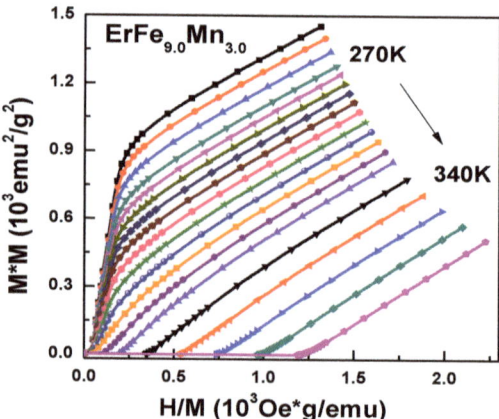

Figure 7. Arrott curve of ErFe$_{9.0}$Mn$_{3.0}$.

The Maxwell relation was used to calculate the isothermal magnetic entropy change in the samples from the isothermal magnetization curves at different temperatures, as shown in Figure 8. The calculated maximum value of the magnetic entropy changes when an applied field change of 50 kOe reaches 1.92 J/kg/K. The peak of $-\Delta S_M$ at 312.5 K corresponds to the ferromagnetic to paramagnetic phase transition, because the magnetization changes drastically near the Curie temperature. Although the maximum $-\Delta S_M$ of ErFe$_{9.0}$Mn$_{3.0}$ is not as large as that of some other magnetic refrigerant materials [28], the $|\Delta S_M|$ vs.

T curve of ErFe$_{9.0}$Mn$_{3.0}$ is significantly broader compared with other materials, which is favorable for active magnetic refrigeration. Additionally, the magnetocaloric effect was caused by the second-order phase transition near the Curie temperature, and the thermal hysteresis and magnetic hysteresis during phase transition were both very small, which has benefits in the practical application of the material.

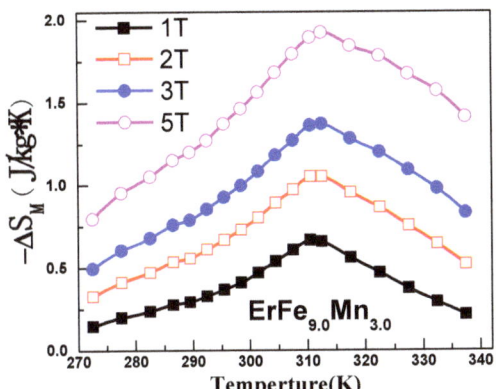

Figure 8. Isothermal magnetic entropy change with temperature for ErFe$_{9.0}$Mn$_{3.0}$.

4. Conclusions

ThMn$_{12}$-type single phase samples with different Fe/Mn ratios were prepared by arc melting and heat treatment, and the magnetic phase diagrams of ErFe$_x$Mn$_{12-x}$-series ($7.0 \leq x \leq 9.0$) samples were obtained by magnetic measurement. At low temperatures, samples with $x < 7.4$ and $x > 8.4$ exhibited ferromagnetism, and ferromagnetism and antiferromagnetism coexisted in samples with $7.4 \leq x \leq 8.2$, with an FC exchange bias effect. The magnetic interaction between transition metal lattices in ThMn$_{12}$-type structural materials can be changed by substituting non-magnetic Y atoms with rare earth Er atoms with magnetic moments. In this study, Y atoms were completely replaced; in the following study, we will partially replace them to finely modulate the exchange bias effect and the magnetocaloric effect of the materials.

Author Contributions: Conceptualization, P.G. and Y.X.; methodology, P.G. and Y.X.; validation, P.G. and Y.X.; formal analysis, P.G. and Y.X.; investigation, P.G. and Y.X.; resources, P.G. and Y.X.; data curation, P.G. and Y.X.; writing—original draft preparation, P.G. and Y.X.; writing—review and editing, P.G. and Y.X.; visualization, P.G. and Y.X.; supervision, J.G. and X.J.; funding acquisition, J.G. and X.J. All authors have read and agreed to the published version of the manuscript.

Funding: This research was funded by the National Natural Science Foundation of China, grant number 11504348.

Data Availability Statement: The data presented in this study are available in this article.

Conflicts of Interest: The authors declare no conflict of interest.

References

1. Florio, J.V.; Rundle, R.E.; Snow, A.I. Compounds of Thorium with Transition Metals. I. The Thorium-Manganese System. *Acta Cryt.* **1952**, *5*, 449. [CrossRef]
2. Deportes, J.; Givord, D.; Lemaire, R.; Nagaï, H. Magnetic interactions in the R-Mn12 compounds. *Physica B+C* **1977**, *86–88*, 69–70. [CrossRef]
3. Kirchmayr, H.R. Magnetic properties of rare earth—Manganese compounds. *IEEE Trans. Magn.* **1966**, *2*, 493–499. [CrossRef]
4. Yang, Y.C. Structural and magnetic properties of Y(Mn(1−x)Fex)12. *Acta Metall. Sin.* **1981**, *17*, 355–358.
5. Li, H.S.; Coey, J.M.D. *Handbook of Magnetic Materials*; Bushcow, K.H.J., Ed.; Elsevier Science Publishers B.V.: Amsterdam, The Netherlands, 1991; Volume 6, pp. 6–41.
6. Mooij, D.B.; De Buschow, K.H.J. Some novel ternary ThMn12-type compounds. *J. Less Common Met.* **1988**, *136*, 207–215. [CrossRef]

7. Ohashi, K.; Tawara, Y.; Osugi, R.; Tawara, Y. The magnetic and structural properties of R-Ti-Fe ternary compounds. *IEEE Trans. Magn.* **1987**, *23*, 3101–3103. [CrossRef]
8. Wang, X.Z.; Chevalier, B.; Berlureau, T.; Etourneau, J.; Coey, J.M.D.; Cadogan, J.M. Iron-rich pseudobinary alloys with the ThMn12 structure obtained by melt spinning: Gd(FenAl12−n), n = 6, 8. *J. Less Common Met.* **1988**, *138*, 235–240. [CrossRef]
9. Muller, K.H. Magnetic material R, Fe, Mo, (Co) with ThMn12 structure. *J. Appl. Phys.* **1988**, *64*, 249–251. [CrossRef]
10. Okamoto, N.; Nagai, H.; Yoshie, H.; Tsujimura, A.; Hihara, T. The coexistence of ferromagnetic and antiferromagnetic interactions in the GdMn12 compound. *J. Magn. Magn. Mater.* **1987**, *70*, 299–300. [CrossRef]
11. Yang, Y.C.; Kebe, B.; James, W.J.; Deportes, J.; Yelon, W. Structural and magetic properties of Y(Mn1−xFex)12. *J. Appl. Phys.* **1981**, *52*, 2077–2078. [CrossRef]
12. Amako, Y.; Saoka, S.; Yoshie, H.; Nagai, H.; Adachi, K. Antiferromagnetic Ordering Temperature of RMn12−xFex (R = Rare Earth). *J. Phys. Soc. Jpn.* **1995**, *64*, 1860–1861. [CrossRef]
13. Stankiewicz, J.; Bartolome, J.; Morales, M.; Bacmann, M.; Fruchart, D. Resistivity of RMn12−xFex alloys. *J. Appl. Phys.* **2001**, *90*, 5632–5636. [CrossRef]
14. Morales, M.; Bacmann, M.; Wolfers, P.; Fruchart, D.; Ouladdiaf, B. Magnetic properties and interactions in the RMn12−xFex series R(Y, Ho, Er, Nd; x ≤ 9). *Phys. Rev. B* **2001**, *64*, 144426. [CrossRef]
15. Mao, W.H.; Yang, J.B.; Cheng, B.P.; Yang, Y.C. Unusual magnetic properties of Ho(Fe0.6Mn0.4)12. *Solid State Commun.* **1999**, *109*, 655–659. [CrossRef]
16. Dong, S.Z.; Yang, J.; Yang, Y.C. Structure and Magnetic Properties of Y(Fe1−xMnx)12 Compounds and Their Nitrides (x = 0.2 and 0.4). *Solid State Commun.* **1995**, *94*, 809–812. [CrossRef]
17. Stankiewicz, J.; Bartolome, J.; Fruchart, D. Spin Disorder Scattering in Magnetic Metallic Alloys. *Phys. Rev. Lett.* **2002**, *89*, 106602. [CrossRef] [PubMed]
18. Yang, J.B.; Yelon, W.B.; James, W.J.; Cai, Q.S.; Eckert, D.; Handstein, A.; Muller, K.H.; Yang, Y.C. Structural and magnetic properties of RFexMn12−x, (R = Ho,Y). *Phys. Rev. B* **2002**, *65*, 064444. [CrossRef]
19. Stankiewicz, J.; Bartolome, J. Magnetic scattering in RMn12−xFex alloys. *Phys. Rev. B* **2002**, *67*, 092409. [CrossRef]
20. Pique, C.; Abad, E.; Blanco, J.A.; Burriel, R.; Fernandez-Diaz, M.T. Interplay between competing exchange interactions and magnetocrystalline anisotropies in YFexMn12−x: The magnetic phase diagram. *Phys. Rev. B* **2005**, *71*, 174422. [CrossRef]
21. Shelyapina, M.G.; Morales, M.; Bacmann, M.; Baudelet, F.; Fruchart, D.; Giorgetti, C.; Hlil, E.K.; Krill, G.; Wolfers, P. Magnetic properties of RMn12−xFex type compounds: I. X-ray magnetic circular dichroism study of the ErMn12−xFex series with x = 0, 7, 8 and 9. *J. Alloys Compd.* **2004**, *368*, 84–93. [CrossRef]
22. Pique, C.; Blanco, J.A.; Burriel, R.; Abad, E.; Artigas, M.; Fernandez-Diaz, M.T. Influence of 3d-4f interactions in the magnetic phases of RFexMn12−x (R = Gd, Tb, and Dy) compounds: Coexistence of ferromagnetism and antiferromagnetism at different crystallographic sites. *Phys. Rev. B* **2007**, *75*, 224424. [CrossRef]
23. Xia, Y.H.; Wu, R.; Zhang, Y.F.; Liu, S.Q.; Du, H.L.; Han, J.Z.; Wang, C.S.; Chen, X.P.; Xie, L.; Yang, Y.C.; et al. Tunable giant exchange bias in the single-phase rare-earth–transition-metal intermetallics YMn12−xFex with highly homogenous intersublattice exchange coupling. *Phys. Rev. B* **2017**, *96*, 064440. [CrossRef]
24. Nogués, J.; Schuller, I.K. Exchange bias. *J. Magn. Magn. Mater.* **1999**, *192*, 203–232. [CrossRef]
25. Nogués, J.; Sort, J.; Langlais, V.; Skumryev, S.; Suriñach, S.; Muñoz, J.S.; Baró, M.D. Exchange bias in nanostructures. *Phys. Rep.* **2005**, *422*, 65–117. [CrossRef]
26. Giri, S.; Patra, M.; Majumdar, S. Exchange bias effect in alloys and compounds. *J. Phys. Condens. Matter* **2011**, *23*, 073201. [CrossRef] [PubMed]
27. Rodríguez-Carvajal, J. Recent advances in magnetic structure determination by neutron powder diffraction. *Physica B* **1993**, *192*, 55–69. [CrossRef]
28. Zarkevich, N.A.; Zverev, V.I. Viable Materials with a Giant Magnetocaloric Effect. *Crystals* **2020**, *10*, 815. [CrossRef]

Article

A Self-Assembly of Single Layer of Co Nanorods to Reveal the Magnetostatic Interaction Mechanism

Hongyu Du [†], Min Zhang [†], Ke Yang, Baohe Li * and Zhenhui Ma *

Department of Physics, Beijing Technology and Business University, Beijing 100048, China; duhongyu2019@163.com (H.D.); tian123zmm@126.com (M.Z.); yangke_0205@163.com (K.Y.)
* Correspondence: libh@btbu.edu.cn (B.L.); mazh@btbu.edu.cn (Z.M.)
† These authors contributed equally to this work.

Abstract: In this work, we report a self-assembly method to fabricate a single layer of Co nanorods to study their magnetostatic interaction behavior. The Co nanorods with cambered and flat tips were synthesized by using a solvothermal route and an alcohol–thermal method, respectively. Both of them represent hard magnetic features. Co nanorods with cambered tips have an average diameter of 10 nm and length of 100 nm with coercivity of 6.4 kOe, and flat-tip nanorods with a 30 nm diameter and 100 nm length exhibit a coercivity of 4.9 kOe. They are further assembled on the surface of water in assistance of surfactants. The results demonstrate that the assembly type is dependent on the magnetic induction lines direction. For Co nanorods with flat tips, most of magnetic induction lines are parallel to the length direction, leading to an assembly that is tip to tip. For Co nanorods with cambered tips, they are prone to holding together side by side for their random magnetic induction lines. Under an applied field, the Co nanorods with flat tips can be further aligned into a single layer of Co nanorods. Our work gives a possible mechanism for the magnetic interaction of Co nanorods and provides a method to study their magnetic behavior.

Keywords: nanomagnets; Co nanorods; solvothermal route; alcohol–thermal method; magnetic interaction

1. Introduction

Nanomagnets, with strong magnetic properties and small volume, have been considered to be the key materials to magnetic and electronic devices that exhibit important applications in artificial intelligence, intelligent robots, wind turbines, and electromobiles [1–12]. Compared with rare-earth-based nanomagnets, the rare-earth-free nanomagnets earn more attention for their high chemical stability and low cost [13–16]. However, these rare-earth-free nanomagnets present small coercivity due to their relatively low magnetocrystalline anisotropy. One strategy to fabricate one-dimensional nanowires or nanorods by the special methods can resolve above problem, since the direction along nanorods length has the lowest demagnetizing field [17–19]. As a result, the shape anisotropy can be summed to magnetocrystalline anisotropy, leading to these nanorods with a larger coercivity than spherical particles. When an external field is applied in these particles, they can orient along the external field direction. Therefore, the system of magnetic nanorods along the applied field will exhibit an enhanced coercivity, high remanence (M_r), and large energy product.

Hexagonal structured Co has a magnetocrystalline anisotropy constant (K_1 = 440 kJ/m^3), and its bulk has a high saturation magnetization (M_s = 160 emu/g) [20]. Therefore, in the last few decades, Co was always considered to be a soft magnetic material since its magnetocrystalline anisotropy is far smaller than those of some well-known hard magnetic structures, such as samarium−cobalt (SmCo$_5$) (17,200 kJ/m^3) and neodymium iron boron (Nd$_2$Fe$_{14}$B) (4900 kJ/m^3) [21–24]. Recently, the Co nanowires have been a potential candidate for rare-earth-free nanomagnets because the Co nanowires with their length direction along [002] (the easy magnetization axis direction) can obtain a large coercivity. To obtain such shape-anisotropic Co nanowires, a conventional approach via electrochemical deposition of Co into a porous alumina template was employed, which yielded an enhanced coercivity from an aligned Co nanowire array, but it is lower than the theoretical value because of the polycrystalline structure [25]. Recently, a solution-phase synthesis was used to reduce a cobalt salt using an organic polyalcohol. The resulted Co nanowires possessed a single crystal structure, and their lengths (50–300 nm) and diameters (5–30 nm) can be tuned to achieve high performance [26–29]. As a consequence, the aligned Co nanomagnet fabricated by nanowires exhibits an ultrahigh coercivity up to 10 kOe, which exceeded the theoretical value of bulk Co material (7.6 kOe) [27]. The high coercivity originated from a sum of magnetocrystalline anisotropy field (7.6 kOe) and shape anisotropy field (9 kOe) [27].

It is worth noting that the magnetic alignment is a key parameter to obtain large coercivity and high M_r. To obtain an aligned nanomagnet, these as-prepared Co nanowires or nanorods have to be mixed with epoxy and aligned under an applied field [26,27]. In such a nanomagnet, the aligned Co nanowires were stacked layer by layer, and the magnetostatic interaction among the Co layers will have an obvious influence on alignment and magnetic performance. However, the single layer of aligned Co nanorods was very hard to prepare by using the current methods; however, it was valuable to study the magnetostatic interaction of nanomagnets.

Here, we developed a novel method to obtain a single layer of aligned Co nanorods. First, we synthesized different shaped Co nanorods by solvothermal route (the nanorods with flat tips) and alcohol–thermal method (the nanorods with cambered tips), respectively. These nanorods were further aligned by using a self-assembly strategy. As a result, a well-aligned single layer of nanorods was obtained under an applied field. Their self-assembly behaviors were further studied without an applied field. It was found that the nanorods with flat tips are prone to aligning tip to tip, and the nanorods with cambered tips are easy to hold together side by side. Furthermore, we analyzed the magnetic interaction mechanism of Co nanorods with different tips to demonstrate their assembly behaviors. Our work provides an important reference to study the magnetostatic interaction of shape anisotropic nanomagnets.

2. Experiments and Methods

2.1. Chemicals and Materials

All chemicals were used without further purifications. Cobalt chloride hexahydrate ($CoCl_2 \cdot 6H_2O$, 98%, M_w = 237.93), sodium laurate ($C_{11}H_{23}COONa$, ≥99%, M_w = 222.30), oleylamine (OAm, 70%, M_w = 267.49), oleic acid (OA, 90%, M_w = 282.46), 2-butanediol (BEG, 98%, M_w = 90.12), Ruthenium chloride (III) hydrate ($RuCl_3 \cdot xH_2O$, 99.98%, M_w = 207.43), hexadecylamine (HDA, 98%, M_w = 241.46), hexane (≥99%, M_w = 86.18), methylbenzene (=99.8%, M_w = 82.14), and chloroform (≥99.8%, M_w = 119.38) were purchased from Sigma-Aldrich, St. Louis, MO, USA.

2.2. The Synthesis of Cobalt (II) Laurate

The Cobalt (II) laurate (Co ($C_{11}H_{23}COO$) $_2$) was prepared by mixing $C_{11}H_{23}COONa$ and $CoCl_2$. In a typical reaction, 9.3366 g of $C_{11}H_{23}COONa$ was dissolved in 30 mL of deionized water, with a mechanical stirrer, in three-neck flask, which was heated to 60 °C, using a water bath. Then 5.000 g of $CoCl_2$ was dissolved in 10 mL deionized water in another flask, forming $CoCl_2$ aqueous solution. The latter was added into former system, dropwise, with vigorous stirring. The mixture was stirred for a further 30 min and kept at 60 °C to obtain a purple precipitate, which was centrifuged at 8000 rpm for 8 min, and washed three times with 50 mL deionized water and methanol, respectively. The as-prepared precipitate was further dried in an air oven at 60 °C.

2.3. The Synthesis of Cobalt Nanorods Using a Solvothermal Route

The Co nanorods were first synthesized by reducing Co-laurate in BEG. In a typical synthesis, 2.0700 g of Co-laurate, 0.5810 g of HDA, and 0.0037 g of $RuCl_3$ were dissolved in 60 mL of BEG in a three-neck flask, which was heated to 90 °C for 30 min to obtain a uniform solution. Then the solution was transferred into a Teflon enclosure (100 mL) and was further heated to 250 °C for 60 min, with a heating rate of 15 °C/min for a solvothermal route. After cooling to room temperature, the black magnetic product was precipitated and washed by toluene by centrifuging at 6000 rpm for 5 min at least 3 times.

2.4. The Synthesis of Cobalt Nanorods Using an Alcohol–Thermal Method

In a representative synthesis, 0.6113 g of Co-laurate, 0.0077 g of $RuCl_3$, and 0.4640 g of NaOH were added to 15.5 mL of BEG. Under mechanical stirring, the reaction mixture was heated to 100 °C for 30 min to obtain the uniform solution. Then the solution was further heated to 175 °C, with a temperature ramping rate of 7 °C/min, and was maintained at this temperature for 30 min. After cooling to room temperature, the black powder was recovered by centrifugation at 6000 rpm for 5 min and washed with ethanol for 3 times. For short Co nanorods, the mechanical stirring rate was 100 r/min; and for the long Co nanorods, the rate was set to 50 r/min.

2.5. The Self-Assembly of Co Nanorods

To obtain single-layer Co nanorods, we used OA and OAm, with a volume ratio of 1:3, as the surfactants. First, 0.0010 g of Co nanorods was dispersed 20 mL of above the mixed solution to achieve the ligand exchange under ultrasonic shock for 24 h. Then these nanorods were recovered by centrifugation and washed with hexane 3 times. Then the centrifuged Co nanorods were dispersed again in 20 mL methylbenzene, using ultrasonic shock. One or two droplets of Co nanorods/methylbenzene were dropped on the smooth surface of the water, and the system was sealed by using glass sheets to make the methylbenzene evaporate slowly for 4 h. A TEM grid was used to collect the self-assembly samples. For the magnetic field assisted self-assembly, a magnet was closed the water surface during the evaporation of organic solution. The other surfactants, HDA and PVP, and the other solutions, hexane and chloroform, were employed to replace the OA/OAm and methylbenzene, respectively. However, they failed to get the well-aligned Co layers.

2.6. Characterization

The Co nanorods' structure was studied by X-ray diffraction (XRD, D/MAX 2200 PC, Rigaku Corporation, Tokyo, Japan) with Cu-K_α radiation (λ = 0.15418 nm). The microstructure and morphology of the samples were analyzed by transmission electron microscopy (TEM, Philips CM 20, Philips, Amsterdam, Netherlands). HRTEM was performed on JEM-2100F (Japan Electronics jeol, Tokyo, Japan). The magnetic properties were measured at room temperature, using a vibrating sample magnetometer (VSM) under a maximum applied field of 30 kOe. Moreover, all of the magnetic properties were measured by using the random aligned Co rods.

3. Results and Discussions

3.1. The Self-Assembly of Co Nanorods with Cambered Tips

We first synthesized Co nanorods with cambered tips by solvothermal route by using a hydrothermal reactor, as shown in Figure 1. The XRD pattern (Figure 1A) demonstrates that the solvothermal route yields hexagonal-structured Co particles. All broad peaks have a good match with standard Co (JCPDS No. 01-1278) pattern. There are obvious differences among the width of the diffraction peaks (100), (002), and (101). The broader (002) plane indicates the smaller size perpendicular to [002] than other directions, suggesting the shape of the anisotropic particles that were obtained. The TEM results further confirm the successful synthesis of shape anisotropic Co nanoparticles. It can be seen from Figure 1B,C that the solvothermal reaction generates the uniform Co nanorods, with their diameter of about 10 nm and length of 100–150 nm. These as-prepared Co nanorods have obvious cambered tips and are prone to being distributed side by side. The microstructure of the Co nanorod was further observed by using HRTEM, as shown in Figure 1D. Along the length direction, the interplanar spacing of 0.20 nm was observed, matching well with the hcp-Co (002) plane, thus indicating that the length direction is consistent with the [002] direction. As we known, the [002] direction is the easy magnetization axis (c-axis). Therefore, such shape can contribute a large shape anisotropy, which can combine with magnetocrystalline anisotropy and give a high coercivity.

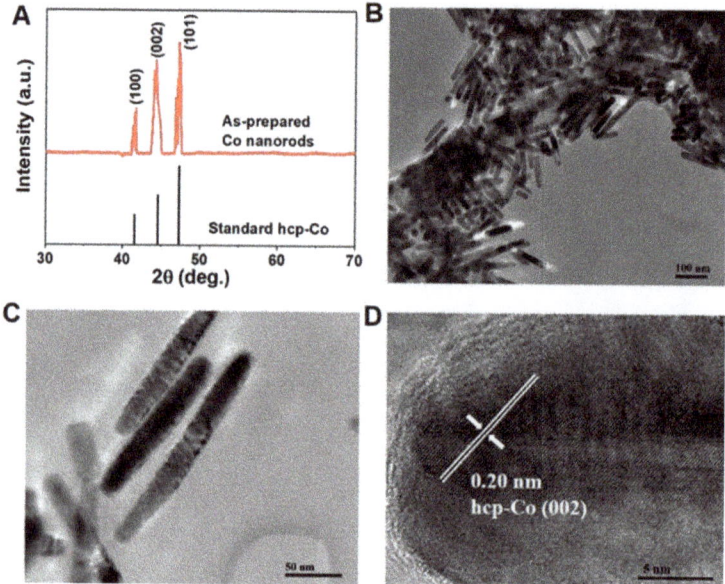

Figure 1. The Co nanorods with cambered tips by solvothermal route: (**A**) XRD pattern compared with standard hcp-Co (JCPDS No. 01-1278); (**B**,**C**) TEM images with different magnification; (**D**) HRTEM image.

We tried to make these Co nanorods self-assemble by the molecular force of surfactant without any magnetic field. The OA and OAm were employed as the surfactant, and the results are shown in Figure 2. From the TEM images in Figure 2A–C, we can see that the Co nanorods were well aligned into an assembly. Unfortunately, these Co nanorods are not aligned along a particular direction. These self-assembly presents radial, and the direction is from the center to all round. This phenomenon can be explained by the fact that some aggregates of Co nanorods become a hard magnetic core, similar to a spherical magnet, which yields a strong radial magnetic field, making other nanorods align along the magnetic field (Figure 2B). We can further observe that these nanorods are aligned side by side (Figure 2C), which may be caused by the magnetostatic interactions among nanorods. We also tried to use other surfactants to achieve the self-assembly, but they failed to align well due to the strong magnetic interaction. This result demonstrates that the magnetic interactions among nanorods are far stronger than the molecular interactions of surfactants. We measured the magnetic hysteresis loop of random Co nanorods at room temperature (Figure 2D), which exhibits a large coercivity of 6.4 kOe and high M_s of 148.3 emu/g under a 2.5 kOe field. Once these nanorods are aligned into a nanomagnet, using a magnetic field, their coercivity will exceed 10 kOe, being an excellent candidate for strong magnetic materials [30].

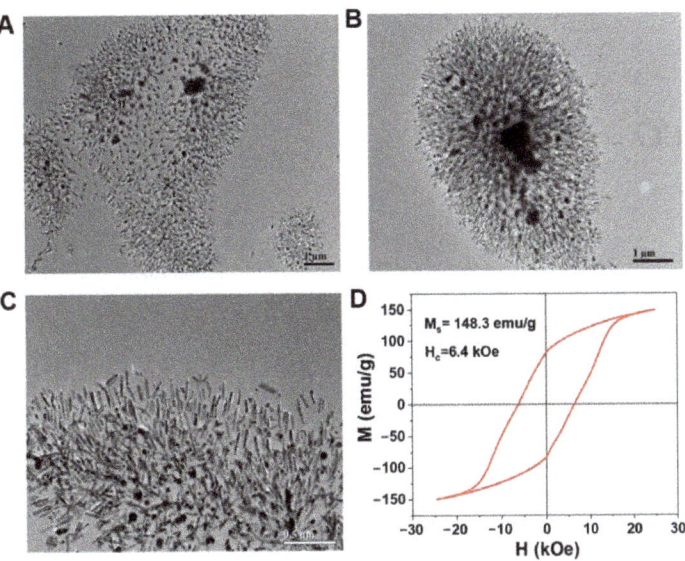

Figure 2. (**A–C**) The TEM images of self-assembly Co nanorods with cambered tips by solvothermal route; (**D**) Magnetic hysteresis loop of Co nanorods at room temperature.

3.2. The Self-Assembly of Co Nanorods with Flat Tips

As a comparison, we also synthesized Co nanorods with flat tips by using an alcohol–thermal method, using a reflux set, as shown in Figure 3. The XRD pattern (Figure 3A; the samples were prepared with a stirring rate of 100 r/min) can correspond to hexagonal structure Co very well. Moreover, we also can find the different peaks width for the (100), (002), and (101) plane, thus indicating that the as-prepared Co particles have a shape anisotropy. Different from Co nanorods obtained from solvothermal method, the diffraction peaks of Co (002) planes exhibit an obvious enhancement when compared with the standard peaks. This may suggest that these Co nanorods have a self-orientation effect. Figure 3B,C shows the TEM images of Co nanorods. The uniform Co nanorods display a bamboo-like shape with an average diameter of ~30 nm and length of ~100 nm. It is interesting that these Co nanorods have larger tips than their body, and their tips exhibit a flat plane. Without any magnetic field or surfactant, several Co nanorods are self-oriented tip to tip, forming a bamboo-like shape, which is totally different from these Co nanorods with cambered tips. We further synthesized a kind of longer Co nanorods by using a lower stirring rate of 50 r/min, as shown in Figure 3D. With the decrease in the stirring rate, there is no obvious change for the diameter of as-prepared Co nanorods, but their length does increase to 200 from 100 nm. Certainly, these longer Co nanorods still represent larger tips and smaller bodies, and they also have a self-orientation effect by going tip to tip. This demonstrates that the mechanical stirring can interrupt the growth of nanorods from the length direction; this can be explained by the fact that the growth of long Co nanorods requires Co^{2+} continuously feeding. With the increase of stirring rate, homogenizing the solution can hinder local increment of Co^{2+} concentration in the solution [31,32]. As a result, the growth of Co nanorods from the length direction was interrupted, forming multiple shorter nanorods.

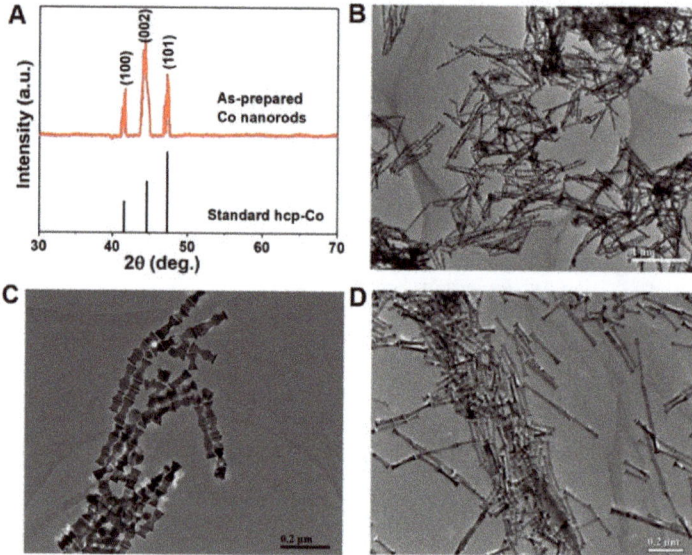

Figure 3. The Co nanorods with flat tips by alcohol–thermal method. (**A**) XRD pattern compared with standard hcp-Co (JCPDS No. 01-1278); (**B**,**C**) TEM images of Co nanorods with 100 r/min stirring rate; (**D**) TEM image of Co nanorods with 50 r/min stirring rate.

The magnetic hysteresis loop of random Co nanorods with the length of 100 nm was measured at room temperature, using VSM. As shown in Figure 4, these Co nanorods exhibit a coercivity of 4.9 kOe and M_s of 146.5 emu/g, at an applied field of 2.5 kOe. Compared with Co nanorods with cambered tips, these Co nanorods with flat tips exhibit a lower magnetic performance. This may be ascribed to two reasons. On the one hand, the ratio of length to diameter (L/D) is a key factor to the coercivity due to the shape anisotropy. In general, the higher ratio will lead to larger coercivity [33]. In our work, the Co nanorods with cambered tips have an L/D of ~10, while the L/D value for Co nanorods with flat tips is only ~4. On the other hand, the Co nanorods with flat tips have smaller coercivity than cambered tips, even though they have the similar L/D. Because Co nanorods with flat tips have more defects (stacking fault) between the tips and bodies, this can lead to a high demagnetizing field, and it causes the low coercivity [34].

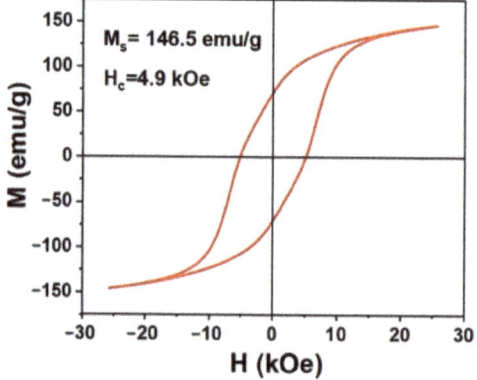

Figure 4. Magnetic hysteresis loop of Co nanorods with flat tips at room temperature.

The coercive mechanism of Co nanorods is totally different from small-sized Co nanoparticles or clusters. As we know, the magnetic particles have the largest coercivity at the key size from the multiple-domain to single-domain structure, and the key size is about 30 nm for Co particles [17]. Thus, our Co nanorods with a diameter of about 10 nm and length of 100–150 nm have been display a relatively higher coercivity than small-sized Co particles (several nanometers). Furthermore, these Co nanorods have been confirmed to be single-domain structures in previous work [27]. Moreover, for Co clusters with a small size, they are superparamagnetic and, thus, have no coercivity at room temperature, since their magnetocrystalline anisotropy at this size cannot compete with thermal disturbance [35]. More important, the coercivity of hcp-Co nanorods is mainly originated from shape anisotropy rather than magnetocrystalline anisotropy. In other words, the high L/D leads to their large coercivity, which can be confirmed by many reports [33].

These as-prepared Co nanorods with flat tips were further assembled on the water solution surface by using OA and OAm as the surfactant. Without any external field, driven by the molecular force of surfactant and magnetostatic force, these Co nanorods tend to align tip to tip and form long nanochains, but there is obvious space between aligned nanochains (Figure 5A). Although this alignment cannot reach our expectation, most of as-formed Co nanochains do present consistent direction, which is an obvious improvement compared to the Co nanorods with cambered tips. We also can observe the well-aligned Co nanorods in some view of TEM, as shown in Figure 5B. In these regions, these nanorods first form straight nanochains, and then these nanochains assemble into a single layer. Unfortunately, these regions are isolated from each other, making it hard to form high quality single-layer self-assembly. This result may be caused by magnetostatic interaction, which is discussed in the next section in detail. To obtain high-quality Co nanorods assembly, an external field was applied during the self-assembly process. The results are shown in Figure 5C,D. According to the XRD pattern in Figure 5C, the assembled Co nanorods under external field only display a single diffraction peak around 44.5°, which can be indexed to hcp-Co (002) plane. As we know, the easy axis of Co nanorods is the [002] direction, which is also the length direction. When these nanorods were aligned, all of Co nanorods displayed the same direction. As a result, the X-ray only detected the (002) plane. The TEM image in Figure 5D demonstrates that these Co nanorods with ~100 nm length were well aligned into a single layer along the magnetic field direction, further confirming the XRD result. We further aligned Co nanorods with a ~200 nm length under the same conditions, but it failed to obtain good alignment; this can be explained by the fact that the stronger properties of Co nanorods make them unfavorable to align due to the strong interaction with each other.

3.3. The Discussion of Magnetostatic Interaction Mechanism

During the self-assembly process, there is a competition between molecular interaction force and magnetostatic force. The former drives these nanorods to align well, while the latter makes them random to reduce the magnetostatic energy. In our work, the magnetostatic force was obviously stronger than the molecular interaction force. Therefore, we should analyze their behavior from the view of magnetostatic interaction.

We analyzed the magnetostatic interaction mechanism by using magnetic induction line distribution, as shown in Figure 6. All Co nanorods can be regarded as a tiny magnet at the nanoscale, and the two tips of nanorods correspond to their N–S poles (Figure 6(A1,B1)) [36]. Moreover, the magnetic induction lines at two tips should be perpendicular to their tips.

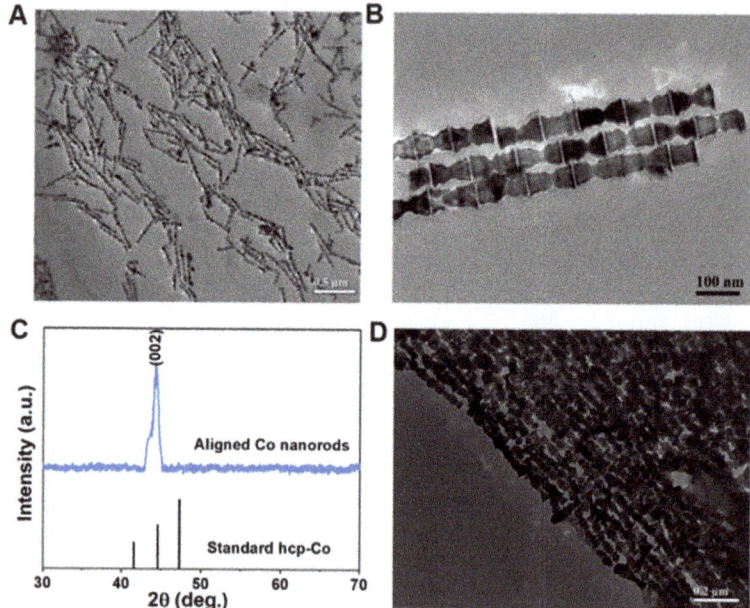

Figure 5. (**A**,**B**) The TEM images of self-assembly Co nanorods with flat tips, without applied field; (**C**) XRD pattern of aligned Co nanorods under magnetic field compared with standard hcp-Co (JCPDS No. 01-1278); (**D**) TEM image of aligned Co nanorods under magnetic field.

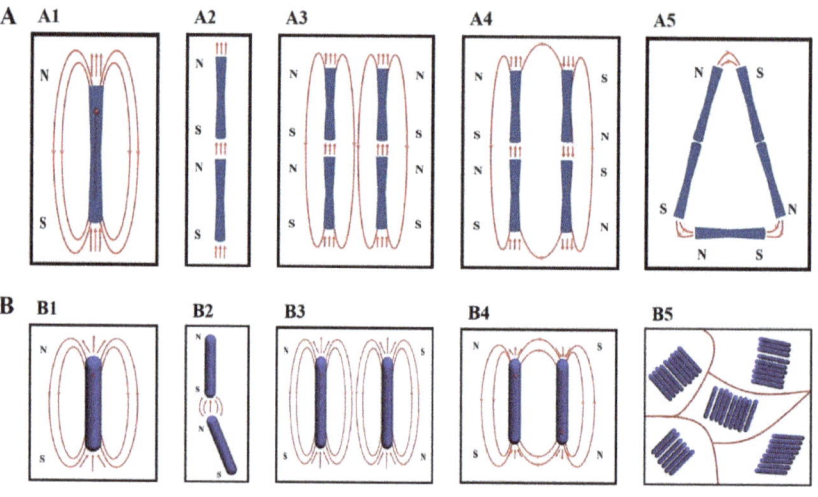

Figure 6. The scheme of magnetostatic interaction mechanism: (**A**) Co nanorods with flat tips and (**B**) Co nanorods with cambered tips.

For the Co nanorods with flat tips, they have larger tips. Therefore, most of magnetic induction lines are perpendicular to the flat tips and, thus, parallel to the [002] direction (Figure 6(A1)). As a result, these Co nanorods are prone to holding together tip to tip since the magnetic induction lines direction for the N pole of one Co nanorod is consistent with the S pole of another Co nanorod (Figure 6(A2)). If these nanochains assembled by Co nanorods had the same direction for N–S poles, they would present mutual exclusion, and, thus, these nanochains would be separated from each other (Figure 6(A3)). If they had the opposite N–S poles, they would hold together again by connecting the third nanochains and rotate at a certain angle to keep the minimum energy (Figure 6(A4,A5)). When an external field was applied, these nanochains can rotate into the magnetic field direction to achieve alignment.

For the Co nanorods with cambered tips, they have smaller tips. Due to the special morphology, they only have very few magnetic induction lines parallel to the c-axis. Most of the magnetic inductions line are random (Figure 6(B1)). Therefore, tip-to-tip assembly is very unstable. It needs to rotate at a certain angle to reduce the energy of system (Figure 6(B2)). On this occasion, these nanorods are prone to holding together side by side for their random magnetic induction lines, since there are some angles between most of magnetic induction lines and nanorods c-axis (Figure 6(B3,B4)). Despite that, the alignment side by side was confined to several nanorods [37]. These clusters assembled by several nanorods with the same direction can be regarded as a "magnetic domain", and the system is similar to a "multidomain magnet" (Figure 6(B5)). To keep the minimum energy of the whole system, these "magnetic domain" directions are distributed randomly. When an external field or self-generated field (some nanorods aggregate similar to a magnet, as seen in Figure 2B) is applied, these "magnetic domains" can rotate into the magnetic field direction to achieve alignment.

4. Conclusions

In summary, we synthesized two kinds of Co nanorods, one with cambered tips and another with flat tips, and achieved a self-assembly of a single layer of Co nanorods to study their magnetostatic interaction behavior. The as-synthesized Co nanorods with cambered tips have an average diameter of 10 nm and length of 100 nm, with coercivity of 6.4 kOe; and flat-tip nanorods with a 30 nm diameter and 100 nm length exhibit a coercivity of 4.9 kOe. These Co nanorods were first assembled on the surface of water, without a magnetic field. It is found that the Co nanorods with cambered tips are prone to aligning side by side, while the Co nanorods with flat tips are easy to hold together tip to tip due to the magnetostatic interaction. Under an applied field, the Co nanorods with flat tips can be further aligned into a single layer of Co nanorods. Furthermore, we studied the magnetostatic interaction mechanism, using magnetic induction lines. Because each nanorod can be considered as a magnet, the flay tips of Co nanorods are similar to the N and S poles of a magnet. Therefore, they are easy to hold together tip to tip since most of magnetic induction lines are parallel to the length direction. For Co nanorods with cambered tips, they are prone to holding together side by side to form clusters for their random magnetic induction lines. Our work provides a method to study magnetic interaction of shape anisotropy.

Author Contributions: Conceptualization, H.D., M.Z. and Z.M.; Data curation, H.D. and Z.M.; Funding acquisition, B.L. and Z.M.; Investigation, H.D., M.Z. and Z.M.; Methodology, H.D., M.Z. and Z.M.; Project administration, H.D., M.Z., B.L. and Z.M.; Resources, B.L. and Z.M.; Supervision, B.L. and Z.M.; Validation, K.Y., B.L. and Z.M.; Visualization, H.D. and M.Z.; Writing—original draft, H.D. and M.Z.; Writing—review & editing, H.D., M.Z., K.Y., B.L. and Z.M. All authors have read and agreed to the published version of the manuscript.

Funding: This work was financially supported by the Capacity Building for Scientific and Technological Innovation Services Project of Beijing Technology and Business University (No. 19008021178), and Beijing Technology and Business University Research Team Construction Project (No. 19008022165).

Conflicts of Interest: The authors declare no conflict of interest.

References

1. Ma, Z.H.; Yue, M.; Liu, H.; Yin, Z.Y.; Wei, K.C.; Guan, H.Q.; Lin, H.H.; Shen, M.Q.; An, S.Z.; Wu, Q.; et al. Stabilizing hard magnetic SmCo$_5$ nanoparticles by n-doped graphitic carbon layer. *J. Am. Chem. Soc.* **2020**, *142*, 8440–8446. [CrossRef] [PubMed]
2. Wang, S.; Xu, J.; Li, W.; Sun, S.; Gao, S.; Hou, Y. Magnetic nanostructures: Rational design and fabrication strategies toward diverse applications. *Chem. Rev.* **2022**, *122*, 5411–5475. [CrossRef] [PubMed]

3. Zhao, P.; Luo, Y.; Yu, D.B.; Peng, H.J.; Yan, W.L.; Wang, Z.L.; Bai, X.Y. Preparation and properties of hot-deformed magnets processed from nanocrystalline/amorphous Nd-Fe-B powders. *Rare Met.* **2021**, *40*, 2033–2039. [CrossRef]
4. Ma, Z.; Tian, H.; Cong, L.; Wu, Q.; Yue, M.; Sun, S. A flame-reaction method for the large-scale synthesis of high-performance Sm_xCo_y nanomagnets. *Angew. Chem.-Int. Ed.* **2019**, *58*, 14509–14512. [CrossRef]
5. Yang, W.; Liang, D.; Kong, X.; Yang, J. Neutron diffraction studies of permanent magnetic materials. *Rare Met.* **2020**, *39*, 13–21. [CrossRef]
6. Shen, B.; Yu, C.; Baker, A.A.; McCall, S.K.; Yu, Y.; Su, D.; Yin, Z.; Liu, H.; Li, J.; Sun, S. Chemical synthesis of magnetically hard and strong rare earth metal based nanomagnets. *Angew. Chem.-Int. Ed.* **2019**, *58*, 602–606. [CrossRef]
7. Ma, Z.; Du, H. Stabilizing interface of $SmCo_5$/Co nanocomposites by graphene shells. *Rare Met.* **2022**, *41*, 1223–1229. [CrossRef]
8. Xu, G.; Lu, H.; Guo, K.; Tang, F.; Song, X. Predictions on the phase constitution of $SmCo_{7-X}M_X$ alloys by data mining. *Nanomaterials* **2022**, *12*, 1452. [CrossRef]
9. Guan, H.Q.; Huang, S.S.; Ding, J.H.; Tian, F.Y.; Xu, Q.; Zhao, J.J. Chemical environment and magnetic moment effects on point defect formations in CoCrNi-based concentrated solid-solution alloys. *Acta Mater.* **2020**, *187*, 122–134. [CrossRef]
10. Zhang, G.; Zhang, Z.; Sun, M.; Yu, Y.; Wang, J.; Cai, S.B. The influence of the temperature on the dynamic behaviors of magnetorheological gel. *Adv. Eng. Mater.* **2020**, 2101680. [CrossRef]
11. Lu, C.J.; Zhou, H.; Li, L.F.; Yang, A.C.; Xu, C.B.; Ou, Z.Y.; Wang, J.Q.; Wang, X.; Tian, F. Split-core magnetoelectric current sensor and wireless current measurement application. *Measurement* **2022**, *188*, 110527. [CrossRef]
12. Li, S.M.; Zhang, S.Q.; Zhao, R.Q. Regulating the electronic and magnetic properties of 1T'-ReS_2 by fabricating nanoribbons and transition-metal doping: A theoretical study. *Nanoscale* **2022**, *14*, 8454–8462. [CrossRef] [PubMed]
13. Cui, J.; Kramer, M.; Zhou, L.; Liu, F.; Gabay, A.; Hadjipanayis, G.; Balasubramanian, B.; Sellmyer, D. Current progress and future challenges in rare-earth-free permanent magnets. *Acta Mater.* **2018**, *158*, 118–137. [CrossRef]
14. Li, H.; Wu, Q.; Yue, M.; Peng, Y.; Li, Y.; Liang, J.; Wang, D.; Zhang, J. Magnetization reversal in cobalt nanowires with combined magneto-crystalline and shape anisotropies. *J. Magn. Magn. Mater.* **2019**, *481*, 104–110. [CrossRef]
15. Jia, Y.; Wu, Y.; Zhao, S.; Zuo, S.; Skokov, K.P.; Gutfleisch, O.; Jiang, C.; Xu, H. L1(0) rare-earth-free permanent magnets: The effects of twinning versus dislocations in Mn-Al magnets. *Phys. Rev. Mater.* **2020**, *4*, 094402. [CrossRef]
16. Crisan, A.D.; Leca, A.; Bartha, C.; Dan, I.; Crisan, O. Magnetism and epsilon-tau phase transformation in MnAl-based nanocomposite magnets. *Nanomaterials* **2021**, *11*, 896. [CrossRef]
17. Ma, Z.; Mohapatra, J.; Wei, K.; Liu, J.P.; Sun, S. Magnetic nanoparticles: Synthesis, anisotropy, and applications. *Chem. Rev.* **2021**. [CrossRef]
18. Proenca, M.P. Multifunctional magnetic nanowires and nanotubes. *Nanomaterials* **2022**, *12*, 1308. [CrossRef]
19. Ma, Z.; Yue, M.; Wu, Q.; Li, C.; Yu, Y. Designing shape anisotropic $SmCo_5$ particles by chemical synthesis to reveal the morphological evolution mechanism. *Nanoscale* **2018**, *10*, 10377–10382. [CrossRef]
20. Mohapatra, J.; Xing, M.; Elkins, J.; Liu, J.P. Hard and semi-hard magnetic materials based on cobalt and cobalt alloys. *J. Alloys Compd.* **2020**, *824*, 153874. [CrossRef]
21. Ma, Z.; Liang, J.; Ma, W.; Cong, L.; Wu, Q.; Yue, M. Chemically synthesized anisotropic $SmCo_5$ nanomagnets with a large energy product. *Nanoscale* **2019**, *11*, 12484–12488. [CrossRef]
22. Ma, Z.; Zhang, T.; Jiang, C. A facile synthesis of high performance $SmCo_5$ nanoparticles. *Chem. Eng. J.* **2015**, *264*, 610–616. [CrossRef]
23. Zhang, D.; Zhu, R.; Yue, M.; Liu, W. Microstructure and magnetic properties of $SmCo_5$ sintered magnets. *Rare Met.* **2020**, *39*, 1295–1299. [CrossRef]
24. Kwon, J.; Lee, D.; Yoo, D.; Park, S.; Cha, H.; Kwon, H.; Lee, J.; Lee, D. Enhancement of magnetic properties of hot pressed/die-upset Dy-free Nd-Fe-B magnets with Cu/Nd coating by wet process. *Rare Met.* **2020**, *39*, 48–54. [CrossRef]
25. Zeng, H.; Zheng, M.; Skomski, R.; Sellmyer, D.J.; Liu, Y.; Menon, L.; Bandyopadhyay, S. Magnetic properties of self-assembled Co nanowires of varying length and diameter. *J. Appl. Phys.* **2000**, *87*, 4718–4720. [CrossRef]
26. Gandha, K.; Mohapatra, J.; Liu, J.P. Coherent magnetization reversal and high magnetic coercivity in Co nanowire assemblies. *J. Magn. Magn. Mater.* **2017**, *438*, 41–45. [CrossRef]
27. Gandha, K.; Elkins, K.; Poudyal, N.; Liu, X.; Liu, J.P. High energy product developed from cobalt nanowires. *Sci. Rep.* **2014**, *4*, 1–5. [CrossRef]
28. Anagnostopoulou, E.; Grindi, B.; Lacroix, L.M.; Ott, F.; Panagiotopoulos, I.; Viau, G. Dense arrays of cobalt nanorods as rare-earth free permanent magnets. *Nanoscale* **2016**, *8*, 4020–4029. [CrossRef]
29. Ener, S.; Anagnostopoulou, E.; Dirba, I.; Lacroix, L.-M.; Ott, F.; Blon, T.; Piquemal, J.-Y.; Skokov, K.P.; Gutfleisch, O.; Viau, G. Consolidation of cobalt nanorods: A new route for rare-earth free nanostructured permanent magnets. *Acta Mater.* **2018**, *145*, 290–297. [CrossRef]
30. Li, C.; Wu, Q.; Yue, M.; Xu, H.; Palaka, S.; Elkins, K.; Liu, J.P. Manipulation of morphology and magnetic properties in cobalt nanowires. *AIP Adv.* **2017**, *7*, 056229. [CrossRef]
31. Patil, S.; Kate, P.R.; Deshpande, J.B.; Kulkarni, A.A. Quantitative understanding of nucleation and growth kinetics of silver nanowires. *Chem. Eng. J.* **2021**, *414*, 128711. [CrossRef]
32. Amirjani, A.; Fatmehsari, D.H.; Marashi, P. Interactive effect of agitation rate and oxidative etching on growth mechanisms of silver nanowires during polyol process. *J. Exp. Nanosci.* **2015**, *10*, 1387–1400. [CrossRef]

33. Li, H.; Wu, Q.; Peng, Y.; Xu, H.; Zhang, J.; Yue, M. Magnetic properties and magnetization reversal in Co nanowires with different morphology. *J. Magn. Magn. Mater.* **2019**, *469*, 203–210. [CrossRef]
34. Mrad, K.; Schoenstein, F.; Nong, H.T.T.; Anagnostopoulou, E.; Viola, A.; Mouton, L.; Mercone, S.; Ricolleau, C.; Jouini, N.; Abderraba, M.; et al. Control of the crystal habit and magnetic properties of Co nanoparticles through the stirring rate. *Crystengcomm* **2017**, *19*, 3476–3484. [CrossRef]
35. Jamet, M.; Wernsdorfer, W.; Thirion, C.; Mailly, D.; Dupuis, V.; Melinon, P.; Perez, A. Magnetic anisotropy of a single cobalt nanocluster. *Phys. Rev. Lett.* **2001**, *86*, 4676–4679. [CrossRef]
36. Ma, Z.; Liu, H.; Yue, M. Magnetically recyclable Sm_2Co_{17}/Cu catalyst to chemoselectively reduce the 3-nitrostyrene into 3-vinylaniline under room temperature. *Nano Res.* **2019**, *12*, 3085–3088. [CrossRef]
37. Wu, Q.; Cong, L.; Yue, M.; Li, C.; Ma, Z.; Ma, X.; Wang, Y. A unique synthesis of rare-earth-Co-based single crystal particles by "self-aligned" Co nano-arrays. *Nanoscale* **2020**, *12*, 13958–13963. [CrossRef]

Article

Theoretical Study on the Electronic Structure and Magnetic Properties Regulation of Janus Structure of M'MCO$_2$ 2D MXenes

Panpan Gao [1,2], Minhui Song [1], Xiaoxu Wang [1], Qing Liu [1,2,3], Shizhen He [1], Ye Su [1,*] and Ping Qian [1,2,*]

1. Beijing Advanced Innovation Center for Materials Genome Engineering, University of Science and Technology Beijing, Beijing 100083, China; gaopanpan@ustb.edu.cn (P.G.); mhuisong@163.com (M.S.); wangxx@dp.tech (X.W.); liuqing6903@163.com (Q.L.); hsz0430@163.com (S.H.)
2. School of Mathematics and Physics, University of Science and Technology Beijing, Beijing 100083, China
3. Department of Physics, National University of Singapore, Singapore 117551, Singapore
* Correspondence: suyechina@163.com (Y.S.); qianping@ustb.edu.cn (P.Q.)

Abstract: Motivated by the recent successful synthesis of Janus monolayer of transition metal (TM) dichalcogenides, MXenes with Janus structures are worthy of further study, concerning its electronic structure and magnetic properties. Here, we study the effect of different transition metal atoms on the structure stability and magnetic and electronic properties of M'MCO$_2$ (M' and M = V, Cr and Mn). The result shows the output magnetic moment is contributed mainly by the d orbitals of the V, Cr, and Mn atoms. The total magnetic moments of ferromagnetic (FM) configuration and antiferromagnetic (AFM) configuration are affected by coupling types. FM has a large magnetic moment output, while the total magnetic moments of AFM2's (intralayer AFM/interlayer FM) configuration and AFM3's (interlayer AFM/intralayer AFM) configuration are close to 0. The band gap widths of VCrCO$_2$, VMnCO$_2$, CrMnCO$_2$, V$_2$CO$_2$, and Cr$_2$CO$_2$ are no more than 0.02 eV, showing metallic properties, while Mn$_2$CO$_2$ is a semiconductor with a 0.7071 eV band gap width. Janus MXenes can regulate the size of band gap, magnetic ground state, and output net magnetic moment. This work achieves the control of the magnetic properties of the available 2D materials, and provides theoretical guidance for the extensive design of novel Janus MXene materials.

Keywords: janus; MXenes; magnetic properties; DFT

1. Introduction

Over the past decade, the two-dimensional (2D) materials have received significant interest since the discovery of graphene [1–3]. Compared with bulk materials, more atoms on the surface of 2D materials are exposed, which is caused by reduced dimensionality. This improves the utilization rates of atoms and makes regulation of band structure and electronic properties easier, thus enabling MXenes to exhibit novel physical and chemical properties [4–6]. Recently, a new family of 2D transition metal (TM) carbides and nitrides, MXenes, has received more and more attention [7,8]. MXenes have the general formula M$_{n+1}$X$_n$T$_x$, where M stands for early TM, X represents C or N, T$_x$ indicates the surface functional groups O, OH, or F, and n = 1, 2, or 3 [9,10]. Usually, MXenes are synthesized by selective etching A layers (A is an element from the A-group 13 or 14) in the MAX phase, using hydrofluoric acid (HF) solutions [11–13]. Sue to their excellent electrical [14], optical [15–18], and mechanical properties [19], MXenes have been widely applied in electronic devices [20], catalysis [21–24], magnetic storage [25], energy conversion, and storage systems [26–28]. Thus far, more than 30 kinds of MXenes have been synthesized in experiments, and more kinds of materials have been theoretically predicted [29,30].

The electronic and optical properties of MXenes with symmetrical configuration have been extensively studied. Previous theoretical investigations have shown that, without surface functionalization, Cr_2C is half metallic and ferromagnetic (FM) configuration [31], V_2C exhibits metallic and antiferromagnetic (AFM) configuration [32], and $Ti_{n+1}C_n$ and $Ti_{n+1}N_n$ (n = 1–9) show magnetic configuration [33]. While functionalized MXenes alter magnetism, others, such as Cr_2CX_2 (X=OH, O and F), V_2CX_2 (X=F, OH), and Ti_2CO_2, are semiconductors [31–35]. Structural symmetry is a key factor in determining the electronic properties of 2D materials [36,37]. If structural symmetry is broken, it is desirable for 2D materials to have electronic and magnetic properties.

Inspired by the successful synthesis of Janus monolayers of TM dichalcogenides [38], MXenes with Janus structures are worth studying further, especially concerning their electronic structures and magnetic properties. Janus refers to MXenes that break the symmetry through asymmetric surface functional groups or different types of TM elements [39]. A previous report has theoretically studied the electronic and magnetic properties of Janus MXenes; it indicated that, by selecting an appropriate terminal group of upper and lower surfaces, the band gap of Janus MXenes can be successfully adjusted to different regions [40]. Therefore, we consider that different TM atoms may also regulate the charge and chemical environment around the atom, which causes Janus MXenes to exhibit significantly different electronic and magnetic properties.

In this paper, using first-principles calculations, we employed $M'MCO_2$ (M' and M stands for V, Cr, and Mn) configurations to investigate the effect of different types of TM on the structure and the magnetic and electronic properties under the same functional groups. We constructed different magnetic configurations (nonmagnetic (NM), FM, and AFM) for each $M'MCO_2$ structure, researched their magnetic properties, and screened out the magnetic ground state. Then, we studied the electronic structure of the magnetic ground state. The results showed that Janus MXenes can adjust the band gap, the magnetic ground state, and the net output magnetic moments, which is a very good control method. Due to its asymmetric structure, Janus MXenes can flexibly control the magnetism of a system by applying small electric fields. This work provides theoretical guidance for the realization of the magnetic controllability of MXene materials.

2. Materials and Methods

All calculations were carried out using the Vienna ab initio simulation package (VASP), based on density functional theory (DFT) [41,42]. The generalized gradient approximation (GGA), with the Perdew–Burke–Ernzerhof (PBE) functional, was used for the exchange and correlation functional [43,44]. Interactions between electrons and nuclei were described by the projector augmented wave (PAW) method [45]. A plane wave kinetic energy cutoff 600 eV was employed. The convergence criteria of total energy and atomic force for each atom were set to 10^{-5} eV per unit cell and 10^{-4} eV/Å, respectively.

To account for the energy of localized 3D orbitals of TM atoms properly, the Hubbard "U" correction was employed within the rotationally invariant DFT + U approach proposed [46]. The spin-polarized DFT + U correction [47,48] was applied to strongly correlated Cr, V, and Mn atoms with the typical U = 4 eV value. The specific U value does not change the predicted magnetic ordering nor the easy axis determination [49,50]. The cutoff kinetic energy for plane waves was set to 600 eV. Considering the van der Waals interaction between layers, the Becke–Jonson attenuation DFT + D3 method was performed for empirical correction.

A vacuum spacing of 20 Å along the $M'MCO_2$ normal was used to avoid the interactions caused by the periodic boundary condition. The Brillouin zone (BZ) was sampled using $11 \times 11 \times 1$ Γ-centered, k-point Monkhorst-Pack grids for the calculations of relaxation and electronic structures for NM, FM, and AFM1 primitive cells. Additionally, k-mesh was decreased to $6 \times 12 \times 1$ for $2 \times 1 \times 1$ AFM2 and AFM3 supercells. In the static self-consistent calculation, k-point grid sampling of $20 \times 20 \times 1$ was used for the primitive cell, and k-point grid sampling of $12 \times 24 \times 1$ was used for the $2 \times 1 \times 1$ supercell.

3. Results

3.1. Stable Structures of Janus MXenes

The monolayer M_2C MXene is a centered honeycomb structure with P3m1 symmetry, in which the 2D hexagonal C atom is sandwiched between two hexagonal M atoms. There are four possible configurations for O atoms absorbed on the M atom [51]: (a) O atoms located right above the M atoms (top sites); (b) O atoms located at the hollow sites of adjacent C atoms (hcp sites); (c) O atoms located at the hollow sites of contralateral M atoms (fcc sites); (d) on the one side, O atoms are at fcc sites, and on the other side, O atoms are at the hcp sites. According to previous research by Tan [52] and Wang [53], (b) configuration is stable for $CrMnCO_2$ and Cr_2CO_2 and (c) configuration is stable for $VCrCO_2$, $VMnCO_2$, V_2CO_2, and Mn_2CO_2. So, we selected those configurations for the following calculations. Figure 1 shows the structures of symmetric V_2CO_2, Cr_2CO_2, and Mn_2CO_2. The arrangement of atoms observed from the top and bottom is the same. Figure 2 shows the structures of Janus MXenes $VCrCO_2$, $VMnCO_2$, and $CrMnCO_2$. It can be seen that the arrangement of atoms seen from the top and bottom is different, so the symmetry of $VCrCO_2$, $VMnCO_2$, and $CrMnCO_2$ is lower than that of V_2CO_2, Cr_2CO_2, and Mn_2CO_2, which is consistent with the symmetry of their space group. The basic information of their lattice parameters is shown in Table 1.

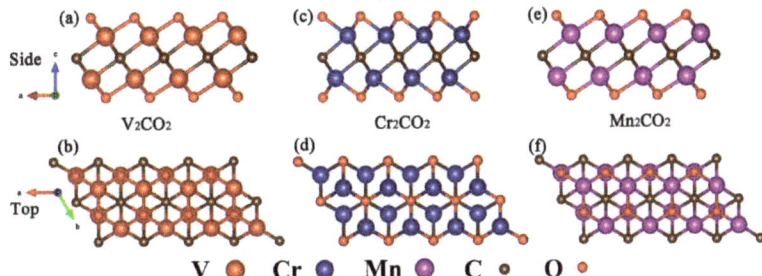

Figure 1. V_2CO_2, Cr_2CO_2, and Mn_2CO_2 structures: (a,c,e) and (b,d,f) are the side and top views of V_2CO_2, Cr_2CO_2, and Mn_2CO_2 structures, respectively.

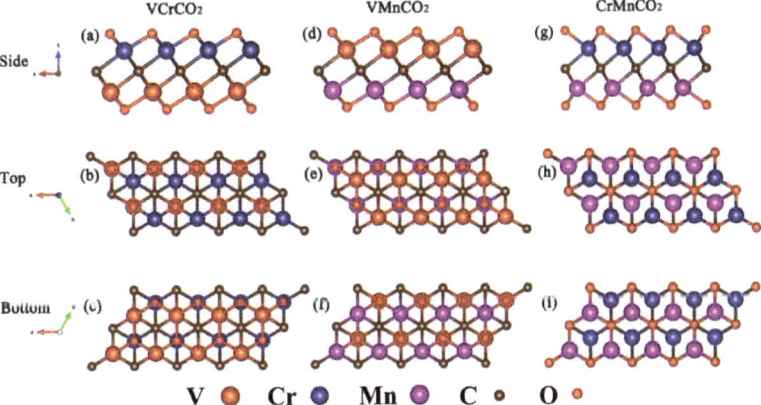

Figure 2. $VCrCO_2$, $VMnCO_2$, and $CrMnCO_2$ structures: (a,d,g) and (b,e,h) and (c,f,i) are the side, top, and bottom views of $VCrCO_2$, $VMnCO_2$, and $CrMnCO_2$ structures, respectively.

Table 1. The lattice parameters of M'MCO$_2$.

	VCrCO$_2$	VMnCO$_2$	CrMnCO$_2$	V$_2$CO$_2$	Cr$_2$CO$_2$	Mn$_2$CO$_2$
Symmetry Group	P3m1 (C_{3V-1})	P3m1 (C_{3V-1})	P3m1 (C_{3V-1})	P$\bar{3}$m1 (D_{3d-3})	P$\bar{3}$m1 (D_{3d-3})	P$\bar{3}$m1 (D_{3d-3})
a/Å	2.88	2.89	2.66	2.88	2.68	2.87
b/Å	2.88	2.89	2.66	2.88	2.68	2.87
c/Å	21.81	21.81	21.81	21.81	21.81	21.81
α	90°	90°	90°	90°	90°	90°
β	90°	90°	90°	90°	90°	90°
γ	120°	120°	120°	120°	120°	120°

Many compounds of V, Cr, and Mn are magnetic [39]. We calculated the total energy of the non-spin-polarized system and the spin-polarized system, respectively, by using standard DFT method: the result is shown in Table 2. It can be seen that, except V$_2$CO$_2$, the total energy of the spin polarization is lower than that of the non-spin polarization. When taking spin polarization into account, obvious magnetic moment can be observed in magnetic atoms. Therefore, the ground states of VCrCO$_2$, VMnCO$_2$, CrMnCO$_2$, Cr$_2$CO$_2$, and Mn$_2$CO$_2$ must be magnetic, while the ground state of V$_2$CO$_2$ is NM. Although the ground states of V$_2$CO$_2$ are NM, the difference between the NM and the magnetic state is very small (about 0.0002 eV). When considering the spin polarization in the system, the V atom has about 1 μB/atom sized magnetic moment. Under certain conditions, NM may become magnetic, so it is necessary to study its magnetic properties.

Table 2. Total energy of spin-polarized and non-spin-polarized systems. NM stands for non-spin-polarized system, magnetic stands for spin-polarized systems, the unit of total energy is eV/u.c (unit cell).

	VCrCO$_2$	VMnCO$_2$	CrMnCO$_2$	V$_2$CO$_2$	Cr$_2$CO$_2$	Mn$_2$CO$_2$
NM	−44.322	−42.926	−42.488	−45.597 (5)	−43.804	−40.701
Magnetic	−44.502	−43.451	−42.499	−45.597 (3)	−43.882	−41.431

Considering the magnetism of the TM atoms, we employed the FM and AFM order for each M'MCO$_2$ configuration, as shown in Figure 3. For FM configurations, the magnetic moments of M atoms are parallel, while for AFM configurations, the magnetic moments are antiparallel with each other. According to the different coupling kinds between atoms, we constructed three different AFM configurations, named AFM1, AFM2, and AFM3. AFM1 configuration is characterized by intralayer FM coupling and interlayer AFM coupling; AFM2 configuration is characterized by intralayer AFM coupling and interlayer FM coupling; AFM3 configuration is characterized by AFM coupling of atoms both intralayer and interlayer. The initial models of NM, FM, and AFM configurations are completely the same, but due to the different coupling models of magnetic atoms, after structure optimization, different configurations will have different lattice parameters, total energy, electronic structure, and other aspects. Compared to the initial models, the lattice constants of NM hardly change, while the corresponding FM and AFM are slight increased. Meanwhile, FM and AFM configurations symmetries are also reduced, which is related to the coupling between magnetic atoms after spin polarization, as shown in Tables S1–S6.

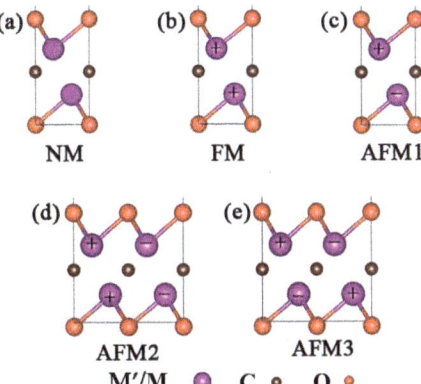

Figure 3. M'MCO$_2$ magnetic state configurations: (**a**–**e**) represent NM, FM, AFM1, AFM2, and AFM3 configurations, respectively. In M'/M atoms, "+" represents spin-up and "−" represents spin-down.

3.2. Magnetic Properties

The spin polarization can be corrected by adopting the DFT + U method, but it will introduce additional situ coulomb interaction energy. However, U is not added in the case of the non-spin-polarized system, which means that the total energy of the two system do not have comparability. Therefore, in the following analysis, we study the electronic structures and magnetic properties of the ground states of VCrCO$_2$, VMnCO$_2$, CrMnCO$_2$, V$_2$CO$_2$, Cr$_2$CO$_2$, and Mn$_2$CO$_2$ configurations under spin polarization, regardless of NM. The calculated total energy is shown in Table 3. The coupling effect between magnetic configurations is different, making the corresponding energy and other properties different. The total energy difference between different magnetic states is very small, which means that the ground states of AFM and FM are unstable in a specific environment. For each M'MCO$_2$ configuration, the configuration with the lowest energy is its most stable configuration state—the magnetic ground state; moreover, this is the focus of the present analysis and study. From the value, we can obtain that the magnetic ground states of VCrCO$_2$, VMnCO$_2$, and Cr$_2$CO$_2$ are FM, and that the ground state energies are −37.321 eV/u.c, −36.936 eV/u.c, and −35.488 eV/u.c, respectively. The magnetic ground state of CrMnCO$_2$ is AFM3, and the ground state energy is −35.707 eV/u.c. The ground state of V$_2$CO$_2$ is NM, the magnetic ground state is AFM1, and the magnetic ground state energy is −38.020 eV/u.c. The magnetic ground state of Mn$_2$CO$_2$ is AFM2, and the ground state energy is −35.989 eV/u.c. In previous studies on the magnetic properties of MXenes, Mohammad Khazaei [35] and Tan [54] calculated the magnetic ground state of Cr$_2$CO$_2$ as FM, while Hu [32] obtained the magnetic ground state of V$_2$CX$_2$ (X=F, OH) as AFM. The conclusion of these studies is consistent with our calculated results. Additionally, we can conclude that, when replacing TM atoms, symmetric MXenes V$_2$CO$_2$, Cr$_2$CO$_2$, and Mn$_2$CO$_2$ become Janus MXenes VCrCO$_2$, VMnCO$_2$, and CrMnCO$_2$, respectively; the magnetic ground states will change, and Janus MXenes can regulate the magnetic ground states.

Table 3. The total energy of FM and AFM structures of M'MCO$_2$, the unit of total energy is eV/u.c (unit cell).

	FM	AFM1	AFM2	AFM3
VCrCO$_2$	−37.321	−37.215	−37.223	−37.223
VMnCO$_2$	−36.936	−36.921	−36.921	−36.886
CrMnCO$_2$	−34.778	−34.349	−34.743	−35.707
V$_2$CO$_2$	−38.013	−38.020	−37.911	−37.918
Cr$_2$CO$_2$	−35.488	−34.794	−35.263	−35.365
Mn$_2$CO$_2$	−35.934	−35.913	−35.989	−35.746

The magnetic moments of all magnetic configurations of each M'MCO$_2$ are summarized in Tables S7–S12. Additionally, we drew the curves of the magnetic moments under different magnetic configurations and M'MCO$_2$ configurations, as shown in Figure 4. From the contribution of atomic species to the magnetic moments, we can find that the magnetic moments of C and O are close to 0 and the magnetic moment is mainly contributed by the TM atoms V, Cr, and Mn. In addition, the curves satisfied $\mu(Mn) > \mu(Cr) > \mu(V)$ because of the different electron numbers—Mn has one more electron than Cr, and Cr has one more electron than V. In the FM configurations, the order of total magnetic moment is $\mu(Mn_2CO_2) > \mu(MnCrCO_2) > \mu(Cr_2CO_2) > \mu(VMnCO_2)$ $\mu(VCrCO_2) > \mu(V_2CO_2)$. The large the specific gravity of Mn in the configuration, the greater the net magnetic moment; the large the specific gravity of V, the smaller the net magnetic moment, which agrees with $\mu(Mn) > \mu(Cr) > \mu(V)$. Because the magnetic moment of V atom is small, when the TM atom in MXene is V, the stable configuration tends to be nonmagnetic. This means that the ground state of V$_2$CO$_2$ is nonmagnetic, while the ground states of VCrCO$_2$, VMnCO$_2$, CrMnCO$_2$, Cr$_2$CO$_2$, and Mn$_2$CO$_2$ are magnetic. As the magnetic order changes, the magnetic moments of V, Cr, and Mn have little change, and it basically maintains a horizontal trend within the range of 0.50 μB. It indicates that the type of TM atom is a decisive factor for the magnetic moment; moreover, the environment of atoms and the coupling mode between atoms have little effect on the magnetic moment.

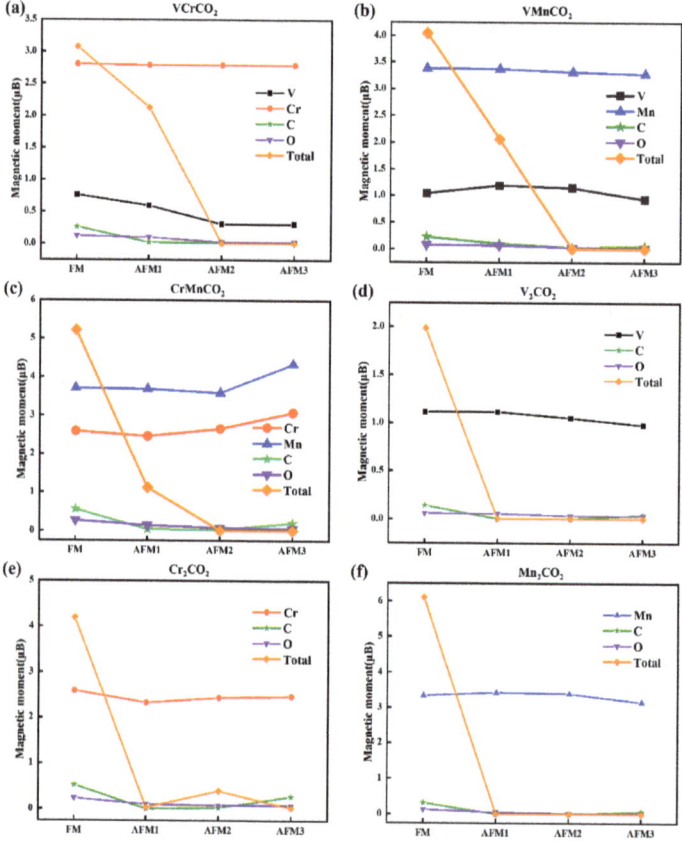

Figure 4. Magnetic moment curves under different M'MCO$_2$ configurations. (**a**) VCrCO$_2$ configuration; (**b**) VMnCO$_2$ configuration; (**c**) CrMnCO$_2$ configuration; (**d**) V$_2$CO$_2$ configuration; (**e**) Cr$_2$CO$_2$ configuration; (**f**) Mn$_2$CO$_2$ configuration.

Meanwhile, the total net magnetic moment is greatly affected by the magnetic configuration. It is clear that FM shows obvious magnetic moment, in which the magnetic moment of magnetic atoms is in the same direction. The total magnetic moment is similar to the algebraic sum of magnetic moments of magnetic atoms, so it has a large magnetic moment output. As for AFM1, Janus MXenes show obvious magnetic moment, while the net magnetic moment of symmetric MXenes is almost 0. Given that the intralayer atoms are composed of FM coupling, the different magnetic moments of the top and bottom atoms cannot completely cancel the Janus MXene; therefore, it has net magnetic moments and exhibits ferromagnetism. In contrast to the symmetric MXenes with the same top and bottom atoms, the magnetic moment can be completely cancelled out, so the net magnetic moment is 0 and it exhibits anti-ferromagnetism. The net magnetic moments of AFM2 and AFM3 are almost 0, a value which does not show magnetic moment externally. The reason is that both the intralayer coupling is AFM and the adjacent electrons have opposite spin directions, which causes the net magnetic moment of top and bottom layers to be 0, a value which does not show magnetic moment externally. We discover that the higher the symmetry of the magnetic moment, the lower the total energy, and the more stable this configuration will be. Besides, there is modulation between different TM atoms in Janus MXenes. When V atoms are replaced with Cr in V_2CO_2, the magnetic moment of the V atoms becomes smaller; meanwhile, when Mn atoms are replaced with Cr in Mn_2CO_2, the magnetic moment of the Mn atoms becomes large—the modulation effects of Cr atoms on V_2CO_2 and Mn_2CO_2 are different. As for the Cr_2CO_2 configuration, replacing the Cr atoms with V or Mn all will cause the magnetic moment of the Cr atoms to become large. Therefore, we conclude that Janus MXenes can manipulate the size of magnetic moment.

3.3. Electronic Properties

For the above magnetic ground states, we further explored their electronic properties, and calculated their band structures and densities of state (DOS), respectively. Figure 5 shows the band structures of each M'MCO$_2$. Additionally, we found that the spin-up and spin-down curves of Mn_2CO_2, $CrMnCO_2$, and V_2CO_2 almost completely coincide, which is consistent with AFM configurations; while the spin-up and spin-down curves of $VCrCO_2$, $VMnCO_2$, and Cr_2CO_2 split, which are almost the only spin-up curves near the Fermi level—this finding is consistent with FM configurations. Therefore, we found that the configuration of Mn_2CO_2, $CrMnCO_2$, and V_2CO_2 are FM, and the configuration of $VCrCO_2$, $VMnCO_2$, and Cr_2CO_2 are FM.

As for M'MCO$_2$ with AFM configurations, near the Fermi level, the valence band and the conduction band of Mn_2CO_2 are clearly separated, and no band curve crosses the Fermi level. Moreover, the top valence band is near the point M (0.622), and the bottom conduction band is near the point K (1.358), illustrating that Mn_2CO_2 is an indirect band gap semiconductor, a finding that is in keeping with the results of Zhou [55]. Meanwhile, both $CrMnCO_2$ and V_2CO_2 have band curves crossing the Fermi level, and their top valence band and bottom conduction band are located at the point Γ, with a small band gap width, showing metallic character. With regard to M'MCO$_2$, with FM configurations, it can be seen from Figure 5a,b,e that the red curves (spin-down) are distributed on both sides of the Fermi level, with large band gap widths, while the blue curves (spin-up) are densely distributed, crossing the Fermi level. The electronic states with spin-up make the band gap narrow—these are metallic materials. On the whole, we concluded that, when M is replaced with M'—causing non-Janus MXene to become Janus MXene—the band gap width was greatly reduced and the conductivity became better; therefore, Janus MXenes can regulate the band gap.

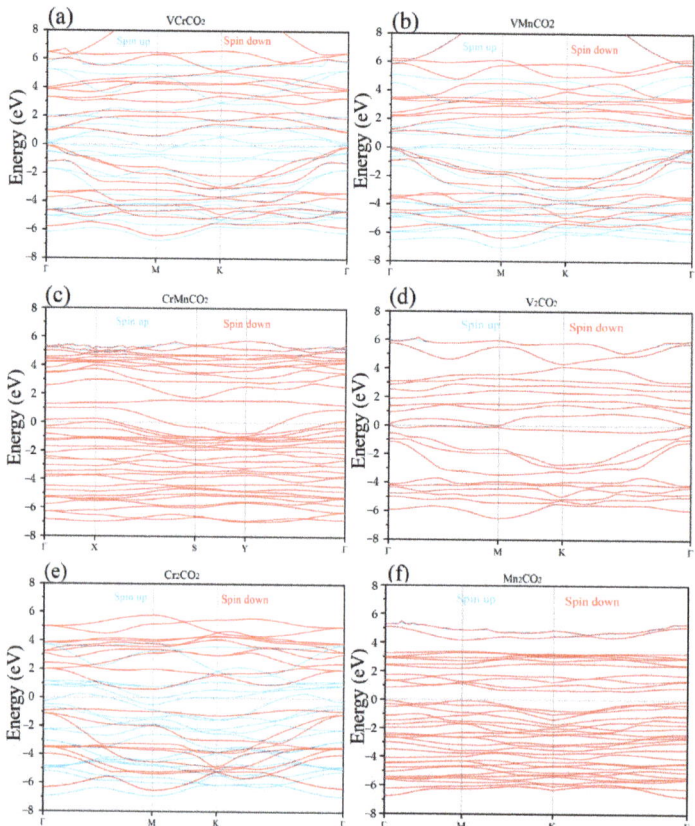

Figure 5. The band gap of M′MCO$_2$. (**a**) VCrCO$_2$; (**b**) VMnCO$_2$; (**c**) CrMnCO$_2$; (**d**) V$_2$CO$_2$; (**e**) Cr$_2$CO$_2$; (**f**) Mn$_2$CO$_2$.

To facilitate the description of the modulation action of Janus MXenes, we plotted the density of state (DOS) of the magnetic ground state of M′MCO$_2$, as shown in Figure 6. There are a large number of electrons near the Fermi level for VCrCO$_2$, VMnCO$_2$, CrMnCO$_2$, V$_2$CO$_2$, and Cr$_2$CO$_2$, indicating that they have metallic properties. However, there are almost no electrons at the Fermi level for Mn$_2$CO$_2$, which can be considered as the forbidden band, and the band width is 0.7071 eV, indicating that it is a semiconductor. In addition, we found that the PDOS of the d orbitals of V, Cr, and Mn contributed most of the TDOS. For symmetric MXenes, the PDOS of d orbitals of V, Cr, and Mn are almost consistent with the TDOS of V$_2$CO$_2$, Cr$_2$CO$_2$, and Mn$_2$CO$_2$ in the energy range of −1~4 eV, −3~6 eV, and 2~6 eV, respectively. Meanwhile, for Janus MXenes, the TDOS are the synergy of d orbitals of different TM atoms. The d orbitals of the V atoms contributed most of the electrons in the following energy ranges: 0~2 eV for VCrCO$_2$; −1~2 eV and 4~6 eV for VMnCO$_2$. The d orbitals of the Cr atoms contributed most of the electrons in the following energy ranges: −2~0 eV and 2~6 eV for VCrCO$_2$, −3~4 eV for CrMnCO$_2$. The d orbitals of the Mn atoms contributed most of the electrons in the following energy ranges: −6~-1 eV and 2~4 eV for VMnCO$_2$, and −5~−3 eV and 4~6 eV for CrMnCO$_2$. The sum of the d orbitals of V, Cr, and Mn is almost equal to the TDOS, showing that the magnetism of the system is mainly derived from the d orbital electrons of magnetic atoms.

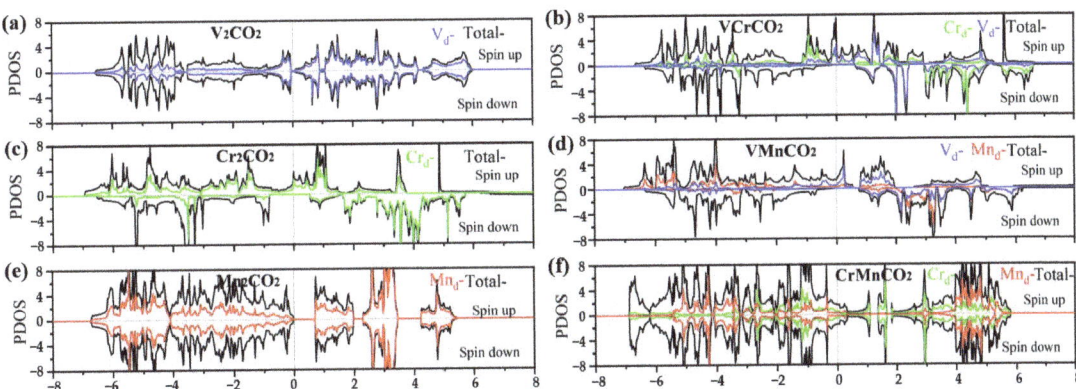

Figure 6. DOS of M'MCO$_2$. (**a**) V$_2$CO$_2$; (**b**) VCrCO$_2$; (**c**) Cr$_2$CO$_2$; (**d**) VMnCO$_2$; (**e**) Mn$_2$CO$_2$; (**f**) CrMnCO$_2$.

Since the magnetic moment of the system is proportional to the area integral of the upper and lower curves, the greater the difference between the spin-up and spin-down, the greater the split, and the greater the net magnetic moment. The upper curves and lower curves of V$_2$CO$_2$ and Mn$_2$CO$_2$ are nearly symmetric and the net magnetic moment is almost 0, which is consistent with the magnetic ground state of AFM. The magnetic ground state of Cr$_2$CO$_2$ is FM, and its upper and lower curves have a certain split, where the peak position energy of the upper curve is lower, and the peak position energy of the lower curve is higher. The area integral of DOS of the upper curve and the lower curve is close, within the energy range of −7~−3 eV, the spin-up electron state occupies the dominant position in the range of −3~2 eV, and the spin-down electron state occupies the dominant position in the range of 2~6 eV, which causes the FM configuration to have 4.184 µB/atom net magnetic moment.

When a V atom in V$_2$CO$_2$ is replaced by Cr or Mn atoms—becoming Janus MXene VCrCO$_2$ or VMnCO$_2$, with the magnetic ground state becoming FM—the upper and lower curves are clearly split. The main electron state is spin-up in the range of −2~2 eV and spin-down in the range of 2~6 eV. Below −2 eV, the peak of spin-up moves to low energy and the peak of spin-down moves to high energy. When a Cr atom is replaced with an Mn atom in Cr$_2$CO$_2$—becoming Janus MXene CrMnCO$_2$, with the magnetic ground state becoming AFM3—the upper and lower curves are symmetric and the net magnetic moment is 0. In conclusion, we can judge the magnetic configuration of a system by the magnetic moment of the atoms, the energy band, and the DOS. Janus MXenes can regulate the band gap width, the magnetic ground state, and the net magnetic moment; this is a great adjusting method.

4. Conclusions

We investigated the magnetic properties and electron structures of M'MCO$_2$ with different magnetic configurations. The following results were found: the magnetic ground states of VCrCO$_2$, VMnCO$_2$, and Cr$_2$CO$_2$ are FM; the magnetic ground state of CrMnCO$_2$ is AFM2; the magnetic ground state of Mn$_2$CO$_2$ is AFM2; the magnetic ground state of V$_2$CO$_2$ is AFM1 and its ground state is NM. The band gap widths of VCrCO$_2$, VMnCO$_2$, CrMnCO$_2$, V$_2$CO$_2$, and Cr$_2$CO$_2$ are no more than 0.02 eV, showing metallic properties; meanwhile, Mn$_2$CO$_2$ is a semiconductor, with a 0.7071 eV band gap width. Moreover, we determined the magnetic configuration of the systems through the magnetic moment, the energy band, and the DOS. Further analysis showed that Janus MXenes can adjust the band gap, magnetic ground state, and output net magnetic moment, resulting in smaller band gap widths and better electrical conductivities when compared with corresponding materials. These theoretical results provide guidance for further experimental verification and electronic device application.

Supplementary Materials: The following are available online at https://www.mdpi.com/article/10.3390/nano12030556/s1, Tables S1–S6: the lattice parameters of $VCrCO_2$, $VMnCO_2$, $CrMnCO_2$, V_2CO_2, Cr_2CO_2, and Mn_2CO_2, Tables S7–S12: the magnetic moment of $VCrCO_2$, $VMnCO_2$, $CrMnCO_2$, V_2CO_2, Cr_2CO_2, and Mn_2CO_2.

Author Contributions: The study was planned and designed by P.G. and P.Q. DFT calculations were performed by P.G. and M.S. The manuscript was prepared by P.G., M.S., X.W., Q.L., S.H., and Y.S. All authors discussed the results and commented on the manuscript together. All authors have read and agreed to the published version of the manuscript.

Funding: This research was funded by National Key Research and Development Program of China(2021YFB3802100), National Natural Science Foundation of China (Grant No.51971031), National Natural Science Foundation of China (51801010), Guangdong Province Key Area R&D Program (2019B010940001), and Guangdong Provincial Key Laboratory of Meta-RF Microwave.

Data Availability Statement: The datasets generated during and/or analyzed during the current study are available from the corresponding author.

Conflicts of Interest: The authors declare no conflict of interest.

References

1. Novoselov, K.S.; Geim, A.K.; Morozov, S.V.; Jiang, D.; Katsnelson, M.I.; Grigorieva, I.V.; Dubonos, S.V.; Firsov, A.A. Two-dimensional gas of massless Dirac fermions in graphene. *Nature* **2005**, *438*, 197–200. [CrossRef] [PubMed]
2. Zhang, Y.; Tan, Y.W.; Stormer, H.L.; Kim, P. Experimental observation of the quantum Hall effect and Berry's phase in graphene. *Nature* **2005**, *438*, 201–204. [CrossRef]
3. Novoselov, K.S.; Geim, A.K.; Morozov, S.V.; Jiang, D.; Zhang, Y.; Dubonos, S.V.; Grigorieva, I.V.; Firsov, A.A. Electric field in atomically thin carbon films. *Science* **2004**, *306*, 666–669. [CrossRef]
4. VahidMohammadi, A.; Rosen, J.; Gogotsi, Y. The world of two-dimensional carbides and nitrides (MXenes). *Science* **2021**, *372*, 6547. [CrossRef] [PubMed]
5. Shekhirev, M.; Shuck, C.E.; Sarycheva, A.; Gogotsi, Y. Characterization of MXenes at every step, from their precursors to single flakes and assembled films. *Prog. Mater. Sci.* **2021**, *120*, 100757. [CrossRef]
6. Ahsan, M.A.; He, T.; Eid, K.; Abdullah, A.M.; Sanad, M.F.; Aldalbahi, A.; Alvarado-Tenorio, B.; Du, A.; Puente Santiago, A.R.; Noveron, J.C. Controlling the Interfacial Charge Polarization of MOF-Derived 0D–2D vdW Architectures as a Unique Strategy for Bifunctional Oxygen Electrocatalysis. *ACS Appl. Mater. Interfaces* **2022**, *14*, 3919–3929. [CrossRef]
7. Naguib, M.; Kurtoglu, M.; Presser, V.; Lu, J.; Niu, J.; Heon, M.; Hultman, L.; Gogotsi, Y.; Barsoum, M.W. Two-dimensional nanocrystals produced by exfoliation of Ti 3AlC 2. *Adv. Mater.* **2011**, *23*, 4248–4253. [CrossRef]
8. Naguib, M.; Mochalin, V.N.; Barsoum, M.W.; Gogotsi, Y. 25th anniversary article: MXenes: A new family of two-dimensional materials. *Adv. Mater.* **2014**, *26*, 992–1005. [CrossRef]
9. Hu, T.; Li, Z.; Hu, M.; Wang, J.; Hu, Q.; Li, Q.; Wang, X. Chemical Origin of Termination-Functionalized MXenes: Ti3C2T2 as a Case Study. *J. Phys. Chem. C* **2017**, *121*, 19254–19261. [CrossRef]
10. Li, L. Lattice dynamics and electronic structures of Ti3C2O2 and Mo2TiC2O2 (MXenes): The effect of Mo substitution. *Comput. Mater. Sci.* **2016**, *124*, 8–14. [CrossRef]
11. Mashtalir, O.; Naguib, M.; Mochalin, V.N.; Dall'Agnese, Y.; Heon, M.; Barsoum, M.W.; Gogotsi, Y. Intercalation and delamination of layered carbides and carbonitrides. *Nat. Commun.* **2013**, *4*, 1716. [CrossRef] [PubMed]
12. Ren, S.; Feng, R.; Cheng, S.; Wang, Q.; Zheng, Z. Synergistic Catalytic Acceleration of MXene/MWCNTs as Decorating Materials for Ultrasensitive Detection of Morphine. *Electroanalysis* **2021**, *33*, 1471–1483. [CrossRef]
13. Feng, X.; Yu, Z.; Long, R.; Li, X.; Shao, L.; Zeng, H.; Zeng, G.; Zuo, Y. Self-assembling 2D/2D (MXene/LDH) materials achieve ultra-high adsorption of heavy metals Ni2+ through terminal group modification. *Sep. Purif. Technol.* **2020**, *253*, 117525. [CrossRef]
14. Hantanasirisakul, K.; Gogotsi, Y. Electronic and Optical Properties of 2D Transition Metal Carbides and Nitrides (MXenes). *Adv. Mater.* **2018**, *30*, 1804779. [CrossRef] [PubMed]
15. Jiang, X.; Kuklin, A.V.; Baev, A.; Ge, Y.; Ågren, H.; Zhang, H.; Prasad, P.N. Two-dimensional MXenes: From morphological to optical, electric, and magnetic properties and applications. *Phys. Rep.* **2020**, *848*, 1–58. [CrossRef]
16. Xia, F.; Wang, H.; Xiao, D.; Dubey, M.; Ramasubramaniam, A. Two-dimensional material nanophotonics. *Nat. Photonics* **2014**, *8*, 899–907. [CrossRef]
17. Yao, H.; Zhang, C.; Wang, Q.; Li, J.; Yu, Y.; Xu, F.; Wang, B.; Wei, Y. Novel two-dimensional layered mosi2z4 (Z = p, as): New promising optoelectronic materials. *Nanomaterials* **2021**, *11*, 559. [CrossRef]

18. Shah, S.A.A.; Sayyad, M.H.; Khan, K.; Sun, J.; Guo, Z. Application of mxenes in perovskite solar cells: A short review. *Nanomaterials* **2021**, *11*, 2151. [CrossRef]
19. Wyatt, B.C.; Rosenkranz, A.; Anasori, B. 2D MXenes: Tunable Mechanical and Tribological Properties. *Adv. Mater.* **2021**, *33*, 2007973. [CrossRef]
20. Fiori, G.; Bonaccorso, F.; Iannaccone, G.; Palacios, T.; Neumaier, D.; Seabaugh, A.; Banerjee, S.K.; Colombo, L. Electronics based on two-dimensional materials. *Nat. Nanotechnol.* **2014**, *9*, 768–779. [CrossRef]
21. Li, J.S.; Wang, Y.; Liu, C.H.; Li, S.L.; Wang, Y.G.; Dong, L.Z.; Dai, Z.H.; Li, Y.F.; Lan, Y.Q. Coupled molybdenum carbide and reduced graphene oxide electrocatalysts for efficient hydrogen evolution. *Nat. Commun.* **2016**, *7*, 11204. [CrossRef] [PubMed]
22. Pandey, M.; Thygesen, K.S. Two-Dimensional MXenes as Catalysts for Electrochemical Hydrogen Evolution: A Computational Screening Study. *J. Phys. Chem. C* **2017**, *121*, 13593–13598. [CrossRef]
23. Liu, A.; Liang, X.; Ren, X.; Guan, W.; Gao, M.; Yang, Y.; Yang, Q.; Gao, L.; Li, Y.; Ma, T. Recent Progress in MXene-Based Materials: Potential High-Performance Electrocatalysts. *Adv. Funct. Mater.* **2020**, *30*, 2003437. [CrossRef]
24. Nguyen, V.H.; Nguyen, B.S.; Hu, C.; Nguyen, C.C.; Nguyen, D.L.T.; Dinh, M.T.N.; Vo, D.V.N.; Trinh, Q.T.; Shokouhimehr, M.; Hasani, A.; et al. Novel architecture titanium carbide (Ti3C2Tx) mxene cocatalysts toward photocatalytic hydrogen production: A mini-review. *Nanomaterials* **2020**, *10*, 602. [CrossRef] [PubMed]
25. Yang, J.; Zhang, S.; Wang, A.; Wang, R.; Wang, C.K.; Zhang, G.P.; Chen, L. High magnetoresistance in ultra-thin two-dimensional Cr-based MXenes. *Nanoscale* **2018**, *10*, 19492–19497. [CrossRef] [PubMed]
26. Bonaccorso, F.; Colombo, L.; Yu, G.; Stoller, M.; Tozzini, V.; Ferrari, A.C.; Ruoff, R.S.; Pellegrini, V. Graphene, related two-dimensional crystals, and hybrid systems for energy conversion and storage. *Science* **2015**, *347*, 1246501. [CrossRef] [PubMed]
27. Wu, X.; Wang, Z.; Yu, M.; Xiu, L.; Qiu, J. Stabilizing the MXenes by Carbon Nanoplating for Developing Hierarchical Nanohybrids with Efficient Lithium Storage and Hydrogen Evolution Capability. *Adv. Mater.* **2017**, *29*, 1607017. [CrossRef]
28. Shinde, P.; Patil, A.; Lee, S.C.; Jung, E.; Jun, S.C. Two-dimensional MXenes for electrochemical energy storage applications. *J. Mater. Chem. A* **2022**, *10*, 1105–1149. [CrossRef]
29. Pan, J.; Lany, S.; Qi, Y. Computationally Driven Two-Dimensional Materials Design: What Is Next? *ACS Nano* **2017**, *11*, 7560–7564. [CrossRef]
30. Frey, N.C.; Wang, J.; Vega Bellido, G.I.; Anasori, B.; Gogotsi, Y.; Shenoy, V.B. Prediction of Synthesis of 2D Metal Carbides and Nitrides (MXenes) and Their Precursors with Positive and Unlabeled Machine Learning. *ACS Nano* **2019**, *13*, 3031–3041. [CrossRef]
31. Lee, Y.; Cho, S.B.; Chung, Y.C. Tunable indirect to direct band gap transition of monolayer Sc2CO2 by the strain effect. *ACS Appl. Mater. Interfaces* **2014**, *6*, 14724–14728. [CrossRef]
32. Hu, J.; Xu, B.; Ouyang, C.; Yang, S.A.; Yao, Y. Investigations on V2C and V2CX2 (X = F, OH) monolayer as a promising anode material for Li Ion batteries from first-principles calculations. *J. Phys. Chem. C* **2014**, *118*, 24274–24281. [CrossRef]
33. Xie, Y.; Kent, P.R.C. Hybrid density functional study of structural and electronic properties of functionalized Tin + 1Xn (X = C, N) monolayers. *Phys. Rev. B-Condens. Matter Mater. Phys.* **2013**, *87*, 235441. [CrossRef]
34. Si, C.; Zhou, J.; Sun, Z. Half-Metallic Ferromagnetism and Surface Functionalization-Induced Metal-Insulator Transition in Graphene-like Two-Dimensional Cr2C Crystals. *ACS Appl. Mater. Interfaces* **2015**, *7*, 17510–17515. [CrossRef]
35. Khazaei, M.; Arai, M.; Sasaki, T.; Chung, C.Y.; Venkataramanan, N.S.; Estili, M.; Sakka, Y.; Kawazoe, Y. Novel electronic and magnetic properties of two-dimensional transition metal carbides and nitrides. *Adv. Funct. Mater.* **2013**, *23*, 2185–2192. [CrossRef]
36. Li, F.; Wei, W.; Zhao, P.; Huang, B.; Dai, Y. Electronic and Optical Properties of Pristine and Vertical and Lateral Heterostructures of Janus MoSSe and WSSe. *J. Phys. Chem. Lett.* **2017**, *8*, 5959–5965. [CrossRef]
37. Yang, Q.; Tan, C.; Meng, R.; Jiang, J.; Liang, Q.; Sun, X.; Yang, D.; Chen, X. The intriguing electronic and optical properties modulation of hydrogen and fluorine codecorated silicene layers. *Appl. Surf. Sci.* **2017**, *398*, 73–80. [CrossRef]
38. Lu, A.Y.; Zhu, H.; Xiao, J.; Chuu, C.P.; Han, Y.; Chiu, M.H.; Cheng, C.C.; Yang, C.W.; Wei, K.H.; Yang, Y.; et al. Janus monolayers of transition metal dichalcogenides. *Nat. Nanotechnol.* **2017**, *12*, 744–749. [CrossRef] [PubMed]
39. Frey, N.C.; Bandyopadhyay, A.; Kumar, H.; Anasori, B.; Gogotsi, Y.; Shenoy, V.B. Surface-Engineered MXenes: Electric Field Control of Magnetism and Enhanced Magnetic Anisotropy. *ACS Nano* **2019**, *13*, 2831–2839. [CrossRef]
40. He, J.; Lyu, P.; Sun, L.Z.; Morales García, Á.; Nachtigall, P. High temperature spin-polarized semiconductivity with zero magnetization in two-dimensional Janus MXenes. *J. Mater. Chem. C* **2016**, *4*, 6500–6509. [CrossRef]
41. Kresse, G.; Furthmüller, J. Efficient iterative schemes for ab initio total-energy calculations using a plane-wave basis set. *Phys. Rev. B-Condens. Matter Mater. Phys.* **1996**, *54*, 11169. [CrossRef]
42. Joubert, D. From ultrasoft pseudopotentials to the projector augmented-wave method. *Phys. Rev. B-Condens. Matter Mater. Phys.* **1999**, *59*, 1758–1775. [CrossRef]
43. Perdew, J.P.; Chevary, J.A.; Vosko, S.H.; Jackson, K.A.; Pederson, M.R.; Singh, D.J.; Fiolhais, C. Atoms, molecules, solids, and surfaces: Applications of the generalized gradient approximation for exchange and correlation. *Phys. Rev. B* **1992**, *46*, 6671. [CrossRef]
44. Perdew, J.P.; Wang, Y. Accurate and simple analytic representation of the electron-gas correlation energy. *Phys. Rev. B* **1992**, *45*, 13244. [CrossRef] [PubMed]
45. Blöchl, P.E. Projector augmented-wave method. *Phys. Rev. B* **1994**, *50*, 17953–17979. [CrossRef] [PubMed]
46. Hubbard, J. Electron correlations in narrow energy bands. *Proc. R. Soc. Lond. Ser. A Math. Phys. Sci.* **1963**, *276*, 238–257. [CrossRef]

47. Jain, A.; Hautier, G.; Ong, S.P.; Moore, C.J.; Fischer, C.C.; Persson, K.A.; Ceder, G. Formation enthalpies by mixing GGA and GGA + U calculations. *Phys. Rev. B-Condens. Matter Mater. Phys.* **2011**, *84*, 045115. [CrossRef]
48. Zhou, F.; Cococcioni, M.; Marianetti, C.A.; Morgan, D.; Ceder, G. First-principles prediction of redox potentials in transition-metal compounds with LDA + U. *Phys. Rev. B-Condens. Matter Mater. Phys.* **2004**, *70*, 1–8. [CrossRef]
49. Kumar, H.; Frey, N.C.; Dong, L.; Anasori, B.; Gogotsi, Y.; Shenoy, V.B. Tunable Magnetism and Transport Properties in Nitride MXenes. *ACS Nano* **2017**, *11*, 7648–7655. [CrossRef]
50. Frey, N.C.; Kumar, H.; Anasori, B.; Gogotsi, Y.; Shenoy, V.B. Tuning Noncollinear Spin Structure and Anisotropy in Ferromagnetic Nitride MXenes. *ACS Nano* **2018**, *12*, 6319–6325. [CrossRef] [PubMed]
51. Chen, Z.; Huang, S.; Huang, B.; Wan, M.; Zhou, N. Transition metal atoms implanted into MXenes (M2CO2) for enhanced electrocatalytic hydrogen evolution reaction. *Appl. Surf. Sci.* **2020**, *509*, 145319. [CrossRef]
52. Tan, Z.; Fang, Z.; Li, B.; Yang, Y. First-Principles Study of the Ferromagnetic Properties of Cr2CO2 and Cr2NO2 MXenes. *ACS Omega* **2020**, *5*, 25848–25853. [CrossRef] [PubMed]
53. Wang, X.; Wang, C.; Ci, S.; Ma, Y.; Liu, T.; Gao, L.; Qian, P.; Ji, C.; Su, Y. Accelerating 2D MXene catalyst discovery for the hydrogen evolution reaction by computer-driven workflow and an ensemble learning strategy. *J. Mater. Chem. A* **2020**, *8*, 23488–23497. [CrossRef]
54. Tan, C.; Cao, X.; Wu, X.J.; He, Q.; Yang, J.; Zhang, X.; Chen, J.; Zhao, W.; Han, S.; Nam, G.H.; et al. Recent Advances in Ultrathin Two-Dimensional Nanomaterials. *Chem. Rev.* **2017**, *117*, 6225–6331. [CrossRef] [PubMed]
55. Zhou, Y.; Zu, X. Mn2C sheet as an electrode material for lithium-ion battery: A first-principles prediction. *Electrochim. Acta* **2017**, *235*, 167–174. [CrossRef]

Article

Comprehensive Study of the Current-Induced Spin–Orbit Torque Perpendicular Effective Field in Asymmetric Multilayers

Baoshan Cui [1,2,3,†], Zengtai Zhu [1,2,†], Chuangwen Wu [1,4,†], Xiaobin Guo [5], Zhuyang Nie [2,6], Hao Wu [1,7,*], Tengyu Guo [1], Peng Chen [1], Dongfeng Zheng [1], Tian Yu [6,7], Li Xi [3], Zhongming Zeng [8], Shiheng Liang [4], Guangyu Zhang [1,2], Guoqiang Yu [1,2,*] and Kang L. Wang [7]

1. Songshan Lake Materials Laboratory, Dongguan 523808, China; cuibaoshan@sslab.org.cn (B.C.); zhuzengtai@sslab.org.cn (Z.Z.); 202111105010070@stu.hubu.edu.cn (C.W.); guotengyu@sslab.org.cn (T.G.); chenpeng@sslab.org.cn (P.C.); zhengdongfeng@sslab.org.cn (D.Z.); gyzhang@iphy.ac.cn (G.Z.)
2. Beijing National Laboratory for Condensed Matter Physics, Institute of Physics, Chinese Academy of Sciences, Beijing 100190, China; 2017141531037@stu.scu.edu.cn
3. Key Laboratory for Magnetism and Magnetic Materials of Ministry of Education, School of Physical Science and Technology, Lanzhou University, Lanzhou 730000, China; xili@lzu.edu.cn
4. Faculty of Physics and Electronic Science, Hubei University, Wuhan 430062, China; shihengliang@hubu.edu.cn
5. School of Physics & Optoelectric Engineering, Guangdong University of Technology, Guangzhou 510006, China; guoxb@gdut.edu.cn
6. College of Physics, Sichuan University, Chengdu 610064, China; work_tian@scu.edu.cn
7. Department of Electrical Engineering, University of California, Los Angeles, CA 90095, USA; wang@ee.ucla.edu
8. Nanofabrication Facility, Suzhou Institute of Nano-Tech and Nano-Bionics, Chinese Academy of Sciences, Suzhou 215123, China; zmzeng2012@sinano.ac.cn
* Correspondence: wuhao1@sslab.org.cn (H.W.); guoqiangyu@iphy.ac.cn (G.Y.)
† These authors contributed equally to this work.

Abstract: The spin–orbit torques (SOTs) in the heavy metal (HM)/ferromagnetic metal (FM) structure hold promise for next-generation low-power and high-density spintronic memory and logic applications. For the SOT switching of a perpendicular magnetization, an external magnetic field is inevitable for breaking the mirror symmetry, which is not practical for high-density nanoelectronics applications. In this work, we study the current-induced field-free SOT switching and SOT perpendicular effective field (H_z^{eff}) in a variety of laterally asymmetric multilayers, where the asymmetry is introduced by growing the FM layer in a wedge shape. We show that the design of structural asymmetry by wedging the FM layer is a universal scheme for realizing field-free SOT switching. Moreover, by comparing the FM layer thickness dependence of (H_z^{eff}) in different samples, we show that the efficiency ($β − H_z^{eff}/J$, J is the current density) is sensitive to the HM/FM interface and the FM layer thickness. The sign of β for thin FM thicknesses is related to the spin Hall angle ($θ_{SH}$) of the HM layer attached to the FM layer. β changes its sign with the thickness of the FM layer increasing, which may be caused by the thickness dependence of the work function of FM. These results show the possibility of engineering the deterministic field-free switching by combining the symmetry breaking and the materials design of the HM/FM interface.

Keywords: spin–orbit torque; perpendicular magnetic anisotropy; perpendicular effective field; zero-field switching

Current-induced spin–orbit torque (SOT) provides an energy-efficient and fast way to electrically manipulate the magnetization [1–4] and dynamics of spin textures (such as chiral domain wall (DW) [5–9] and magnetic skyrmions [10–12], etc.) in the heavy metal (HM)/ferromagnetic metal (FM) multilayers. In such a structure, an in-plane current (*I*)

flowing through the HM layer is converted to a pure spin current (J_s) due to the spin Hall effect [1,13,14] and/or interfacial Rashba effect [15]. The J_s injects into the adjacent FM layer and thus exerts the SOTs. To enable the SOT-driven perpendicular magnetization switching, an external magnetic field is inevitable to break the mirror symmetry [2], which is impractical for high-density nanoelectronics applications. Until 2014, the field-free SOT switching of a perpendicular magnetization was achieved by introducing a laterally asymmetric structure [16], providing a new pathway to realize all-electric deterministic switching. After that, many other strategies have been proposed for realizing field-free SOT switching [17–31]. For the case of laterally asymmetric structure, the field-free SOT switching is driven by the current-induced out-of-plane effective magnetic field (H_z^{eff}). The magnitude and sign of H_z^{eff} determine the switching efficiency and switching polarity at zero external field, respectively. However, the key factors that affect the magnitude and sign of H_z^{eff} are still elusive.

In this work, we aim to explore the key factors that affect the current-induced H_z^{eff} and the resulting field-free SOT switching in a variety of laterally asymmetric structures. We find that the H_z^{eff} is generally introduced in various laterally asymmetric structures. By comparing the FM thickness dependence of the efficiency (β) of H_z^{eff} (i.e., $\beta = H_z^{eff}/J$, where J is the current density), we show that β is closely related to the HM/FM interface and the FM layer thickness. Our results advance the understanding of the current-induced out-of-plane effective magnetic field in the laterally asymmetric structures.

The film stacks consisting of (i) Ta(5)/Gd(1)/CoFeB(w)/MgO(2), (ii) Pt(5)/CoFeB(w)/MgO(2), (iii) IrMn(5)/CoFeB(w)/MgO(2), (iv) Ta(5)/CoFeB(w)/MgO(2), (v) Ta(5)/Mo(1)/CoFeB(w)/MgO(2), and (vi) W(5)/CoFeB(w)/MgO(2) (thickness in nm) were prepared by magnetron sputtering at room temperature on Si substrates capped with a 100 nm thermal oxide under a base pressure of $<1 \times 10^{-8}$ Torr. The CoFeB layer was grown by the oblique sputtering method and hence has a wedge-sharp structure (w). The CoFeB layer thickness (denoted as t_{CoFeB}) varies from 0.50 nm to 1.20 nm within the lateral length of ~5 cm. It is worth noting that we calibrate the wedged thickness in a large lateral scale, therefore, the several nm-scale thickness difference can be detected precisely. The other layers were uniformly grown by rotating the substrate during the deposition. The stacks were annealed at 250 °C for 30 min to enhance the perpendicular magnetic anisotropy (PMA). The basic magnetic properties of the different samples are similar, therefore, only the results of the Ta(5)/Gd(1)/CoFeB(w)/MgO(2) multilayer are presented. The schematic illustration of the Ta/Gd/CoFeB/MgO structure is shown in Figure 1a. The films were patterned into Hall bar devices with the dimension of 130 × 20 µm² (see Figure 1b) via standard photolithography and dry etching techniques for anomalous Hall effect (AHE) and magneto-optical Kerr effect (MOKE) microscopy measurements. For the Hall bar device, there could be some thickness variation, however, the wedged trend should be kept.

Figure 1c shows the AHE loops of the devices with a series of t_{CoFeB}, in which the R_H and H_z are the Hall resistance and out-of-plane external magnetic field, respectively. The sharp-square loops indicate the existence of a PMA for the devices. The dynamics of the domain wall driven by H_z for the whole Hall bar device with t_{CoFeB} = 0.70 nm is shown in Figure 1d. In image ①, the red dotted line shows the current channel of the Hall bar device. At first, a large H_z along +z direction was applied to saturate the sample, and the picture was chosen as the reference as shown in image ①. As H_z increases in the −z direction and reaches the switching field, a reversed domain is nucleated at the bottom edge of the device (see image ②). As the field increases, the domain expands to the whole Hall bar device, as shown in images ③–⑤. These results show that the switching is accomplished by domain nucleation and the domain wall motion. We also measured the perpendicular anisotropy energy density K_u ($K_u = \mu_0 H_k M_s / 2$). M_s is the saturation magnetization, which is measured by the superconducting quantum interference device (SQUID) and has a magnitude of ~710 emu/cm³, μ_0 is the vacuum permeability. H_k is the effective anisotropy field, which is measured by the in-plane AHE loops, as shown in Figure 1e. It is known that R_H is only proportional to the z-axis component of magnetization (*M*) in a system with PMA. As the in-plane magnetic field (H_x) increases, *M* will be rotated from the z direction (easy axis) to the x direction (hard axis). Consequently, there is a reduction of the R_H at high fields, as shown in Figure 1e. Figure 1f summarizes the t_{CoFeB} dependence of K_u. K_u increases first when t_{CoFeB} < 0.77 nm, which has been attributed to the change of the CoFeB/MgO interface (i.e., the interfacial anisotropy) caused by B diffusion [32]. With further increasing the CoFeB thickness, K_u starts to decrease since the PMA has an interfacial origin [33].

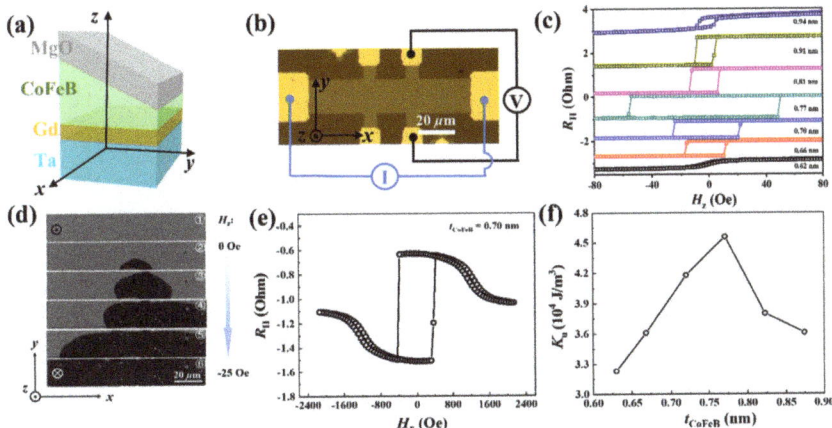

Figure 1. (a) Sketch of the multilayer stack of Ta(5)/Gd(1)/CoFeB(*w*)/MgO(2) (layer thickness in nm). (b) Hall bar device and the measurement configuration. (c) AHE loops for the devices with different CoFeB thicknesses under a current of *I* = 1 mA. (d) Representative MOKE images of the perpendicular magnetic field induced domain wall motion for the whole Hall bar device with t_{CoFeB} = 0.70 nm. (e) Hall resistance (R_H) as a function of the in-plane magnetic field (H_x). (f) The CoFeB thickness dependence of perpendicular magnetic anisotropy energy (K_u).

Next, we show the field-free SOT switching in the devices with different t_{CoFeB} performed by the Keithley 2612A source/measure unit. In the measurements, the writing pulses with a width of 1 ms were injected along the device channel. To avoid applying a zero-writing current, the step number was set as 101 for scanning the writing pulse between −35 (−30) mA and 35 (30) mA. After each writing pulse, the Hall resistance was measured by a reading current of 3 mA. Figure 2a shows the field-free SOT switching for devices of 0.66 nm < t_{CoFeB} < 0.91 nm. The current density (J_e) is calculated by assuming a uniform current distribution across the film stack. As a reference, we have measured the magnetization switching curve in the absence of external magnetic fields loop in the structure without the wedge structure, as shown in Figure 2a. One can see that there is no deterministic SOT-induced magnetization switching. The previous work has demonstrated that the field-free SOT switching was driven by the current-induced H_z^{eff} that originates from the lateral asymmetry [16]. In detail, the wedged layer is deposited at an oblique angle with respect to the substrate surface (along the *y* axis) without rotating the substrate, so it is grown in a tilted direction away from the substrate normal. Consequently, it breaks the mirror symmetry with respect to the *x–z* plane and allows for the creation of a built-in effective electric field (*E*) along the *y* axis. Consequently, a current induced H_z^{eff} is expected due to the Rashba spin–orbit coupling (SOC), which is expressed by $H_z^{eff} = \alpha \cdot (p \times E)$ [15,34]. Here, *α* is the Rashba SOC constant depends on the materials, and *p* represents the electron's momentum. If H_z^{eff} is larger than the coercivity, the magnetization switching can be achieved, although the thickness gradient is very tiny. Figure 2b shows the SOT switching loops under different H_x for the device with t_{CoFeB} = 0.81 nm. We found that the switching polarity changes from a clockwise mode to an anticlockwise mode when the field is increased to $H_x = 50$ Oe. In this case, the switching is not dominated by the H_z^{eff} anymore, and the conventional damping-like SOT dominates the switching with the assistance of the in-plane magnetic field. It is worth noting that the switching polarity under a large positive H_x is consistent with the case in the Ta/CoFeB/MgO system [2]. Nevertheless, the Gd has a positive spin Hall angle (θ_{SH}) [35], which is opposite to Ta. In this regard, we conclude that the conventional damping-like SOT originates from the spin current that is generated in the Ta layer and diffuses through the Gd layer even it is partially compensated by the spin current in the Gd layer.

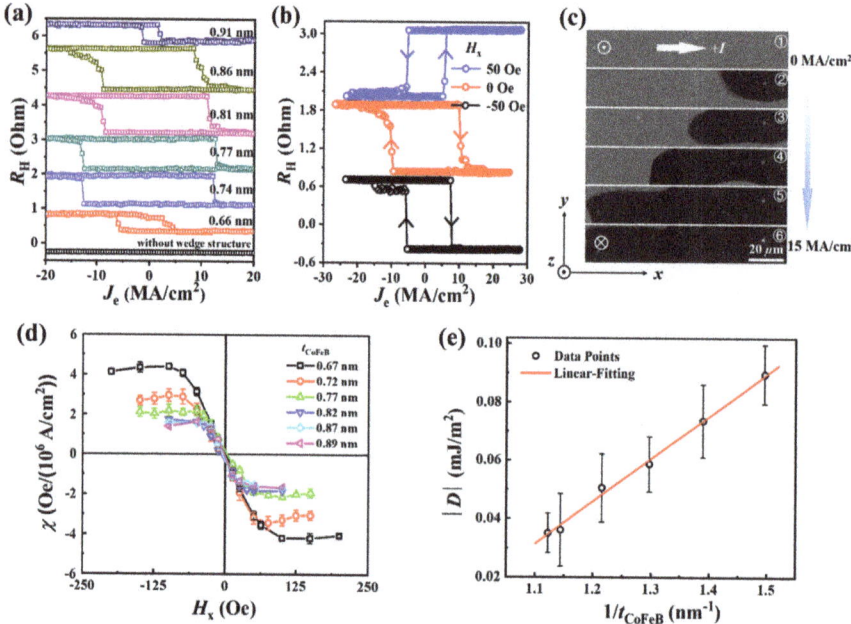

Figure 2. (a) Field-free SOT switching loops for the devices with different t_{CoFeB}. (b) SOT switching at zero field and in-plane magnetic fields of $H_x = \pm 50$ Oe for the device with $t_{CoFeB} = 0.81$ nm. (c) Representative MOKE images of pulsed current–driven magnetization switching for the whole Hall bar device with $t_{CoFeB} = 0.70$ nm. (d) The measured SOT efficiency χ as a function of the in-plane magnetic field for different t_{CoFeB}. (e) The relationship between the estimated DMI constant $|D|$ and $1/t_{CoFeB}$.

During the current-driven magnetization switching, the MOKE measurements were performed simultaneously. Figure 2c shows the MOKE images of SOT switching by using the same Hall bar device as shown in Figure 1d. During the measurements, a large H_z along +z direction was first applied to saturate the magnetization, and the picture was chosen as the reference, as shown in image ①. Then, the current pulses were applied to drive the magnetization switching. Interestingly, the nucleation position is different from that in Figure 1d. For the SOT-driven switching, the initial reversal of domain occurs on the right side of the device channel, as shown in image ②, which is likely due to the presence of Dzyaloshinskii–Moriya interaction (DMI) [36,37] that tilts the magnetization at the device boundary [38], lowering the SOT switching barrier for the right edge and thus to induce the domain nucleation. As the current increases, the DW is subsequently driven to the left side of the device channel. For the observed DW, the neighboring magnetizations on its two sides point ↑↓. When a current is applied along +x axis, the spin orientation (σ) of the net spin current is along +y direction. The dimpling-like field (H_{DL}) as a driven force of the DW can be expressed as $H_{DL} = m \times \sigma$, pointing respectively along +z or −z direction when the magnetic moment (*m*) in the DW along +x (→) or −x (←) directions. Here, "→" and "←" refer to the in-plane component of *m* in the center of DW. Obviously, the current along +x drives the DW to move along −x direction, namely, an effective H_{DL} along −z direction is generated. Thus, the magnetic configuration in the domain wall is ↑←↓, i.e., a left-handed chirality. The magnitude of DMI can be obtained by measuring the AHE loops of switching a Hall cross with combined H_x and H_z under a series of DC current densities [39,40]. Figure 2d summarizes the SOT efficiency (χ) as a function of H_x. The saturated field with the maximum χ corresponds to the effective DMI field ($|H_{DMI}|$). Then, the DMI exchange constant ($|D|$) can be obtained by using $|D| = \mu_0 M_s \Delta |H_{DMI}|$ [41], where Δ is the DW width and is related to exchange stiffness constant $A \approx 1.5 \times 10^{-11}$ J/m and K_u, with the form of $\Delta = (A/K_u)^{1/2}$ [6,39]. The value of the $|D|$ scales linearly with the inverse of t_{CoFeB}, as shown in Figure 2e, indicating its interfacial origin.

In the following, we extract the current-induced H_z^{eff}. Figure 3a shows the hysteresis AHE loops under currents with opposite polarities. The AHE loop shifts to the left under a negative current, indicating the existence of a perpendicular effective field along +z direction (H_z^+). Similarly, a positive current generates an effective perpendicular field along the −z direction (H_z^-). The averaged perpendicular effective field H_z^{eff} can be obtained by $H_z^{eff} = (H_z^- - H_z^+)/2 = \beta J$, where β and J refer to the efficiency of H_z^{eff} and the current density, respectively. Figure 3b shows the t_{CoFeB} dependence of β, where β decreases firstly and changes its sign at $t_{CoFeB} \approx 0.66$ nm, after that β increases negatively. The sign change of β is consistent with our previous work [16]. It is worth noting that the β at 0.63 nm < t_{CoFeB} < 0.66 nm shows small magnitudes, which are likely responsible for the partial switching and same polarity of switching, as shown in Figure 2a.

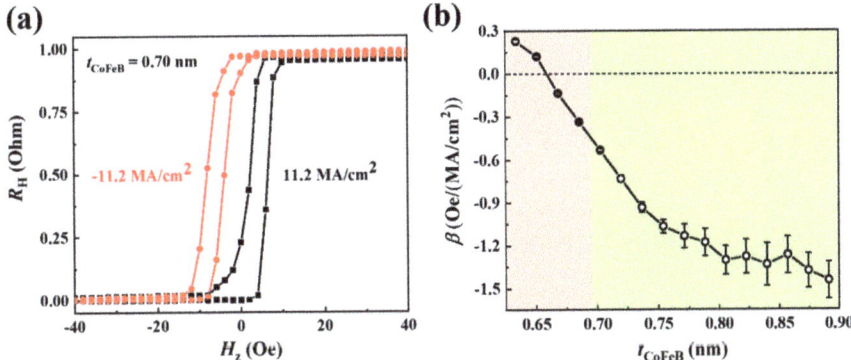

Figure 3. Current-induced out-of-plane effective magnetic fields measured using loops-shift methods. (a) Out-of-plane hysteresis loops under opposite current polarities for t_{CoFeB} = 0.70 nm. (b) β as a function of CoFeB thickness, where the full field-free SOT switching only can be found in the green region.

To explore the key factors that affect the β, samples ii-vi were measured. Figure 4 summarizes the CoFeB thickness dependence of β for these samples. We found that H_z^{eff} exists in all laterally asymmetric structures, indicating this is a universal phenomenon. The Pt/CoFeB, IrMn/CoFeB, and Gd/CoFeB samples have a similar thickness dependence, while Ta/CoFeB, Mo/CoFeB, and W/CoFeB samples show an opposite dependence. To better compare these samples, the β, θ_{SH}, and $D_s = D \cdot t_{FM}$ are extracted for all the samples, as shown in Table 1. We note that some parameters are obtained from the literatures. We found that the interfacial HM layer attached to the FM has a significant relation to the β. First, the sign of β for the thin CoFeB region is determined by the sign of θ_{SH} of the interfacial HM layer attached to the FM layer. The β is positive (negative) for the interfacial HM that has a positive (negative) θ_{SH}. For example, the β values are positive when the interfacial HM are Pt, IrMn, and Gd, which have a positive θ_{SH}. Similarly, β values are negative when the interfacial HM are Ta, Mo, and W, which have a negative θ_{SH}. Consequently, β is negative for Ta/CoFeB, Mo/CoFeB, and W/CoFeB samples. As the thickness of CoFeB increases, β changes its sign at thick CoFeB side except for the Pt/CoFeB and IrMn/CoFeB samples. Our previous work pointed out that the sign of β likely depends on the work functions of interfacial HM and FM [34]. Thus, the sign-change of β in our systems may be ascribed to the thickness dependence of work function of CoFeB layer. The sign-change was not observed for Pt and IrMn samples, which may be attributed to the fact that the PMA regions in these two systems are narrow and the sign reversal thickness is not reached. For all the samples, the magnitude of β strongly depends on the CoFeB thickness, which may be caused by the thickness dependences of the work function of FM layer, K_u, interfacial DMI, and the oblique deposition induced crystal structure asymmetry of CoFeB layer. Further work is required to elucidate the microscopic origin of the thickness dependence.

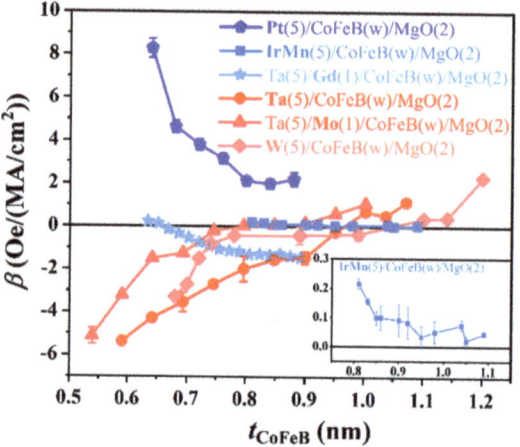

Figure 4. The summarized β as a function of wedged CoFeB thickness in X/CoFeB systems, where X are Pt, IrMn, Gd, Ta, Mo, and W. The inset shows the enlarged IrMn/CoFeB case.

Table 1. Room temperature β, θ_{SH}, and D_s in this work.

HM	β (Oe/(10^6 A/cm^2))		θ_{SH}	D_s (10^{-15} J/m)
	Thin CoFeB	Thick CoFeB		
Pt	8.3	2.2	0.05~0.15 [42]	−965 [43]
IrMn	0.2	0	0.057 [44]	−172 [45]
Gd	0.24	−1.44	0.04 [35]	−146
Ta	−5.5	1.2	−0.05~−0.35 [2,46]	36 [43]
Mo	−5.1	1.1	−0.003 [47]	490 [47]
W	−3.5	2.5	−0.14~−0.49 [48]	73 [43]

In conclusion, we have demonstrated that the current-induced SOT perpendicular effective field is universal for a variety of laterally asymmetric multilayers with a wedged FM layer. The efficiency β is sensitive to the HM/FM interface and the FM layer thickness. The sign of β in a laterally asymmetric structure at thin FM thickness position is determined by the sign of the θ_{SH} of interfacial HM layer attached to the FM layer. As the thickness of FM increases, the sign reversal of β is observed, which may be related to the thickness dependence of the work function of FM. Our work advances the understanding of the out-of-plane effective field in the laterally asymmetric device and provides a pathway in engineering the perpendicular effective field. However, additional advantages may be added to the field-free SOT devices. For example, Pt usually introduces a large DMI [3], IrMn provides antiferromagnetic coupling [20,44], Mo improves the sample's thermal stability [49], and W has a larger θ_{SH} [48].

Author Contributions: Investigation, formal analysis, data collection, B.C., Z.Z. (Zengtai Zhu), C.W., X.G., Z.N., T.G., P.C. and D.Z.; data analysis, data interpretation, B.C., Z.Z. (Zengtai Zhu), C.W., X.G., Z.N., T.G., P.C., D.Z. and T.Y.; conceptualization, methodology, supervision, H.W., L.X., Z.Z. (Zhongming Zeng), S.L., G.Z., G.Y. and K.L.W.; writing—original draft preparation, B.C., Z.Z. (Zengtai Zhu), C.W., H.W. and G.Y.; writing—review and editing, B.C., Z.Z. (Zengtai Zhu), C.W., H.W. and G.Y. All authors have read and agreed to the published version of the manuscript.

Funding: This work was supported by the National Key Research and Development Program of China (Grants No. 2021YFB3601300), the Guangdong Basic and Applied Basic Research Foundation (Grant No. 2020A1515110553), the Beijing Natural Science Foundation (Grant No. Z190009), the National Natural Science Foundation of China (NSFC, Grant Nos. 11874409, 11904056, 91963201, 5167109), the Science Center of the National Science Foundation of China (No. 52088101), the K. C. Wong Education Foundation (Grant No. GJTD-2019-14), the China Postdoctoral Science Foundation (Grant No. 2020M670499), the 111 Project (Grant No. B20063), and Guangzhou basic and applied basic research project (No. 202102020053). T.Y. acknowledges the International Visiting Program for Excellent Young Scholars of SCU. H.W. acknowledges the start-up funding from Songshan Lake Materials Laboratory (Y1D1071S511).

Data Availability Statement: All the data present in this paper will be made available upon reasonable request. Please contact the corresponding author for further information.

Conflicts of Interest: The authors declare no conflict of interest.

References

1. Miron, I.M.; Garello, K.; Gaudin, G.; Zermatten, P.-J.; Costache, M.V.; Auffret, S.; Bandiera, S.; Rodmacq, B.; Schuhl, A.; Gambardella, P. Perpendicular switching of a single ferromagnetic layer induced by in-plane current injection. *Nature* **2011**, *476*, 189–193. [CrossRef] [PubMed]
2. Liu, L.; Pai, C.-F.; Li, Y.; Tseng, H.; Ralph, D.; Buhrman, R. Spin-torque switching with the giant spin Hall effect of tantalum. *Science* **2012**, *336*, 555–558. [CrossRef] [PubMed]
3. Liu, L.; Lee, O.; Gudmundsen, T.; Ralph, D.; Buhrman, R. Current-induced switching of perpendicularly magnetized magnetic layers using spin torque from the spin Hall effect. *Phys. Rev. Lett.* **2012**, *109*, 096602. [CrossRef] [PubMed]
4. Kim, J.; Sinha, J.; Hayashi, M.; Yamanouchi, M.; Fukami, S.; Suzuki, T.; Mitani, S.; Ohno, H. Layer thickness dependence of the current-induced effective field vector in Ta/CoFeB/MgO. *Nat. Mater.* **2013**, *12*, 240–245. [CrossRef] [PubMed]
5. Yamanouchi, M.; Chiba, D.; Matsukura, F.; Ohno, H. Current-induced domain-wall switching in a ferromagnetic semiconductor structure. *Nature* **2004**, *428*, 539. [CrossRef] [PubMed]
6. Metaxas, P.; Jamet, J.; Mougin, A.; Cormier, M.; Ferré, J.; Baltz, V.; Rodmacq, B.; Dieny, B.; Stamps, R. Creep and flow regimes of magnetic domain-wall motion in ultrathin Pt/Co/Pt films with perpendicular anisotropy. *Phys. Rev. Lett.* **2007**, *99*, 217208. [CrossRef] [PubMed]
7. Parkin, S.S.; Hayashi, M.; Thomas, L. Magnetic domain-wall racetrack memory. *Science* **2008**, *320*, 190–194. [CrossRef]
8. Miron, I.M.; Moore, T.; Szambolics, H.; Buda-Prejbeanu, L.D.; Auffret, S.; Rodmacq, B.; Pizzini, S.; Vogel, J.; Bonfim, M.; Schuhl, A. Fast current-induced domain-wall motion controlled by the Rashba effect. *Nat. Mater.* **2011**, *10*, 419. [CrossRef]
9. Ryu, K.-S.; Thomas, L.; Yang, S.-H.; Parkin, S. Chiral spin torque at magnetic domain walls. *Nat. Nanotechnol.* **2013**, *8*, 527. [CrossRef]
10. Nagaosa, N.; Tokura, Y. Topological properties and dynamics of magnetic skyrmions. *Nat. Nanotechnol.* **2013**, *8*, 899–911. [CrossRef]
11. Jiang, W.; Upadhyaya, P.; Zhang, W.; Yu, G.; Jungfleisch, M.B.; Fradin, F.Y.; Pearson, J.E.; Tserkovnyak, Y.; Wang, K.L.; Heinonen, O.; et al. Blowing magnetic skyrmion bubbles. *Science* **2015**, *349*, 283. [CrossRef]
12. Jiang, W.; Zhang, X.; Yu, G.; Zhang, W.; Wang, X.; Jungfleisch, M.B.; Pearson, J.E.; Cheng, X.; Heinonen, O.; Wang, K.L.; et al. Direct observation of the skyrmion Hall effect. *Nat. Phys.* **2016**, *13*, 162–169. [CrossRef]
13. Dyakonov, M.; Perel, V. Current-induced spin orientation of electrons in semiconductors. *Phys. Lett. A* **1971**, *35*, 459–460. [CrossRef]
14. Hirsch, J.E. Spin hall effect. *Phys. Rev. Lett.* **1999**, *83*, 1834. [CrossRef]
15. Bychkov, Y.A.; Rashba, E.I. Properties of a 2D electron gas with lifted spectral degeneracy. *JETP Lett.* **1984**, *39*, 78–81.

16. Yu, G.; Upadhyaya, P.; Fan, Y.; Alzate, J.G.; Jiang, W.; Wong, K.L.; Takei, S.; Bender, S.A.; Chang, L.T.; Jiang, Y.; et al. Switching of perpendicular magnetization by spin–orbit torques in the absence of external magnetic fields. *Nat. Nanotechnol.* **2014**, *9*, 548–554. [CrossRef]
17. You, L.; Lee, O.; Bhowmik, D.; Labanowski, D.; Hong, J.; Bokor, J.; Salahuddin, S. Switching of perpendicularly polarized nanomagnets with spin orbit torque without an external magnetic field by engineering a tilted anisotropy. *Proc. Natl. Acad. Sci. USA* **2015**, *112*, 10310–10315. [CrossRef]
18. Oh, Y.W.; Chris Baek, S.H.; Kim, Y.M.; Lee, H.Y.; Lee, K.D.; Yang, C.G.; Park, E.S.; Lee, K.S.; Kim, K.W.; Go, G.; et al. Field-free switching of perpendicular magnetization through spin–orbit torque in antiferromagnet/ferromagnet/oxide structures. *Nat. Nanotechnol.* **2016**, *11*, 878–884. [CrossRef]
19. Fukami, S.; Zhang, C.; DuttaGupta, S.; Kurenkov, A.; Ohno, H. Magnetization switching by spin–orbit torque in an antiferromagnet–ferromagnet bilayer system. *Nat. Mater.* **2016**, *15*, 535. [CrossRef]
20. Lau, Y.C.; Betto, D.; Rode, K.; Coey, J.M.; Stamenov, P. Spin–orbit torque switching without an external field using interlayer exchange coupling. *Nat. Nanotechnol.* **2016**, *11*, 758–762. [CrossRef]
21. Cai, K.; Yang, M.; Ju, H.; Wang, S.; Ji, Y.; Li, B.; Edmonds, K.W.; Sheng, Y.; Zhang, B.; Zhang, N.; et al. Electric field control of deterministic current-induced magnetization switching in a hybrid ferromagnetic/ferroelectric structure. *Nat. Mater.* **2017**, *16*, 712–716. [CrossRef]
22. Ma, Q.; Li, Y.; Gopman, D.B.; Kabanov, Y.P.; Shull, R.D.; Chien, C.L. Switching a Perpendicular Ferromagnetic Layer by Competing Spin Currents. *Phys. Rev. Lett.* **2018**, *120*, 117703. [CrossRef] [PubMed]
23. Wang, M.; Cai, W.; Zhu, D.; Wang, Z.; Kan, J.; Zhao, Z.; Cao, K.; Wang, Z.; Zhang, Y.; Zhang, T.; et al. Field-free switching of a perpendicular magnetic tunnel junction through the interplay of spin–orbit and spin-transfer torques. *Nature Electronics* **2018**, *1*, 582–588. [CrossRef]
24. Liu, Y.; Zhou, B.; Zhu, J.-G. Field-free Magnetization Switching by Utilizing the Spin Hall Effect and Interlayer Exchange Coupling of Iridium. *Sci. Rep.* **2019**, *9*, 325. [CrossRef] [PubMed]
25. Chang, M.; Yun, J.; Zhai, Y.; Cui, B.; Zuo, Y.; Yu, G.; Xi, L. Field free magnetization switching in perpendicularly magnetized Pt/Co/FeNi/Ta structure by spin orbit torque. *Appl. Phys. Lett.* **2020**, *117*, 142404. [CrossRef]
26. Cao, Y.; Sheng, Y.; Edmonds, K.W.; Ji, Y.; Zheng, H.; Wang, K. Deterministic Magnetization Switching Using Lateral Spin–Orbit Torque. *Adv. Mater.* **2020**, *32*, 1907929. [CrossRef] [PubMed]
27. Wei, J.; Wang, X.; Cui, B.; Guo, C.; Xu, H.; Guang, Y.; Wang, Y.; Luo, X.; Wan, C.; Feng, J.; et al. Field-Free Spin–Orbit Torque Switching in Perpendicularly Magnetized Synthetic Antiferromagnets. *Adv. Funct. Mater.* **2021**, *32*, 2109455. [CrossRef]
28. Wu, H.; Nance, J.; Razavi, S.A.; Lujan, D.; Dai, B.; Liu, Y.; He, H.; Cui, B.; Wu, D.; Wong, K.; et al. Chiral Symmetry Breaking for Deterministic Switching of Perpendicular Magnetization by Spin–Orbit Torque. *Nano Lett.* **2021**, *21*, 515–521. [CrossRef]
29. Zheng, Z.; Zhang, Y.; Lopez-Dominguez, V.; Sanchez-Tejerina, L.; Shi, J.; Feng, X.; Chen, L.; Wang, Z.; Zhang, Z.; Zhang, K.; et al. Field-free spin–orbit torque-induced switching of perpendicular magnetization in a ferrimagnetic layer with a vertical composition gradient. *Nat. Commun.* **2021**, *12*, 4555. [CrossRef]
30. Chen, R.; Cui, Q.; Liao, L.; Zhu, Y.; Zhang, R.; Bai, H.; Zhou, Y.; Xing, G.; Pan, F.; Yang, H.; et al. Reducing Dzyaloshinskii-Moriya interaction and field-free spin–orbit torque switching in synthetic antiferromagnets. *Nat. Commun.* **2021**, *12*, 3113. [CrossRef]
31. Xie, X.; Zhao, X.; Dong, Y.; Qu, X.; Zheng, K.; Han, X.; Han, Y.; Fan, Y.; Bai, L.; Chen, Y.; et al. Controllable field-free switching of perpendicular magnetization through bulk spin–orbit torque in symmetry-broken ferromagnetic films. *Nat. Commun.* **2021**, *12*, 2473. [CrossRef] [PubMed]
32. Liu, T.; Cai, J.W.; Sun, L. Large enhanced perpendicular magnetic anisotropy in CoFeB/MgO system with the typical Ta buffer replaced by an Hf layer. *AIP Adv.* **2012**, *2*, 032151. [CrossRef]
33. Ikeda, S.; Miura, K.; Yamamoto, H.; Mizunuma, K.; Gan, H.D.; Endo, M.; Kanai, S.; Hayakawa, J.; Matsukura, F.; Ohno, H. A perpendicular-anisotropy CoFeB-MgO magnetic tunnel junction. *Nat. Mater.* **2010**, *9*, 721–724. [CrossRef] [PubMed]
34. Cui, B.; Wu, H.; Li, D.; Razavi, S.A.; Wu, D.; Wong, K.L.; Chang, M.; Gao, M.; Zuo, Y.; Xi, L.; et al. Field-Free Spin–Orbit Torque Switching of Perpendicular Magnetization by the Rashba Interface. *ACS Appl. Mater. Interfaces* **2019**, *11*, 39369–39375. [CrossRef] [PubMed]
35. Reynolds, N.; Jadaun, P.; Heron, J.T.; Jermain, C.L.; Gibbons, J.; Collette, R.; Buhrman, R.A.; Schlom, D.G.; Ralph, D.C. Spin Hall torques generated by rare-earth thin films. *Phys. Rev. B* **2017**, *95*, 064412. [CrossRef]
36. Dzyaloshinsky, I. A thermodynamic theory of "weak" ferromagnetism of antiferromagnetics. *Sov. Phys. JETP* **1957**, *5*, 1259. [CrossRef]
37. Moriya, T. Anisotropic Superexchange Interaction and Weak Ferromagnetism. *Phys. Rev.* **1960**, *120*, 91–98. [CrossRef]
38. Yu, G.; Upadhyaya, P.; Shao, Q.; Wu, H.; Yin, G.; Li, X.; He, C.; Jiang, W.; Han, X.; Amiri, P.K.; et al. Room-Temperature Skyrmion Shift Device for Memory Application. *Nano Lett.* **2017**, *17*, 261–268. [CrossRef]
39. Pai, C.-F.; Mann, M.; Tan, A.J.; Beach, G.S.D. Determination of spin torque efficiencies in heterostructures with perpendicular magnetic anisotropy. *Phys. Rev. B* **2016**, *93*, 144409. [CrossRef]
40. Cui, B.; Yu, D.; Shao, Z.; Liu, Y.; Wu, H.; Nan, P.; Zhu, Z.; Wu, C.; Guo, T.; Chen, P.; et al. Néel-type elliptical skyrmions in a laterally asymmetric magnetic multilayer. *Adv. Mater.* **2021**, *33*, 2006924. [CrossRef]
41. Thiaville, A.; Rohart, S.; Jué, É.; Cros, V.; Fert, A. Dynamics of Dzyaloshinskii domain walls in ultrathin magnetic films. *Europhys. Lett.* **2012**, *100*, 57002. [CrossRef]

42. Lee, J.W.; Oh, Y.-W.; Park, S.-Y.; Figueroa, A.I.; van der Laan, G.; Go, G.; Lee, K.-J.; Park, B.-G. Enhanced spin–orbit torque by engineering Pt resistivity in Pt/Co/AlO$_x$ structures. *Phys. Rev. B* **2017**, *96*, 064405. [CrossRef]
43. Ma, X.; Yu, G.; Tang, C.; Li, X.; He, C.; Shi, J.; Wang, K.L.; Li, X. Interfacial Dzyaloshinskii-Moriya interaction: Effect of 5d band filling and correlation with spin mixing conductance. *Phys. Rev. Lett.* **2018**, *120*, 157204. [CrossRef]
44. Wu, D.; Yu, G.; Chen, C.-T.; Razavi, S.A.; Shao, Q.; Li, X.; Zhao, B.; Wong, K.L.; He, C.; Zhang, Z.; et al. Spin–orbit torques in perpendicularly magnetized Ir$_{22}$Mn$_{78}$/Co$_{20}$Fe$_{60}$B$_{20}$/MgO multilayer. *Appl. Phys. Lett.* **2016**, *109*, 222401. [CrossRef]
45. Ma, X.; Yu, G.; Razavi, S.A.; Sasaki, S.S.; Li, X.; Hao, K.; Tolbert, S.H.; Wang, K.L.; Li, X. Dzyaloshinskii-Moriya interaction across an antiferromagnet-ferromagnet interface. *Phys. Rev. Lett.* **2017**, *119*, 027202. [CrossRef]
46. Sagasta, E.; Omori, Y.; Vélez, S.; Llopis, R.; Tollan, C.; Chuvilin, A.; Hueso, L.E.; Gradhand, M.; Otani, Y.; Casanova, F. Unveiling the mechanisms of the spin Hall effect in Ta. *Phys. Rev. B* **2018**, *98*, 060410. [CrossRef]
47. Chen, T.-Y.; Chan, H.-I.; Liao, W.-B.; Pai, C.-F. Current-induced spin–orbit torque and field-free switching in Mo-based magnetic heterostructures. *Phys. Rev. Appl.* **2018**, *10*, 044038. [CrossRef]
48. Zhang, C.; Fukami, S.; Watanabe, K.; Ohkawara, A.; DuttaGupta, S.; Sato, H.; Matsukura, F.; Ohno, H. Critical role of W deposition condition on spin–orbit torque induced magnetization switching in nanoscale W/CoFeB/MgO. *Appl. Phys. Lett.* **2016**, *109*, 192405. [CrossRef]
49. Wu, D.; Yu, G.; Shao, Q.; Li, X.; Wu, H.; Wong, K.L.; Zhang, Z.; Han, X.; Khalili Amiri, P.; Wang, K.L. In-plane current-driven spin–orbit torque switching in perpendicularly magnetized films with enhanced thermal tolerance. *Appl. Phys. Lett.* **2016**, *108*, 212406. [CrossRef]

MDPI
St. Alban-Anlage 66
4052 Basel
Switzerland
Tel. +41 61 683 77 34
Fax +41 61 302 89 18
www.mdpi.com

Nanomaterials Editorial Office
E-mail: nanomaterials@mdpi.com
www.mdpi.com/journal/nanomaterials

www.ingramcontent.com/pod-product-compliance
Lightning Source LLC
LaVergne TN
LVHW070719100526
838202LV00013B/1125